Silvopasture

ALSO BY STEVE GABRIEL

Farming the Woods
(with Ken Mudge)

Praise for *Silvopasture*

"With farmland getting scarcer and the climate getting warmer, we must figure out novel approaches to growing food on less land with a smaller environmental footprint. Getting meat, firewood, lumber, mushrooms, berries, nuts, and other crops off the same piece of land will be even more important. One of the best approaches to that is silvopasturing—combining food animals with tree crops. Author Steve Gabriel gives us a well-organized, practical guide to this centuries-old approach of land management."

—**Rebecca Thistlethwaite**, author of
Farms with a Future and *The New Livestock Farmer*

"A heartfelt, humble, and hope-filled account of the need for people to embrace one another and the landscapes we inhabit, told through the invaluable language of silvopasture. This book is packed with information and practical examples for anyone interested in the benefits of trees and grazing for the health of soil, plants, herbivores, and human beings."

—**Fred Provenza**, professor emeritus,
Department of Wildland Resources, Utah State University

"There comes a time when modern messengers are needed to bring new life to ancient practices so that they can be utilized by the masses to transform society. In this foundational book, Steve Gabriel masterfully blends years of hard work, listening, and studying to present the complex subject in simple yet meaningful ways. *Silvopasture* provides much needed support and inspiration for anyone interested in becoming part of the solution to our climate, ecological, economic, and health challenges."

—**Vail Dixon**, Simple Soil Solutions

"There is rapidly increasing interest among European farmers in combining trees and livestock, but until now there's been no really good book to guide them that explores all the benefits, as well as the tricky management decisions, involved in silvopasture systems. This is where Steve Gabriel's book, appearing at exactly the right moment, comes in. Covering both tree management (stocking, species, etc.) and animal management (fencing, shelter, and breeds) as well as the ecology of the interactions between the two, this book should become the first port of call for farmers needing that extra information and confidence to take the step toward becoming agroforesters."

—**Martin Crawford**, director, Agroforestry Research Trust

"To practice silvopasture grazing successfully we must recognize the complexity of the farm ecosystem as a whole, and understand the needs of the many different parts of the system: livestock; understory plants, including grasses, legumes, and forbs; trees and other woody plants; and, of course, the soil itself, which must be maintained in good health. This book is an excellent resource to gain the essential knowledge needed to manage silvopasture well."

—**Sarah Flack**, author of *The Art and Science of Grazing*

Silvopasture

A Guide to Managing
Grazing Animals, Forage Crops, and
Trees in a Temperate Farm Ecosystem

STEVE GABRIEL

Foreword by Eric Toensmeier

Chelsea Green Publishing
White River Junction, Vermont | London, UK

Project Manager: Alexander Bullett
Acquisitions Editor: Fern Marshall Bradley
Project Editor: Benjamin Watson
Copy Editor: Laura Jorstad
Proofreader: Nanette Bendyna
Indexer: Linda Hallinger
Designer: Melissa Jacobson
Page Composition: Abrah Griggs

Cover photographs, *clockwise from top left*, by Jen Gabriel, Steve Gabriel,
Johnaapw/123RF.com, and Uberprutser/Wikimedia.

Printed in the United States of America.
First printing May, 2018.
10 9 8 7 6 5 4 3 2   21 22 23 24

Library of Congress Cataloging-in-Publication Data
Names: Gabriel, Steve, 1982– author.
Title: Silvopasture : a guide to managing grazing animals, forage crops, and trees in a temperate farm ecosystem / Steve Gabriel.
Description: White River Junction, Vermont : Chelsea Green Publishing, [2018] | Includes bibliographical references and index.
Identifiers: LCCN 2018000802 | ISBN 9781603587310 (pbk.)
Subjects: LCSH: Silvopastoral systems.
Classification: LCC S494.5.S95 G33 2018 | DDC 333.76/14—dc23
LC record available at https://lccn.loc.gov/2018000802

Chelsea Green Publishing
85 North Main Street, Suite 120
White River Junction, VT 05001
(802) 295-6300
www.chelseagreen.com

This book is dedicated to those who mentor and teach
in a deep way, by being their most authentic self.

To Dale Bryner,
who taught me to slow down and listen, honor my gifts,
and to share what I know with the next generation.

and

To Mike DeMunn, Da' Ha' da' nyah,
who invited me to build a relationship with the woods,
and work for the rest of my life, in service to the trees.

Contents

Foreword

There is wide agreement among scientists that climate change mitigation is the most critical human endeavor of the 21st century. Stepping back from the brink of catastrophe involves reassessment and redesign of every aspect of civilization. Agriculture and land use, currently responsible for roughly a quarter of human emissions, are certainly no exception. The challenge of transforming food production means we must develop new forms of food production, and this is particularly true for the livestock sector.

In spring 2015 I joined environmental thought leader Paul Hawken's Project Drawdown, with a mission to model and rank the impact of climate change mitigation solutions across sectors including land use, food, electricity generation, buildings and cities, materials, educating women and girls, and transport. My team's task was to look at both food production and ecosystem management. In the final analysis, silvopasture, a relative unknown, emerged with the most powerful climate impact of all agricultural production solutions. *Drawdown* was a *New York Times* bestseller when it was released in 2017, bringing much attention to the practice of integrating trees with grazing.

Climate change gives us an opportunity to put this solution in the spotlight, which it was already deserving. Silvopasture offers many additional benefits including improved animal welfare due to shade availability, along with increased productivity, wildlife habitat, water quality, and soil organic matter. Though largely overlooked, silvopasture is quite widespread. Drawdown estimates that this practice is already used on 351 million acres globally. Some silvopasture practices, like Spanish *dehesa*, go back thousands of years.

In coming decades, as the world looks for ever more powerful mitigation strategies, silvopasture could and should take center stage in ruminant production wherever rainfall permits tree growth. Mexico is already paying ranchers to shift from degraded pasture to intensive silvopasture—and finding that, once the initial establishment costs are covered, these ranches (which had been subsidized for decades) become profitable on their own. Government payments for environmental services, access to low-interest loans, preferential markets, and premium prices—all of these incentives and more are already being seen. Indeed the USDA Natural Resources Conservation Service (NRCS) currently includes silvopasture as a conservation practice under its Environmental Quality Incentives Program (EQIP)—though it is not yet a funded priority in most states.

Thus, due to its climate and other benefits for people, livestock, and ecosystems, silvopasture's time has come. Yet the practice is unfamiliar to the great majority of farmers and ranchers in the United States and Canada. There are few operations to visit here, and, importantly, there is a huge need for a how-to guide.

Enter Steve Gabriel's *Silvopasture: A Guide to Managing Grazing Animals, Forage Crops, and Trees in a Temperate Farm Ecosystem*. Steve is an ideal person to produce this book. First and foremost he is a silvopasture producer himself. He works for Cornell University agricultural extension, including serving as editor of *Small Farm Quarterly*. And he has already written a how-to agroforestry manual for temperate climates (with Ken Mudge), the fantastic *Farming the Woods*, a guide to forest farming.

Steve provides readers with the information and encouragement they need to succeed. Case studies of working silvopasture systems, research on the benefits of silvopasture for the land, livestock, and more make the case for combining trees, forages, and animals. Planning tools and guidelines assist readers, whether they are starting with pasture, forest, or overgrown bush.

Silvopasture includes valuable planning tools and guidelines that demonstrate Steve's deep commitment to ecological production and producer success. Of special value is the discussion of trade-offs, drawbacks, and when and where *not* to practice silvopasture. Were I a silvopasture producer (and this book sorely tempts me to become one), this book would be the single most important resource I'd use for planning and management.

Eric Toensmeier
author of *The Carbon Farming Solution*

Acknowledgments

This book has so many voices. Not just mine. While I wrote it, I am just a librarian, a curator of the thoughts, ideas, experiments, and research of so many committed people, communities, and institutions. Between the research and case studies in this book, there is the work of at least 300 people in here. Above all, I am thankful for and honored by the privilege and opportunity of trying to bring all these pieces together, in one place. It is, at best, a skeleton of the knowledge and wisdom out there. I want to thank many people, in no particular order:

My work in the world has always been with the support of teachers, colleagues, and friends who are eager to explore the world and question the ways we steward it. Among the most special teachers in my life are the two this book is dedicated to.

To Dale Bryner, thank you. Your invitation to observe the natural world, trust my instincts, and build knowledge from a place of love and respect is something I strive for each day.

To Michael DeMunn, who remains a close teacher, mentor, and friend. Your love of the forest, knowledge of it, and passion to share the deep lessons you have learned are the foundation of my work, to connect as many people as possible to the forested landscape, and to support them in developing a positive relationship to it.

So many wonderful colleagues contributed their own expertise to this work, and I am appreciative of their willingness to share. Thanks to Eric Toensmeier, whose work has helped define the importance of silvopasture in the context of climate change. Thanks to Connor Stedman, who offered his time to discuss many ideas and approaches in this book, and who offered contributions on keyline and tree spacing considerations. And thanks to Jonathan Bates, Kass Urban Mead, Jono

Neiger, and Akiva Silver for writing sidebars that help enrich this text with your own personal experiences.

I was fortunate to get some firsthand accounts of farmers practicing and developing silvopasture, just a few of the many out there. My appreciation to Don Kilpela of Michigan, Chris Fields Johnson of Virginia, Cliff Davis of Tennessee, and Ann Wilhelm from Connecticut, for giving us a snapshot of how you are living this practice on your own farms.

One of the great discoveries of this book was the work of Fred Provenza and staff and students at Utah State University Extension, as well as his work with Michel Meuret in France, which has blown open the door to a deeper understanding of animal behavior. Fred and Michel were both very kind and helpful in reviewing relevant parts of the manuscript and providing photos and clarity. Thanks.

Other institutions and their people that have contributed greatly to this text include Uma Karki and Tuskegee University, Gabriel Pent and John Munsell from Virginia Tech, and Mike Gold and the folks at University of Missouri's Agroforestry Center. These relationships have been built over several years, thanks to the Association for Temperate Agroforestry and the biannual agroforestry conference they organize. Much of the support for this network can be traced back to Kate MacFarland and the National Agroforestry Center staff. My appreciation for your dedication to research, knowledge, and open sharing of information, for the benefit of the farmers we seek to serve.

On the home front, I am blessed to be part of a university extension program that includes some of the foremost thinkers and practitioners of agroforestry. My thanks goes to Peter Smallidge, New York State extension forester, and Joe Orefice, who is now director of the Cornell Maple Program in Lake Placid but has done extensive work to research, educate, and practice silvopasture in his previous roles and as an active farmer in his own right. And I couldn't thank Brett Chedzoy enough, for all the years of discussion, collaboration, support, and inspiration for moving the ideas and education of silvopasture forward.

Thanks to my friend Camilo Nascimento for his drawings and to my friend and sister Jen Gabriel for her wonderful photos. I am honored to have connections to those who are much more artistic than I. Our small community is home to an incredible array of farmers, builders, foresters, craftspeople, and community organizers. To all of them, I am grateful.

Some of the best teachers of all have been the land, our animals (ducks, geese, and sheep), and the time we are afforded to interact with them. I am grateful to Suzanne and Daryl Anderson for giving us a little slice of their land and supporting our dreams. I am thankful for the trees, the water, and the soil we are blessed to work with. Our farm had been a learning journey like all ventures in life, and we could not have come so far without our wonderful apprentices who helped us in so many ways over the seasons: Thanks to Joshua, Dominic, Costa, Sara, Kat, Jaya, Wyatt, Claire, Jonathan, and Shaun.

And finally and most central I give my thanks to my wife, partner, and friend Elizabeth, who probably shouldn't have let me write another book, but has given me the space and encouragement to do so. I love you.

And as a further extension of family, I am grateful to our dogs Sadie and Vida, and to our families: Bob and Susan, Jennifer and Scott, Marcia and Stephen, and Ben and Erica. Their ongoing love and support bring all the abundance we have in our life to fruition.

There are, no doubt, voices missing in this note of thanks—and so I acknowledge those, in both the seen and the unseen beings that have touched my life. Thank you.

About This Book

Just a few notes to begin this text and help get you oriented so you can make best use of it. This book was organized with the intention of supporting readers who are actively developing a silvopasture plan for their farm. Some readers may not be in this situation currently, but can still easily come along for the ride.

While researching and developing this book, I came across a great many gaps in the available information. If you have ever explored the topic of silvopasture, you will likely be well acquainted with this phenomenon. My goal has been to pull together what information is out there (and credible), as well as what my experience has been in working for extension and as a farmer in the agroforestry community for the past decade or so. I aimed to gather all this information and offer a framework for developing a viable silvopasture enterprise.

So I recognize that many parts of this narrative are missing, and you might find yourself unsatisfied with this as you read the book. Take note of those moments. Then please help fill those gaps in the knowledge. People love to poke holes in emerging agricultural practices, focusing on all the deficiencies they can find. I'm inviting you *not* to do this, and instead become involved in the conversation. We need everyone's thoughtful participation.

Geographic Scope

It's dangerous to generalize in agriculture, because so many variables exist in the landscape. To this end, and because the experience I have is limited to the cool temperate zones of North America, the context-specific material in this book is going to be most relevant to these regions, as indicated on the map.

That said, there is much in the assessment and planning process that is useful worldwide, and some of the top tree species (such as willow, mulberry, and poplar) are prevalent in a vast range of climate

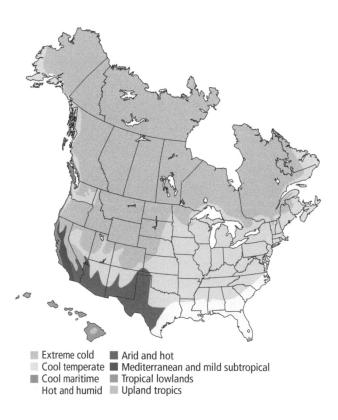

Extreme cold   Arid and hot
Cool temperate   Mediterranean and mild subtropical
Cool maritime   Tropical lowlands
Hot and humid   Upland tropics

Figure 0.1. One version of a climate zone map created by Eric Toensmeier for his book *Perennial Vegetables*. Eric cross-indexed USDA hardiness zones, American Horticulture Society heat zones, and *Sunset* gardening zones for the map. His definition of *cool temperate* is hardiness zones 4 through 7 with warm to hot summers and 30 or more inches of rain, which fits the author's definition of *cool temperate* for this book. Illustration courtesy of Eric Toensmeier.

types. Readers in regions outside the main scope of this book will likely find a lot of good information, though they will need to translate this into their own context, including considerations for animal type and breed, plant and tree species, and environmental factors, which change readily with a change in latitude or ecotype.

How This Book Is Organized

Chapters 1 and 2 aim to define *silvopasture* and lay out the basic tenets, drawing upon our collective history and the known ecology of forests, grasslands, and savanna ecosystems to inform our thinking.

Chapter 3 is all about the animals. We look at how a deeper understanding of animal sentience and behavior leads to better management, and discuss the considerations for different types of animals, their breeds, and their needs on-site to live good lives.

Chapter 4 focuses on applying silvopasture to existing woodlots, and to landscapes dominant with woody vegetation. There is an opportunity to begin mapping the land you are working with, and utilize some assessment tools to develop a good plan.

Chapter 5 looks at the opposite end of silvopasture—its application to open pastureland—by looking at good grazing practices and engaging in activities to assess the land and consider how trees can best be incorporated. Here we also name some of the most promising silvopasture trees, which offer some of the best attributes based on what we are after.

Chapter 6 brings the assessment work from chapters 4 and 5 together to help you develop a complete plan, articulate overall goals, and develop a grazing map, plan, and budget. We look at opportunities to get help, and discuss how marketing silvopasture and your farm narrative are key parts of success.

I encourage you to follow along and, if useful, work through the planning tools and suggested activities. This would result in the start of a good plan for silvopasture by the time you finish reading.

You'll Need Other Books and Resources

The nature of silvopasture is cross-disciplinary, and includes aspects of forestry, animal husbandry, grazing, ecology, and agroforestry. With the goal of keeping the text focused on patterns, process, and systems, we simply couldn't get every detail in here. You will want to get more specific knowledge on how to raise specific types of animals, on the wide range of forestry applications, and on approaches to managed grazing. Throughout the book, there are links and references to supplemental material that is also required reading along your path to developing a silvopasture practice.

References and Citations

There are more than 250 citations in this book, linking the text to the credible sources of ideas, figures, and information. The references serve to link you to a long list of resources for further reading and exploration. Many of the materials are available as free downloads online, though one of the challenges to this is that links change over time. The materials that we think readers will most benefit from reading are posted at www.silvopasturebook.com under resource library, to ensure that you can access them over time. Don't hesitate to contact the author if you cannot locate a resource.

Thanks for reading. I hope that the work put into the book helps clarify and articulate silvopasture for you in a way that helps you meet your goals and does good for our precious world.

Prologue

We are living in a time of great confusion and controversy around the ways livestock are raised in the United States. This is a heated debate, where each side seems to only dig their heels in more, resting on overly simplistic arguments that are recycled over and over. On the side of those who advocate for the total elimination of meat and dairy products from the human diet, there is a judgment that there is no possible way it can be done right. On the extreme opposite side are grazing advocates, who say that getting animals back on pasture will not only mitigate the problem but even reverse all the problems of climate change.

As usual, the truth lies somewhere in between and is nuanced, with many dependent variables. For instance, it's definitely true that, in the United States, people need to eat less meat. It's also true that at certain densities, on certain types of land, with well-designed grazing systems, animals can be benign or even regenerative in their effect on the land and environment. Silvopasture, with the addition of trees and forests to the conversation, offers one of the clearer paths to climate-friendly farming.

At the root of this controversy is a lot of talk that seems to blame the livestock and animal agriculture, not the humans who designed it. At the root of our poor design of farming systems—regardless of whether they produce meat, grain, or vegetables—is the reductionist and isolationist approach, which reduces complex systems into overly simplistic terms. *This is good, and this is bad*, goes the narrative. In order to change our behavior, however, we need to examine our own personal assumptions, as well as those of our communities and society as a whole.

Complex problems won't be solved with simple answers. Not everyone will choose to be a vegan, and not all farmers will invest the time and energy into grass-based grazing operations. It all comes down to perspective. When I first learned about the way the vast majority of meat and milk is produced, I was appalled. I continue to feel this way today. The approach of confining animals and feeding them formulated rations of grain grown on thousands of acres of land focuses on one goal only: to make as much product, as cheaply as possible. As a result, animals are not given any dignity, massive inputs of water and energy are required, and manure becomes a toxic waste problem. Not to mention that the farmer can barely make a living, and the consumer's health suffers from the poor quality of the food.

My newfound awareness of this dominant approach led me to become a vegetarian for six years, a process I now see as an exploration of my relationship to food, where I really began to take notice of how what I ate affected my body. I ultimately found that, for myself, a lack of animal protein had some adverse effects. I choose to live in a cold climate, and I found that meat and dairy were necessary parts of the picture for me if I wanted to source foods from my region and not from far away. Still, I was only going to consume these products if there was a better way to do it—if in fact it could be done in a way that did good and not harm to the world. I wanted to build relationships to the land and to the foods that came from it.

Flash-forward now about 15 years spent exploring this idea, among other interests I had in stewarding land in more responsible ways—a journey that led me through organic farming, permaculture, and then agroforestry. Each discipline has its own positives and negatives, full of individuals who see the nuance in each approach, as well as those who evangelize each as "perfect." What I have come to see is that it's important to be wary of any solution that claims to work for all land and all people. Rather, we need to think about being flexible, adaptive, and very attentive to where we are doing the work.

This book describes the aspects of silvopasture that we know, while also trying to shine a light on the areas that need more research and development. It is by no means a perfect system. It is technical in nature, attempting to bring together the science and art of grazing and forestry, two practices that have been intentionally kept apart. And while it does appear to have some of the best numbers when it comes to forms of farming that sequester carbon and fight climate change, this actually depends on a range of variables that the farmer must continually mull over in his or her head.

What excites me about silvopasture is that it offers an ecosystems approach to farming, where many goals can be achieved side by side, in scenarios that are win-win-win for each of the components. We can raise animals in an environment that encourages them to explore for their food and live a life that is in line with their evolution. We can reforest the land and increase diverse habitats for wildlife. We can build soil health and see the benefits of manure on the landscape. We can increase the quantity and quality of foods on the land for our animals, which results in lower costs and inputs from outside systems. And with good planning, we can see profit from the farm.

It is an extreme privilege to be able to embark on this journey. When I say privilege, I mean to recognize that I have many things in my life that everyone does not have equal access to: a safe home to grow up in, a good education, available land and natural places, and capital to buy land and start a farm. The movement toward ecological farming is leaving many people behind, and it has become a movement of privilege. It doesn't have to be this way, but it does require that we look hard at the big picture, and how our history as a nation set these circumstances in motion.

Agriculture and the history of the United States are intimately connected. We cannot be good farmers without understanding and reckoning with this narrative. Anyone who is farming is doing so on land that was, at some point, stolen from indigenous people,

most likely in terrible and brutal ways. Land was concentrated in the hands of a few, with the early economy of this nation built on selling this stolen property, then working the land with bodies that were stolen and forced into slavery. Trees, forests, soil, and water have also been exploited, valued solely for their economic benefits to those with access and power.

The forest, which was held as a sacred home to many native people, has been devalued for too long. When land was being claimed, colonist settlers were required to "improve" it to stake their claim, where improving land meant clearing all its trees. Old survey maps often labeled forest as "wastelands" and deemed them fit for use however an owner chose. These patterns continue today, where forests are only seen as repositories for timber, and very few foresters and forest owners balance economic value with considerations for wildlife, future forest regeneration, and maintenance of healthy watersheds.

As this nation has shifted from millions of small family farms to larger, more industrial operations, immigrants and people of color largely do the work to produce the food we all eat, while the vast majority of owners and managers on farms are white. Underserved groups with people who want to farm include women, African Americans, Hispanics, and Native Americans (as identified by USDA) as well as many other people of color. The USDA has itself admitted and documented its own history of discrimination to these individuals.

The solution to these myriad problems is not to continue to separate and isolate, but to look at them together. This is known as intersectionality: the study of how parts are related, in which the overlapping issues actually help us see the problem better. The history of the ways in which people have been exploited (and continue to be exploited) as the basis of our food system is similar to the ways we have exploited and abused forests, animals, and the soil. In every case the subject is treated as an object, and thereby devalued so that it can be overtaken. Once you give value, acknowledge that people and living things have rights, and build relationships, you can no longer colonize them.

How does this all tie in to the topic of this book? As we zoom into the details of building a silvopasture, we must also keep the bigger picture in mind. Even if we get it right on our farms, we won't "solve" the larger problems in the system if we can't see how all the parts play a role. Let's name three big problems in the larger food system:

1. Modern agriculture systems degrade land, animal, and people resources, cause pollution, and are a major contributor to climate change.
2. Access to land, farm ownership, and healthy foods is not equitable across race and class. Those historically exploited are left behind in the good food movement.
3. American farmers are getting older and don't have solid plans for transition. And as we lose them, we lose valuable knowledge and skills to continue farming.

Most often each of these issues is looked at and dealt with in isolation, yet not one of them can be solved without also trying to solve the others. It is at the intersection of these issues that true solutions emerge. For instance, some of those people most interested in farming in the future are farmworkers and immigrant farmers, who could be taught about agroforestry practices and then linked to farmers looking to transition out of farming for land access. Another one: Multilayered agroforestry systems could employ people to manage the various "layers"—one person does the animals, another the trees, and a third a byproduct of these activities, such as processing animal hides for sale or tree prunings into biochar.

In seeing these problems, and exploring where they intersect, we can develop meaningful solutions to actually change systems on a large scale. This type of thinking requires that we think not only about our own farms, but about our communities as well. It asks us to challenge our preconceptions and ideas about the value of trees, the intelligence and sentience of animals, and the ways we are leaving others behind as we develop innovative farming methods. But above all this, it leads us to a place closer to where our values lie, and offers a life rich with meaning.

So as we dive into the details of silvopasture, keep these larger questions in mind. Do not expect that any of the answers will be straightforward. Grazing animals have taught me this better than almost anything I've done before in my life.

Good grazing requires constant attention to the present moment. You have to see the grasses, monitor the animals, and make decisions for the next day, the next week, and the next month. The conditions are always changing, and never what you think they will be when you make a plan for how the day will go in the morning.

Silvopasture, as you will see, has great promise as a leading ecological farming technique. It offers a great opportunity to retain existing forests, reforest the land, and raise healthy and happy animals, all on the same piece of land. It also has space to be widely adopted, and provide livelihoods for many people in the process. Those who are up for a journey that is more process and less destination—who want to explore the intersection of trees, animals, and forage—are in for a treat. Thanks for joining me on the journey!

1 What Is Silvopasture?

The foundational concepts of silvopasture challenge our notions of modern agriculture and land use as we know it. For centuries European colonizers of North America have engaged in practices that separate the field from the forest, and even the food from the animal. In silvopasture, trees, animals, and forages for those animals are integrated as a whole system that is greater than just the sum of these parts. The word is a combination of the Latin root word *silvo-* (as in *silviculture* or forestry) and *pasture*, which implies grazing. Such a system offers not only the promise of ecological regeneration of the land but also an economic livelihood, and even the ability to farm extensively while adapting to a changing climate. And as we will learn, planted silvopastures rank among the most effective approaches to sequestering carbon while farming the land and soil.

Silvopasture is not, however, as simple as allowing animals into the woodlot, or planting trees into the pasture. It is, and must be, intentional, steeped in careful observation skills, and flexible to the dynamics of such a complex ecology. It requires a farmer who is proficient in understanding grassland ecology, forestry, and animal husbandry at once. She or he does not need to be an expert in all of these disciplines, but rather familiar enough to make decisions on a wide variety of timescales. A silvopasture system will inevitably look different from year to year, and careful design, creativity, and visioning for the future are all part of the equation.

If we travel away from North America, silvopasture is sometimes just called "farming," whether it's because in dryland climates animals demand shelter from the hot weather to survive, or because of cultural custom. Though this type of mixed farming is common in Europe, South America, and many other regions worldwide, it never arrived with the colonization of the temperate eastern and midwestern parts of the United States—the regions where the conversation in this book is focused. This lack of transfer was likely because the forest ecotypes found by early settlers were so dense, diverse, and vast. During their imperialism, Europeans spent most of their time clearing trees, opening land to the plow. A well-documented fear of the woods[1] meant that harsh lines were drawn between field and forest in the minds of early colonists, as they came to be a dominant force on the American landscape through exploitation of both the land and its longtime inhabitants.

Prior to colonization, native peoples had cultivated a wild ecology that included mixed woodlands, forests, and grasslands. They traditionally hunted wild game for food while cultivating a mosaic of gardens and farms for staple crops. Their main tool—fire—created a mixed woodland in many places, where trees were widely spaced and the concepts of "field" and "forest" were blended on a continuum, much more mixed in their composition. Some crops, such as the three sisters of corn, beans, and squash, were widely cultivated on an annual basis,[2] while others such as black walnut,

Figure 1.1. A honey locust silvopasture at the Virginia Tech research farm. This example shows the potential for animals to forage both grasses and forbs, as well as tree fodders from previously cut trees. Photo by Gabriel Pent.

apple, and peach orchards were a multigenerational community effort.[3]

Today we are left with a legacy of not only the choices of early American settlers to extensively clear land, but also the footprint of modern industrial agriculture, which has largely stripped the soil of nutrients and degraded its structure. Trees are all but gone from the pasture, limited to the occasional hedgerow that a farmer happened to keep. Farm woodlots are an afterthought for productive use, only occasionally visited for timber or firewood harvest. Animals are confined and fed predetermined rations of food imported from places far away. Yet despite this being the dominant farm paradigm, we see the slow emergence of a new type of agriculture, one that re-blends the best that field and forest have to offer. This practice is known as

agroforestry, and silvopasture offers one of the most promising agroforestry practices for this time in history.

While silvopasture as a practice is relatively small in the temperate United States, interest and momentum are growing. Examples of specific systems are what really give us a sense of the possibilities. Just a short list of the varied systems includes:

- A honey locust plantation for shade, pod production, and leaf fodder combined with sheep grazing in Virginia (see figure 1.1).
- Oxen and pigs used to clear forested land in New Hampshire to create space for new market gardens and orchards.
- Turkey used for controlling pests and fertilization on an apple cider and asparagus farm in New York.

Figure 1.2. Sheep grazing under four-year-old black locust trees, one of the most useful and versatile of all trees for silvopasture.

- Sheep who graze the understory of hybrid chestnut and hickory plantings to make for an easier harvest for a nut nursery in Minnesota.
- Cattle maintaining the understory and providing short-term yields (meat) for southern pine plantations in Alabama.

Each of these examples is quite unique and different from the others, yet they all share common goals, components, and philosophies. The systems may take several years to establish, but many farmers see the benefits of this type of production in the longer-term view. Some of these benefits include better support of animal health, more yields off the same acreage, reduced inputs to deal with pests and keep fields mowed, and healthier soil.

There is another big benefit that is not often the first reason for farming in silvopasture but one that will continue to prove critical: climate change mitigation. As described later in this chapter, research shows that mixed systems such as silvopasture sequester significant amounts of carbon from the atmosphere better than forests or grasslands alone—which represents a substantial part of the potential solution to global warming. Equally compelling is the positive aspect of silvopasture to buffer against the unpredictable nature of change: increased rainfall, longer droughts, and more intense storm events that are an inevitable part of our future.

Attitude Determines Success

All this sounds very positive, but how do we get there? It's important to note up front that while we have a lot of knowledge about silvopasture's parts (pasture management, forestry, animal science, and so on), the combining of systems in the temperate United States is still a bit of a grand experiment. As such, it's important to identify potential hazards and pitfalls, and proceed with caution as we design and implement.

One of the largest challenges is to be thinking in the long term—decades and generations—rather than in just months or years. When a vegetable farmer tries out a new technique, he or she gets feedback if it's working (or not), usually that same year or maybe the next. Damage to trees can take upward of a decade to show up, however, and by the time we notice the symptoms, it's often too late to do anything about it.

The aim of this book is to articulate the components and design of silvopasture systems and highlight best management practices so that those interested in working silvopasture into their land-use strategy do so in the best possible way. A combination of research into the practice, and the experiences of farmers on the ground, provides readers with a solid list of dos and don'ts as they translate the information to their own context. By no means is this list complete.

Each silvopasture plot, even within the same farm, is unique. It expresses itself differently depending on the year, season, weather patterns, animal behavior, and choices the farmer makes. Each element has its role to play: The plants translate sunlight into plant mass, then the animals harvest the green parts and produce foods; the humans harvest wood, food, and materials; and the fungi, bacteria, and other microbes balance the soil.

In this great dance, we humans ultimately play the role of determining the fate of each player, a role none of us should take lightly or for granted. The more we as farmers act in response to the clues each element of the farm gives us, the less work we have to do, and the more ecological and sustainable is the system.

Rather than thinking of ourselves as "the deciders" or as having dominion, we are better off seeing our role as that of providers and orchestrators: We provide the needs of each living element in the system, so they may thrive. In doing this we orchestrate the type, frequency, and duration of the interactions among organisms. If we wait too long to move our animals, the trees or forage might suffer. If we don't keep our animals on fresh ground, they might succumb to malnourishment or disease. If we time everything right, and sometimes with some luck, all parties benefit.

To some, conducting this symphony of nature is also known as "work." Many who already raise animals are resistant to rotational management because it's "too much work." And in some ways it is more work—especially in the establishment phase of the process. Yet this book will show how the notion of work is relative, and that in the end it isn't about more or less work, but about what *kind* of work the farmer does.

Silvopasture is not for everyone. It is for someone who reads descriptions of a practice that is part science and part art, and gets excited by it. It is for those farmers who love being part-time ecologists, naturalists, mechanics, and engineers, all wrapped into one. It is for those who are eager to ask questions and not necessarily find out the answer right away. It is for those who find farming to be a lifestyle, and not only a job, though it does need to pay. It is for those who want to grow and change in their thinking and perspective as they form a more intimate understanding of the great wide world we are fortunate enough to inhabit.

Ultimately, those who possess a desire to farm in a way that balances practicality with creativity, determination with flexibility, and planning with adaptation will succeed not only in farming, but in life. The willingness and eagerness to rise in the morning and seek to make the system more efficient, and to better support ecological good, as well as the positive health and well-being of all creatures involved (human and not), is essential. This is paramount not only to a good life, but to the overall survival of our species in a world that is quickly changing.

Defining Silvopasture and Agroforestry

Silva in Latin means "forest" or "woods." *Pasture* comes from the Latin *pastura*, meaning "feeding" or "grazing." For the purposes of this book, the practice of silvopasture is defined as:

The intentional combination of trees, domesticated animals, and forages as a multilayered system where each benefits from its relationship to the others, with multiple yields harvested from the same piece of land.

In other words, the presence of these three groups of living organisms, along with the deliberate and intentional design on the part of the farmer, is what defines an active silvopasture system. This leaves the possibilities wide open for its application—there is not a limit to the practice based on tree spacing, on canopy density, on whether the animals are grazing planted grasses or foraging for insects and roots, or on other factors.

Silvopasture is considered one of six temperate agroforestry practices, five of which have been recognized by the USDA in an effort to help landowners and farmers understand the possible options. Many consider forest gardening to be an important sixth practice, especially for farmers in urban and peri-urban areas. Each practice relates to another, and you could often argue over exactly which practice is being utilized on a particular site. Often multiple practices are employed simultaneously. The named practices are:

Silvopasture. The intentional mixing of trees, animals, and forage.

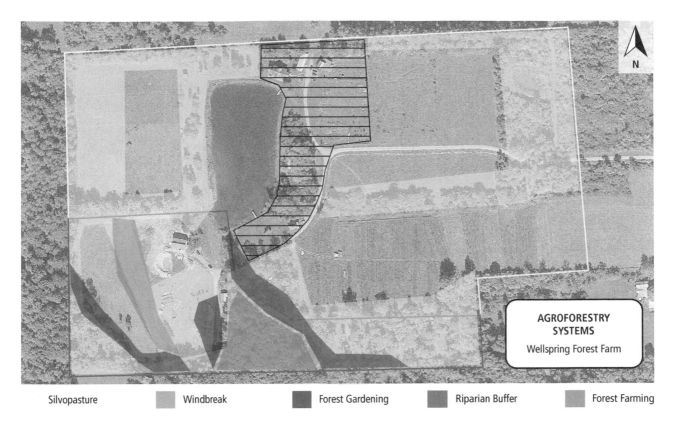

Silvopasture ▮ Windbreak ▮ Forest Gardening ▮ Riparian Buffer ▮ Forest Farming

Figure 1.3. Our farm uses multiple agroforestry practices, many of which overlap in our production systems. Where there are gaps are areas we are actively planning to bring in more trees.

Forest farming. Cultivating non-timber crops under the canopy of an existing forest (for more, see my previous book with Ken Mudge, *Farming the Woods*).

Alley cropping. Growing traditional field or row crops (hay, grain, vegetables) in between rows of trees.

Windbreaks. Growing a multistoried hedge of trees to mitigate the effects of wind.

Riparian buffers. Tree and woody shrub plantings along a water's edge.

Forest gardening. Growing a diverse set of edible and medicinal plants in a way that mimics forest ecology and evolves into a forest.

Each of these practices has its nuances, and it's easy to see how someone could plant trees in their pasture both to benefit livestock and to act as a windbreak, for example. The designation of these practices is merely to encourage thinking around the main functions of each approach. Readers should take note that the lines are thin between each of these, and that there are hundreds of other named agroforestry practices that have been articulated worldwide.

OUR FARM— AND FURTHER BLURRING OF LINES

At Wellspring Forest Farm & School, my wife and I produce mushrooms, maple syrup, pastured lamb, duck eggs, and elderberry extract, all in agroforestry-based systems. The mushrooms grow underneath the canopy of our sugarbush, which provides maple sap and syrup each season (aka forest farming), while our elderberry plantings are part of the design for riparian buffers on two waterways that flow through the 35-acre property (10 owned and 25 leased). We have planted a windbreak of alder, locust, and willow to help buffer the strong winter winds that visit our land.

Interestingly enough, the animals have found their way into almost every inch of the landscape, and thus the practice of silvopasture is both throughout the landscape and intertwined with the other agroforestry systems on the farm. In this way, while silvopasture is a specific practice, it often works well as part of other practices, depending on the timing and season. For instance, we rotate our ducks through the maple grove three to four times a season to reduce slug pressure on the mushrooms. Our sheep rotate through riparian buffer zones during dry times, which minimizes harm to the more sensitive ecosystem while providing good forage during the hotter summer months. So much of what we have found at the farm is that paying attention and being flexible are key. Nature often offers small windows of opportunity for the right conditions to play out, and we need to be ready as farmers to take advantage of all that is occurring at a given point in time.

The Main Components of Silvopasture

Since this is a book primarily about the practice of silvopasture, it is useful at the outset to offer an overview of the key principles and approaches that are part of the system, regardless of the specific species or site context at play. Whether you choose to graze sheep in a Christmas tree farm, move cows through a walnut plantation, or graze chickens through an apple orchard, these elements are universal for successful silvopasture.

1. SILVOPASTURE CAN BE ESTABLISHED IN EXISTING WOODLANDS, OR TREES CAN BE BROUGHT INTO PASTURE.

One of the nice aspects of silvopasture is that you can establish a system on almost any type of land. Of course, establishing it in existing forest is in many ways a very different process than bringing the trees into open pasture. The similarities and differences of establishment in such different contexts will be discussed in detail in chapters 4 and 5.

The only land types on which we might consider avoiding silvopasture are very sensitive areas such as wetlands and healthy, maturing hardwood forests that might be best left to their own process of succession. Given that silvopasture has an experimental aspect to it, working first on more marginal lands is the best way to begin, offering the opportunity to learn with lower stakes. In ecological approaches to farming, it's good to practice precaution, easing into the development of novel systems, all while carefully monitoring for any adverse impacts.

2. ANIMALS ARE MATCHED TO LAND TYPE AND SUCCESSIONAL STAGE.

It's critically important from the outset that the appropriate animal is chosen for a given site in order to reduce the potential of inflicting damage on the landscape. Animals are incredible at what they do, but it cannot be overstated that they have just as much potential to do harm as they do good. While we will cover considerations for animals in depth in chapter 3, here is a short list of potential risks for given species:

Cows. Excessive stocking/duration with their weight could damage soil and tree roots, as well as cause erosion; also, they can easily destroy young trees.

Pigs could root and trample desired vegetation and make a moonscape of your woods or pasture in a very short period of time. Pigs are the most challenging animals to incorporate into silvopasture.

Sheep and goats. Depending on forage type, sheep and goats could overgraze the landscape and/or strip the bark off young trees, killing them.

Poultry could scratch or root down to bare soil and damage roots and plantings.

You can see from the above list that most of the problems can be avoided by doing proper assessment of the land and engaging with the animals to ensure they are moved before they do harm. The key elements of stocking rate, density, and duration in a paddock come into play here; they must be well designed, and redesigned each year, to optimize the system.

In addition to choosing the right type of animal for the system, careful selection of the specific breed is an essential task. Some breeds are able to utilize a wider

Figure 1.4. Silvopasture can be designed from either existing forest or pasture. On the left, cattle graze in an existing woodlot thinned for silvopasture. On the right, cattle graze among established orchard trees. Photos courtesy of Eric Toensmeier (*left*) and USDA-SARE (*right*).

Figure 1.5. Many sheep and goats will eagerly strip the bark of young trees such as this black locust, which can set the trees back or even kill them if the damage is too severe. Still, in many cases such as this, the trees recover even after severe damage so long as the entire bark layer isn't severed.

range of forage and conditions, whereas others are not as willing to be as flexible. Chapter 3 discusses the role animals play in silvopasture systems in much greater detail.

3. ANIMALS ARE ALWAYS ON A ROTATION.

Grasses evolved alongside grazing herbivores, and while it might be surprising, they arguably benefit from being grazed so long as they have a rest period. For wild grazing animals, such rest is achieved when grazing animals need to move on to new places because of the threats predators pose or from seasonal changes to weather and climate. In the context of modern grazing systems, designated paddocks, the farmer, and electric fencing act as the "predator."

When the animal consumes the top of a plant, a proportion of the roots are sloughed off or deposited into the soil, which contributes organic matter content. After plants are grazed, a rest period is critical to their recovery, where the shoots grow back. Overgrazing means these plants take a longer time to recover, while severe overgrazing means the plant might die altogether. Moving animals is also good for them, as they have reduced exposure to disease risks and receive the highest-quality food possible during the season.

The rotational process also benefits the farmer's bottom line, as it's been shown to improve the quality and quantity of forage on the pasture.[4] More intensively

Figure 1.6. Cows on a rotational grazing system on a farm in Virginia. Fencing, whether it's permanent (like this) or temporary, is used to keep livestock off forage until the timing is right. Perhaps this is where "the grass is always greener" comes from? Photo by Jeff Vanuga and courtesy of NRCS/USDA.

managing pasture also allows you to feed more animals on the same amount of land. The clear promise of managed grazing comes in your ability to have more control over food for your animals, on-site, with the potential to increase the value of their feed and thus the number of animals and/or the amount of land and duration of the season that land can be grazed.

How long the organic matter remains in the soil, known as *sequestration*, is another matter altogether. Many rotational grazing proponents laud any practice of rotational grazing as an important way to address climate change. But as always, the details matter. The soil type, climate, and bioregion, along with the variables of management, make it hard to be conclusive, and it's an open debate among scientists.[5] Some studies show net positive effects, while others show that all animal grazing systems are emitters of greenhouse gases, no matter the style or approach.[6]

Regardless of the climate impacts, rotational grazing is essential to a healthy pasture system. This aspect of silvopasture is *non-negotiable*, and is often the biggest hurdle for

adopting the practice, especially by graziers who have been practicing continuous grazing for some time. Fortunately, advances in our knowledge and technology have made rotational grazing easier than ever before. Farmers new to livestock are almost always convinced from the start that rotating their animals is a good thing, though there are many details as to what a rotation can look like, including the size of paddocks, the duration of stay animals have in a given paddock, and so on. These considerations are discussed in chapters 3 and 6 in great detail.

Regardless of the specifics, it is the universal opinion of silvopasture advocates that animals should not be placed in tree-based systems if they will not be managed through rotational grazing.

4. TREES SHOULD MATCH THE SOIL TYPE AND MICROCLIMATE AND HAVE MULTIPLE FUNCTIONS.

You could arguably plant trees for the sole purpose of shading the livestock, but why not aim a bit higher? There are so many choices in the temperate climate for

FORAGE VOCABULARY

Here are a few key phrases we will use throughout this book when referring to grazing practices:

Forage/fodder. Both of these terms are generally all-encompassing for any type of plant an animal is consuming in the pasture, though *fodder* usually refers to tree leaves.

Grasses are monocots and belong to the family Poaceae (formerly Gramineae). Leaves of these herbaceous plants generally appear as blades, with parallel veins. Within grasses there are two distinct types:

Cool-season grasses (C4) thrive during the cooler times of year (spring and late summer/fall), when temps are between 40 and 75 degrees F (4–24 degrees C).

Warm-season grasses (C3) grow best during the heat of the summer, when temps are between 70 and 95 degrees F (21–35 degrees C).

Forbs. The general term *forb* refers to any herbaceous, broadleaf plant (non-grasses) without regard to family classification, and would include legumes, flowers, and other forage plants.

Legumes. Pod-forming plants in the family Fabaceae (or Leguminosae) valued for their ability to fix nitrogen from the atmosphere.

Browse. Plants other than grasses and forbs; these are usually taller and often woody plants, such as trees, shrubs, and vines. This word overlaps with some definitions of *fodder*.

offer both a potential food value to animals and potential wood and timber products down the road.

The goals of the farmer or landowner also come into play, as there is no use planting apple trees, for instance, if you don't want to harvest apples. Some farmers want to establish the lowest-maintenance trees possible. Some want a yield of fruit in 3 to 5 years or of nuts in 5 to 10. And some are happy to plant timber species and wait 50 or more years to harvest. We will discuss options for tree species and weigh the pros and cons of each in chapter 5.

5. FORAGE AND FODDER SHOULD BE DIVERSE AND SUPPORT A RESILIENT FOOD SUPPLY FOR ANIMALS.

One of the largest opportunities in silvopasture is the creation of a wide range of ecotypes, which can support a wider range of grasses, forbs, herbaceous plants, and trees for animal feed. This gives animals a more diverse and healthy diet that is not only nutritious but also medicinal. In essence, the design of a diverse silvopasture offers animals a habitat that might resemble or even exceed their original experience grazing in the wild.

Modern farming has greatly oversimplified the animals' experience of seeking food; in some operations animals only visit the feed bin for grain or hay. This not only offers animals a limited diet in terms of nutrition, but also starves their innate desire to seek out food in the landscape. Behavior specialists argue that this creates boredom in animals, which can lead to disease and to a lower quality in the final product. Ethics and markets also come into play, as animals have an innate right to live a good-quality life, and more and more consumers are lining up with their dollars to support farming practices that promote animal welfare. More on this in chapter 3.

In addition to supporting the overall health and well-being of the animals, a focus on diverse forage also provides an economic incentive for the farmer. More diversity in feeds should reduce the feed bill, and also provide food in lean times, because tree-based systems can often buffer better against long-term drought and even excessive rain. Grasses grow on a bell curve, often peaking in early summer with lower production in July

trees that will do well in even the worst of soils, that provide not only shade but also a number of other possible yields. Of course, the yields will depend on how the trees are managed and are often easier to "control" when you're establishing a silvopasture in open field conditions versus an existing forest.

While you could choose any number of trees to plant into a silvopasture, there is a shorter list of trees that offer specific benefits in the context of silvopasture systems (see chapter 5). These trees are generally fast growing, hardy, and resilient to weather and climate. They often

Figure 1.7. This sheep paddock at Wellspring Forest Farm includes access to willow and black locust fodder, as well as a clover- and orchard-grass-rich pasture. This allows each animal to explore and meet its specific individual dietary needs, which are diverse, like those of people.

and August—unless the forages are shaded and can thus remain better quality for a longer period of time. The invitation to include trees into grazing systems is ultimately one to create more dynamic ecosystems for our animals to explore.

Careful matching of forages to the micro-environments on the farm is the challenge. For example, for most silvopasture in the eastern United States, cool-season grasses are utilized, as they excel in part-sun environments. Warm-season grasses are best for overly sunny or dry areas, or warmer climates. The trees effectively help retain optimal conditions for cool-season grasses throughout the summer months. This, coupled with the careful selection of trees that leaf out at various times and provide a range of shade conditions, can optimize production. For instance, black locust is a great silvopasture species, as it leafs out late in the spring, and when fully leafed out casts

a mild shade, allowing the space underneath to be cool and somewhat shady but not to the point where grasses would be stressed for light. More on forage and fodder considerations in chapter 5.

6. THE SYSTEM IS IDEALLY OPTIMIZED TO STACK INPUTS AND OUTPUTS IN BOTH SPACE AND TIME.

The beauty in silvopasture systems is not in the parts but in the complex whole created by bringing all the parts together. Yet with complexity comes a challenge in management—this is indeed why agriculture in the United States and other industrialized nations has been on a general trend toward more straight rows, single-species monoculture, and rationed feeds. It's easier to do the math. But as we will discuss, the benefits of creating a more complex ecology outweigh the time it takes to design, establish, and manage such a system. Each chapter in this book walks readers through the process, and helps make more sense of the complexities.

Be patient with yourself. Few of us are raised in cultures where we understand a more natural way of farming. Many are interested in the concept of a more complex ecology, yet find themselves overwhelmed and frustrated as they try to comprehend things. It's wise, then, to start small and slow, especially if you are new to one or more of the two main aspects of this practice: grazing and forestry. Draw upon the knowledge of others, and recognize that you're in for a lifetime of learning. Get the foundations of grazing right from the start, then bring in the forestry aspects. The content of this book, along with the case studies of farms actively practicing silvopasture, will help paint a picture of how this can be done.

The Benefits of Silvopasture

Usually, farmers and landowners get into silvopasture to take advantage of one or more of its benefits. In 2014 forestry professor and agroforestry advocate Joe Orefice interviewed 20 farmers in the Northeast who were intentionally practicing silvopasture on their land.[7] His research found that the top four reasons these farmers were using this system were:

1. Shade for livestock.
2. Expanding pasture acreage and diversity.
3. Increased utilization of existing farm woodland.
4. Increased forage availability during midsummer and droughts.

Other less important reasons mentioned by respondents included diversifying livestock diet, overall animal welfare, the management of undesired vegetation, tree health/fertilization, and increased farm aesthetics.

Each landowner will have to determine for him- or herself what the main motivations are for engaging in silvopasture. While we can promote myriad benefits to farms and the land, in the end the ability to manage a system comes down to economics: Can we pay the bills or not? So think about arranging benefits in two categories: those that support the farm, and those that support the larger world community. Here, we will flesh out these benefits in order to get the big picture of silvopasture as one of the more remarkable agricultural systems available to us.

BENEFITS TO THE FARMER AND LAND

Most important to getting more silvopasture practiced on the ground are the opportunities it provides to farmers and land managers. The following can be seen as the foundation of an argument for silvopasture, while many of the secondary benefits are positive effects on the regional and global scale, described in the next section.

1. Increased use of farmland for production.

Whether or not it produces, land costs money to own. There are many acres of land that are in farms, but not used. These are often edges, hedgerows, and forestland. In New York State, for example, 21 percent of farmland is forest, much of which is only periodically visited and used for activities such as hunting, firewood harvesting, and a periodic timber harvest. Much of this forested land is not pristine woods but a mix of young trees, thorny shrubs, and thickets that don't allow easy passage.

Silvopasture presents an opportunity to use these more marginal stands of forest. Animals can benefit from the protective aspects the forest has to offer, while the farmer benefits from better utilization of his or her land, as well as a diverse ecology of habitats to work from in a dynamic and ever-changing climate.

On our small farm, we were amazed at just how much land we were missing out on using on our own property and the acreage we lease. Using Google Earth, we quickly realized that by getting into the acres that were "scrubby" and marginal we could increase grazable land that we both own and lease from 22 to 30 acres, a significant increase.

2. Increased carrying capacity and stocking rate.

While one aspect of being able to raise more animals is the ability to utilize more acres of land, as addressed above, a second is the ability to increase both the quantity and

Figure 1.8. When we started grazing our owned and leased land, we assumed that the only usable space was the open pasture. As we learned about silvopasture, we realized there were about 8 acres in marginal woods, hedges, and edges (in purple) that we could incorporate into our pasture, increasing our grazing land significantly.

the quality of forage on a given acre, thereby increasing the number of animals that the land can handle.

Carrying capacity refers to the number of animals a land base can support in total, while *stocking* has to do with the amount of animals on a given area of land for a given duration of time. Both concepts are based in the foundations of rotational grazing, where a rest period allows forages to regenerate. By excluding animals and allowing this to happen, more food is made available, and *if the food is harvested at the proper time*, it can mean that more animals can be sustained from less land.

Through the combination of good rotational grazing of grasses, forbs, and legumes, along with woody plants, you can dramatically increase the available food on your land for your animals. This is one of the more empowering aspects of silvopasture; the idea that you can always improve the food on your land each season means you have a considerable amount of

control over the ultimate yield and therefore profit, and also the ability to reduce the need for outside inputs. This has important implications for economics, but also for autonomy.

This possibility is specific to the use of ruminant grazers (sheep, goats, cows) and can apply to any rotational grazing system, trees or not. We discuss this further in chapters 5 and 6.

*3. Increased animal comfort
equals improved performance.*

One of the most critical benefits cited for silvopasture is shade. The basic logic is that shade reduces heat stress in animals, an effect that anyone who has driven by a pasture on a hot day will notice. Often, though, it's a scene where the livestock are crammed under the one, lone tree in the pasture, eager to enjoy the meager shade it offers. Those trees, unfortunately, often take such a

Figure 1.9. On a hot day we inevitably find sheep enjoying the shade of the trees, especially as they ruminate and digest the pasture they recently grazed.

beating that they die off in a few years, and with them goes the shade.

As readers of this book will come to understand, what at first seems like a simple cause-effect relationship is in fact much more complex. So when you're looking into research, the benefits of shade are hard to pin down. Some studies show clear results, while others debate this, noting the context-specific nature of the claim.[8] The benefits vary depending on the location, seasonal variation, and type of animal. Studies often use artificial shade, which doesn't really capture the environment trees provide, and only examine the effect of blocking or dissipating sunlight.

Heat stress in animals results not only from direct sunlight hitting animals in pasture, but also from radiated heat from the ground and, most significantly, from the animals' own metabolism.[9] So the real heart of the matter is the type of environment trees offer to animals, in dynamic weather and climate conditions. In this frame, what is likely more accurate to think about is that trees provide shade and shelter, all while moderating the microclimate for animals. Especially in colder climates, what may be more critical to animal performance are the effects of wind and cold stress rather than heat, which may only become overwhelming in uncharacteristically hot summers, at least in the more northern reaches of the cool temperate climate.

In the end trees help moderate the climate from extremes, and with a changing climate, weather is becoming more unpredictable—and more extreme. While it is hard to pin down specifically what benefits come to pass, it's important to give weight to them all, and to consider the benefits of shelter from a variety of elements, in addition to shade.

Trees in silvopasture also offer a mechanism to better distribute animal impact in a paddock. A better

Figure 1.10. Certain breeds are better than others at utilizing and working on undesirable brush. It's remarkably satisfying to let animals loose on a vegetation problem and see them enjoy it while solving it, such as with this honeysuckle.

distribution of trees means that rather than animals clustering in one spot during the hottest part of the day, they will continue to comfortably forage, as well as rest, in various parts of the pasture. Thus their impact can be spread more evenly, and they will potentially forage for longer periods of time, better utilizing the forages available.

4. Improved animal health through diverse diets.

Most meat and dairy products we consume are from animals fed corn, soy, and other grains and residues from the cropping production systems that represent the overwhelming majority of modern farming. The animals we raise on these feeds were not designed to digest them, especially in the case of ruminants (cows, sheep, goats). Further, the type of environment a concentrated animal feeding operation (CAFO) provides is nothing short of disgusting. Animals are unable to exercise their muscles, spend hours on

hard-packed, dusty ground, and stand in their own manure and urine. Disease and stress are rampant. And animals are fed a formulated ration of feed, with little variety from one day to the next. Research efforts to support CAFOs are usually framed around reducing the myriad issues such environments create.[10] At best this scenario can be made "less bad," but never will it be good.

When animals eat from pasture, they explore their environment, seeking out their own balanced diet, which varies from day to day, and throughout the season. They exhibit natural patterns of feeding and rest, and need a lot less intervention in the form of antibiotics or other treatments. Researchers have found that animals can even engage in "self-medication" and develop a customized diet to meet their needs in a diverse landscape.[11] Much more on the importance of this in chapter 3.

5. Cost-effective vegetation control.

Related to benefit 1 above, often land is not utilized fully because of the presence of thick, undesired vegetation. The cost of removing vegetation either mechanically or with chemical application is often prohibitive, or at least not the top priority on the farmer's to-do list. Yet if vegetation can be removed efficiently, it opens up new places and possibilities for the farmer to gain access and income from additional land.

There is also a substantial amount of land that has only recently been abandoned by farming, meaning in just the past 10 to 20 years. These old crop and hay and grazing fields are in the process of transitioning back into forest, though often this is hampered by the range of brushy and thorny species that choke out trees, slow the process, and make accessing the land a challenge, to say the least. The more this interfering vegetation grows, the harder and thus more costly it is to clear it for grazing. Knowing this, many landowners default to at least haying their lands or mowing them occasionally to prevent this from occurring. Once vegetation becomes thick and woody, the difficulty of maintenance greatly increases.

Undesired vegetation can be managed by mechanical, chemical, and biological means, as we will discuss further in chapter 4. Each of these carries a cost, potential negative impacts, and differing time factors. Often, some combination will occur, but animals already being raised as livestock on a farm are arguably the most cost-effective way to transition a landscape over time. This of course depends on the situation, but animal behavior specialists have been able to train livestock to eat weeds with remarkable success.[12] One study at Cornell, focused less on a silvopasture system and more on an in-and-out treatment of undesired vegetation with goats, found that control with goats was competitive with the use of herbicides in terms of cost, especially on smaller-acreage projects.[13]

6. More yields from the same acreage.

While the shade and shelter benefit of trees discussed previously may alone be valuable enough to justify silvopasture, we cannot ignore the potential to go beyond the livestock product to seek out other yields from such a system. Trees offer a number of possible yields we will explore in this book, including feed for animals (known as fodder), firewood, fence posts, hop poles, mushroom logs, polewood, timber, and more. Additionally, edible yields such as tree fruits and nuts are possible, provided we are aware of the regulations and best practices around food safety (see the sidebar "Grazing with Food" in chapter 5 on page 212).

Local demand and markets will be a big driver in the tree choices a farmer makes, as well as the approach to harvesting so that the canopy of the silvopasture remains intact while you harvest a sufficient wood resource. In other words, in most cases it won't work to clear-cut a silvopasture; it's better to selectively harvest stems over time.

A big advantage of silvopasturing is that the animals and their products can provide the farmer with short- to midterm financial gain, which helps buffer the wait time for trees to mature to some sort of marketable product. The 10-plus years needed for trees to get established and grow is often a large barrier to farmers adopting them, but the shade and other benefits can be realized beginning just 3 to 5 years from planting, depending on the species chosen. More on tree selection in chapter 5.

Figure 1.11. Many secondary products increase both the productivity and the yields from the same acreage in silvopasture. Here a roughly 20-year-old black locust stand has been thinned for high-value posts, desired for their rot-resistant qualities. Photo by Brett Chedzoy.

7. Climate change resilience.

The focus on resilience here is different from mitigation, which is discussed on page 17. Resilience has to do with the ability of a system to withstand and recover from changes in the environment. With climate change, animal grazing systems are particularly vulnerable in the temperate climate to increased heavy rainfall, as well as excessive heat and drought. The inclusion of trees and forest-based systems can help buffer impacts from these patterns, as well as shelter animals from intense storms.

Resilience impacts farmers on a day-to-day and seasonal basis, and may be one of the more immediate benefits of implementing silvopasture systems. One of the most striking impacts is in regard to drought conditions, where trees can help maintain soil moisture content as well as protect forages and provide alternative feed, because shrubs and trees tend to do better when the grasses have long since dried up.[14]

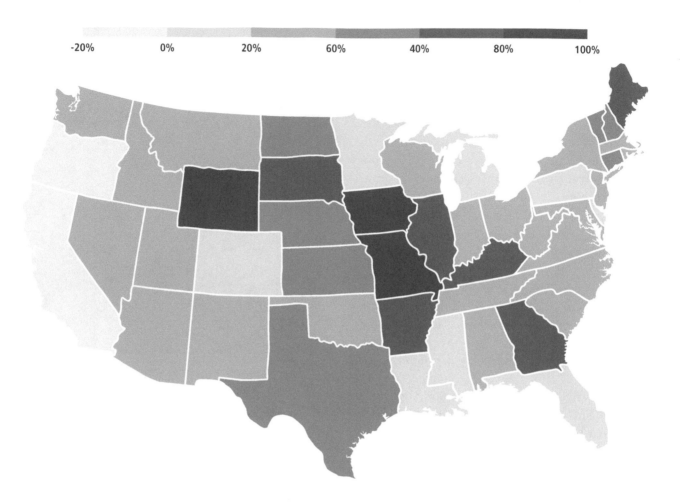

-20%   0%   20%   60%   40%   80%   100%

Figure 1.12. This map, from ClimateCentral.org, shows an increase in the percent change in top 1 percent of events between 1950–1956 and 2007–2016. Adapted from NOAA/Applied Climate Information System, RCC-ACIS.org.

Trees and forests also fare better in heavy and persistent rain events, something much of the United States is projected to see increase due to climate change. Recent analysis shows that for many temperate locations in the US, heavy rain events (downpours) are on the rise, when comparing the average events in 1950–1956 versus 2007–2016[15] (see figure 1.12).

8. Improvements in whole farm viability.

Zooming out to the bigger picture, each of the preceding benefits contributes to an overall improvement in the viability of a farm. The word *viability* refers to the ability for something to work successfully. What is success on the farm? It is when the production systems meet the economic and social goals of the farmer, all while doing no harm or, better yet, improving the ecological health of the farm landscape.

At the end of the day, while farmers are always concerned with economics, they are also farming because they enjoy the work. Farmers are a social group, and share a lifestyle. It's important to recognize that farming isn't just about the money, but also about the type of life it provides for those who participate in it.

So while most of the justification above centers on the ways silvopasture can save time, energy, and resources, it's also a beautiful and vibrant way to farm. Adding trees creates a more interesting place in which to work and live, for both the animals and the farmer.

Providing multiple yields and buffering from the inevitable effects of weather, climate, and markets reduces the farmers' stress. In the end all these factors contribute to an increased likelihood of success for farms, and a better ability to adapt to change over time.

BENEFITS TO THE LARGER COMMUNITY, SOCIETY, AND CULTURE

In addition to the range of benefits silvopasture offers to individual farms, this practice also brings a number of promising benefits to the larger society and global community. While these might not be at the top of a farmer's to-do list for the year, it's important to outline them, as it provides good incentives for government, industry, and society to better support and encourage silvopasture as a practice.

1. Wildlife habitat and forest restoration.

Yet another benefit silvopasture specifically offers is its potential to restore wildlife habitat. Farming itself is largely responsible for the fragmented chunks of forest and hedgerows we see littering the rural landscape today. Silvopasture can erase the stark line that is so often drawn between field and forest. Lands lack structural diversity, which is critical for birds and migrating animals. Ideally, we need grasslands, shrublands, and deep forests to support the widest range of species. The inclusion of silvopasture into the farm landscape can greatly enhance the structural diversity of vegetation, which in turn supports a greater diversity of wildlife.

In bird habitat ecology, the practice is called habitat connectivity. The goal is to reconnect the remnants of habitats that are still intact, whether you're looking at a suburban development or rural farmscape. One strategy is to identify three types: patches, stepping-stones, and corridors.[16] A patch is a substantial section of habitat, and will serve as major residencies for wildlife, whereas stepping-stones are small chunks of connective patches, and corridors are long stretches that connect patches.

Silvopasture can be used as a tool to facilitate this connectivity. And it doesn't just benefit the birds. Native pollinators (see the sidebar on page 18), soil biology, and many other animals also benefit. Paying attention

Figure 1.13. Since we've started planting trees, we've seen a sharp rise in the number and diversity of species of birds in our pastures. In 2017 our sheep had a resident pair of flycatchers, which seemed fond of sitting on their backs and cleaning up flies. The sheep didn't seem to mind.

to this aspect of silvopasture also opens doors to federal funding, which is discussed more in chapter 6.

2. Climate mitigation.

In contrast with the previous discussion of the ways silvopasture can support farm resilience in the face of climate change, mitigation has to do with reducing emissions and especially sequestering carbon, which is the process of removing CO_2 from the atmosphere, thereby reducing the impact of change. Eric Toensmeier's sidebar on page 21 chapter nicely summarizes the science;

SILVOPASTURE AND BEES
By Kass Urban-Mead

What is a wild bee? If you think of a hive of feral honey bees living in a high-up cavity in old-growth woods, or angrily remember wasps ruining a late-summer picnic, you're not alone. However, there are 20,000 species of bees in the world that are neither honey bees nor wasps. Unlike wasps, bees have fuzzy bodies for collecting pollen, and they don't eat other insects. Bees come in all shapes and sizes—some have extra-long antennae, or metallic bodies. There are even species of wild bees that only fly at dusk and dawn, or with the full moon![17]

One species called the squash bee didn't even exist in the United States until squash cultivation expanded across the continent.[18] Wild bees visit wildflowers, and farms: The more different kinds of wild bees there are, the greater the successful fruit-set.[19] Honey bees may even be best thought of, in fact, as managed backup for when an ecosystem is not supporting wild bees.

How Does This Connect to Silvopasture?

Silvopasture is rooted in multifunctionality, and in allowing connections between organisms that don't happen when we separate their component parts. Wild bees live in lots of habitat niches, thriving in complex landscapes: Bee diversity is higher when there are more land-use types; in regularly disturbed forests than in old growth or plantations,[20] and in cities than in monoculture crops.[21] A silvopasture system includes multi-year rotations, intentional disturbances, and, most important, variation in light level across space and time.

Bee Food

More floral diversity means more choices for bees: Different bees use different flowers depending on their body size and shape, and some prefer legumes, while others prefer asters, mustards, or borages. Thinning a forest creates light gaps where blooming flowers will likely join high-quality forage plants. A silvopasture's regular rotational grazing (with long recovery periods) ought to increase floral diversity within each stand, and will definitely create variety in

the stages of plant growth across the whole property. Beyond the *pasture*, the *silvo* part of your system provides bee food, too. A single male red maple tree, for example, can have 1,350 pollen-producing blooms per cubic meter of canopy; a female can have 600 nectar-producing blooms in the same area.[22] That is an immense resource.

But, you ask, aren't maples wind-pollinated? Since we're often trained to think of bees as being symbiotic with flowers, it can be hard to remember that bees only actually visit flowers to *eat* (pollination is incidental). Research shows that when a wind-pollinated tree—such as a willow, maple, oak, even alder and chestnut—produces protein-rich pollen, then wild bees will happily take advantage of them. At least 50 species of wild bees visit willow trees in the early spring (see table 1.1). Higher-protein diets make bigger baby bees,[23] and simple analyses have found that pollen grains of oaks and maples are up to 40 percent protein—as much as an insect-pollinated redbud and, in fact, as much as a steak![24]

Trees and shrubs are also useful for insects in unexpected ways. Some resins produced by wind-pollinated *Populus* and *Betula* are specifically sought out for self-medication by honey bees challenged by a fungal disease,[25] and other trees' and shrubs' nectar alkaloids can help reduce gut parasite infections.[26]

Bee Homes

Finally, a silvopasture doesn't provide just bee food but also bee *habitat*. Much of this habitat happens incidentally, if you don't spray insecticides and leave the corners of your property a bit messy! Coarse woody debris and other rotting logs are nests for shiny green bees, who use abandoned beetle burrows to protect their young. Uncut, pithy stems of sumacs, berry canes, and reeds standing through the winter let leaf-cutter bees, miniature carpenter bees, and mason bees stay protected in the hollowed-out stems. Don't forget hedgerows and forest edges with multispecies shrubs: Not only do many of these flower, but bumblebees nest at their bases and inside grassy tussocks, or

Table 1.1. Tree species hosts for various wild bees

	Protein Content*	Bumble	Mining	Mason	Leaf-Cutter	Long-Horned	Cellophane	Shiny Green	Carpenter	Tiny Carpenter	Sweat	Tiny Sweat	Wool-Carder	Yellow-Masked	Parasitic	Honey
Black Locust		x	x		x						x	x	x			x
Willow	36.8–46.4	x	26+	x			x	x	x	x	x	x			x	
Honey Locust			x									x				
Mulberry																
Poplar			x													x
Tulip Poplar	37.1	x														
Basswood		x	x	x	x						x	x		x	x	
Hawthorn			x	x				x		x	x	x				
Fruit Trees	35–45															
Chestnut		x	x	x	x								x			
Maple	39.4–46.2	x	15+				x				x				x	
Oak	30.6–41.5	x	x	x				x				x				
Redbud	40		x	x			x				x					
Sumac		x	15+			x	x	x	x	x	x			x	x	
Horse Chestnut	26.7	x	x	x							x	x			x	
Wild Cherry	28.5		12+				x					x				
Serviceberry			26+	x			x					x			x	
Mountain Laurel		x	x	x				x			x					
Honeysuckle		x	6+	x	x		x	x	x	x		x				

Sources: AMNH database, personal observation, Giuseppe Tumminello, Laura Russo PhD.
* Source for protein content, from trees of the same genus: Roulston, T. H., and James H. Cane. "The Effect of Pollen Protein Concentration on Body Size in the Sweat Bee *Lasioglossum zephyrum* (Hymenoptera: Apiformes)." *Evolutionary Ecology* 16, no. 1 (2002): 49–65.

in the abandoned burrows of rodents who once found refuge in hedgerow edges. Buffer zones along streams or rivers protect bumblebee queens, who burrow into the soil to overwinter.

Speaking of soil, 60 percent of wild bee species nest in vertical underground tunnels with small branching cells where they lay their eggs. Grazing can compact soil (although frequent pasture rotation can mitigate this), and compact soil allows a soil bee nest not to collapse in on itself. So any grassy or shrubby areas with small, bare soil patches between can be perfect—but usually unrecognized—habitat. Note that these are not the aggressive ground-nesting bees of playground fears—instead, they are nearly all solitary bees, where a single female is in charge of provisioning her brood, and is gentle, as she alone is responsible for the reproductive success of her offspring. Alternatively, extensive rooting by pigs meant to turn the soil could also disturb soil-nesting bees, or destroy bumblebee nests.

All in all, a well-managed silvopasture inherently creates a patchwork of grazing intensities, soil quality, and light regimes. When shrubs and trees are in bloom, protect the flowers from grazers. Keep a diversity of trees to provide continuous pollen sources from the earliest willows and maples of spring through midsummer, and be sure to allow some "untidy" edges and old woodpiles.

Then don't hesitate to add beneficial insects in your calculations of ecosystem health and multifunctionality!

Kass Urban-Mead is a passionate researcher and advocate for wild bees. She is currently a PhD student at Cornell University, and can be reached at kru4@cornell.edu.

his work, along with that of a team of researchers and scientists in the book *Drawdown*, finds that silvopasture is likely one of the top climate-saving forms of agriculture available.[27] In essence this is because such systems utilize both aboveground (trees) and belowground (roots) storage of carbon, and can help build healthy, carbon-rich soils. Another critical point is that, while there are definitely carbon benefits to establishing trees in pasture, it's likely that many forms of converting existing woodland to silvopasture may have a negative effect on emissions.

3. Addressing the need to feed a growing population on less land.

At the meta level, the agricultural challenge of our time is often described as the need to grow more food, on less land, within the context of a changing climate. In many respects the Green Revolution set out to do this, albeit without taking into account the climate change factor. The problem is that increasing tillage led to rapid soil erosion; increasing use of fertilizers and pesticides polluted waterways and people; and technologies like genetic engineering fell short of the stated goal of reducing the impacts of industrial agriculture.[28]

These technological attempts to solve the problems of farming also devalued traditional methods and knowledge that have sustained cultures for generations. While many large corporations continue to promote technological fixes to this problem, many communities are taking a different approach, valuing organic methods, soil health, and cultural diversity as the foundations of good farming.

This "feed the world" paradigm is actually riddled with issues. For starters, with a world population expected

to reach 9.7 billion by 2050, we actually already grow enough food to feed this many people; it's just that many cannot afford it.[29] Economic inequalities and the industrial exploitation of people and land are the major issue. Further, most of what is counted as yield in agriculture is not in fact edible by humans, but grown for use as either biofuel or animal feed. Well over 50 percent of the corn and soy grown in the United States goes to livestock.[30] This leads to the often touted notion that animals are the problem, blamed for all sorts of environmental ills.

Yet some organic methods, and the growing interest in regenerative agriculture approaches, indicate that a number of healthy environmental indicators can be improved while at the same time producing crops.[31] Seen in this light, farmers can be stewards of ecological restoration, which would lead us to want to have farming happen on *more* land, not less. And silvopasture offers one of the most compelling forms of regenerative agriculture, with proven benefits to land and people.

So we might do better to phrase the global need as:

Growing good-quality food for people, in ways that support community sovereignty, on restored lands.

Silvopasture definitely has a role to play in this scenario, as animals are a renewable resource that can support land restoration, keep forests in place, and provide real food and livelihoods for people.

4. A shift from factory-farmed meats.

People love to blame animals for our environmental problems. Livestock have been blamed for major contributions to greenhouse gas emissions,[32] and rightly

SILVOPASTURE:
A POWERFUL TOOL FOR CARBON SEQUESTRATION
By Eric Toensmeier

Climate change may well be the biggest challenge facing humanity this century. Livestock production is a major source of greenhouse gas emissions. Fortunately, some livestock systems fight climate change by removing and storing excess atmospheric carbon dioxide. Silvopasture is among the very best tools we have available for limiting global warming to 1.5 degrees C.

Project Drawdown recently completed and published an extensive analysis of 80 climate change mitigation solutions, across many sectors like energy, food, educating women and girls, buildings and cities, land use, transport, and materials.[33] Silvopasture was ranked as the ninth most powerful solution overall, and as *the most powerful of all agricultural strategies.*

The Drawdown model results show silvopasture's sequestration impact at 31.9 gigatons of carbon dioxide by 2050. This is about double the 16.4-gigaton impact Drawdown calculated for managed grazing, even though our model showed better grazing being adopted five times more widely. This disproportionate impact is due largely to the climate-change-fighting power of trees.

Biosequestration

How does carbon sequestration work? During the process of photosynthesis, plants remove atmospheric carbon dioxide, breaking it apart and using the carbon atoms to create sugars. Many of these sugars are converted into starch, fibers, lignins, and other compounds. Carbon in plant biomass (such as tree trunks and roots) can be stored for the life of the plant, which in the case of trees can be a long time indeed. It's estimated that the biomass of trees is about 50 percent on a dry-weight basis.

But that's not all. Within an hour of photosynthesis, some of that carbon is exuded from plant roots to feed beneficial soil organisms. Over time leaves drop and root hairs die back. Some of their carbon returns to the atmosphere, but some portion becomes part of soil organic matter. This organic matter, or humus, is about 57 to 58 percent carbon on a dry-weight basis.

Some will argue that carbon sequestered in aboveground biomass doesn't really count, as trees can be cut down or burned. But soil carbon can also be easily lost, through a return to poor grazing practices, tillage, or other forms of degradation. Both soil and biomass carbon are considered temporary storages by the International Panel on Climate Change. With good management, both biomass and soil can hold on to carbon for decades or even centuries—long enough to make a big difference in the critical decades ahead.

Annual Sequestration Rates

The annual rate of sequestration in an agroecosystem is measured in tons per hectare per year. This reflects the net gain in carbon in soils and/or biomass. How does the sequestration rate of silvopasture compare with managed grazing?

Typical sequestration rates for managed grazing and improved pasture management range from 0.3 to 0.5 tons per hectare per year. Though some researchers report much higher rates of 3.0 to 4.0 tons/ha for sophisticated practices like intensively managed and adaptive grazing,[34] many grazing scientists have aggressively disputed these results. Higher figures for sophisticated grazing practices like holistic grazing, should they be confirmed, would be very good news for climate change mitigation.

While we wait for the scientists to resolve this issue, it is worth turning our attention to silvopasture. Unlike grazing alone, there is little controversy about silvopasture's sequestration potential. Agroforestry expert P. K. Nair estimates a range of 3.0 to 10.0 tons per hectare per year. Experts from the University of Missouri agroforestry program estimate 6.1 tons per hectare for silvopasture in temperate North America. Project Drawdown meta-analysis calculated a global rate of 4.8 tons per hectare per year.[35] The five temperate studies included actually had a higher average rate of 7.2 tons/ha (unpublished results).

Assuming the more conservative rates for grazing, silvopasture sequesters carbon 6 to 33 times faster. Even with the controversial figures for sophisticated grazing systems, silvopasture sequesters carbon up to three times faster. Managed grazing has received a great deal of press as a mitigation practice, but it would appear that silvopasture deserves even more attention, despite its current low profile.

The USDA has developed a carbon tool called the COMET-Planner. The rates assigned to all practices are extremely conservative, but the rate of silvopasture (at 0.2 to 0.4 t/ha/yr) is still double that of managed grazing alone (0.1 to 0.2 t/ha/yr).

Saturation

Here's the challenge: Agroecosystems don't keep sequestering at these rates forever. After 10 to 50 years, most become "saturated," meaning they have had their fill of carbon. After saturation, sequestration continues, but is offset by roughly equal emissions. Thus, there is a limit to how much carbon can be stored in both soil and aboveground biomass.

What happens to the mitigation impact of managed grazing when pasture soils become saturated with carbon? These lands once again become net emitters via methane.

Can this return to net emissions be avoided? Addition of biochar (a special type of charcoal) to soils of pastures and silvopastures may well be able to raise carbon stocks beyond saturation. Silvopasture offers a second "hack" of saturation—trees can be harvested when they are mature, and used for timber. The carbon in the wood will last the life of the building or furniture it is used in. Meanwhile new trees can be planted, their young growth once again providing high sequestration rates for decades to come.

Carbon Stocks

As with sequestration rates, the total carbon accumulated in saturated soils varies between practices. While grazing scientists argue over sequestration rates, there is little controversy about the total carbon stocks that can be accumulated under improved grazing regimes. The lifetime carbon sequestered in a managed grazing system ranges from 30 to 60 tons per hectare. Soil carbon stocks in silvopasture systems range from 60 to 250 tons per hectare—a substantial increase over grazing alone.[36] Thus, measuring soil carbon alone, silvopasture carbon stocks can be as much as eight times larger than managed grazing, and even the low end of silvopasture stocks matches the high end for grazing alone.

But carbon in aboveground biomass is also present in silvopastures. While estimates of aboveground carbon stocks in temperate silvopastures are lacking, in tropical systems the reported range is an impressive 13 to 92 tons per hectare.[37]

Methane Impacts

Carbon sequestration is not the only relevant process impacting climate change in livestock systems. Ruminants emit methane from both digestion and manure, and nitrous oxide from manure. Both are very powerful greenhouse gases. Improving pasture quality can decrease methane emissions from ruminants a bit, as can consuming high-tannin tree leaves in silvopastures. Figures from the IPCC show methane emissions equivalent to 1.39 tons of carbon dioxide per hectare per year (equal to 0.4 ton of carbon).[ ] These methane emissions offset (counteract) 58 percent of the sequestration from managed grazing. Due to the much higher sequestration rates of silvopasture, only 7 percent of sequestered carbon is offset by methane emissions.

Some advocates of climate-friendly grazing have stated that there is no methane impact from managed grazing, as the methane is consumed by microbes in the soil of healthy pastures.[38] It would be great news if this were true, but there is no scientific evidence whatsoever that this is the case. Allan Savory, developer of holistic grazing, has also wildly overstated the mitigation potential of grazing.[39] In the long term these kinds of claims can damage the legitimacy of grazing, which, though not a silver bullet solution, is certainly a very critical component of any land-based mitigation effort.

Conclusions

Note that the discussion here only applies to tree planting and managed natural tree regeneration in pastures.

Thinning forest to create pastures results in net emissions (though it may still have a better climate impact than alternative economic uses of the same tract of forest).

The majority of the world's grasslands are too dry for trees to grow. In these regions managed grazing is the highest and best climate mitigation practice. However, where there is enough rainfall to support tree growth, silvopasture is clearly the more desirable climate change mitigation practice. That still doesn't mean it is the right choice for any given farm or ranch, but it is deserving of far more attention than it has received to date.

Eric Toensmeier is a senior researcher with Project Drawdown, an appointed lecturer at Yale University, and the author of The Carbon Farming Solution. *For further information visit www.drawdown.org and www.carbonfarmingsolution.com.*

so. Of course it is not the animals' fault, but rather our inability to manage livestock with good methods. Chances are that if you eat meat, the animals were raised in confinement and fed grain as their main food source. This method isolates the animal from its habitat, and both animals and the land suffer as a result. It does not recognize that animals, especially ruminants, are meant to eat grasses and vegetation. It creates dangerous working conditions while turning a fertility resource—manure—into a major pollution problem.

An important hidden cost of confinement operations is the land you don't see in the photos of cattle penned in tight quarters with not a speck of green to be found underfoot. Feedlots are only possible with the large importation of grains, grains that are farmed by tilling up the soil on thousands of acres and, in many cases (especially in the tropics), the clearing of large swaths of forests to make way for grain production. And the effects of confinement practices aren't noticeable just in the context of large feedlots in Colorado or Brazil. They exist in a range of farms all across the continent.

In this paradigm many advocates for the vegetarian or vegan diet say that the problem here is that we humans should be eating the grains ourselves instead of feeding them to animals, which supposedly would greatly reduce the impacts of our food choices. And while it's true that many of us should eat less meat,[40] there are two problems with the overall argument. The first is that these "grains" are not edible by humans, so converting fields growing feed to human foods would require a massive shift in the way agriculture works. The second issue is that a shift from meat to other forms of protein still relies heavily on tillage, which destroys soil structure, releases greenhouse gases, decreases water infiltration, reduces soil biology, and degrades organic matter. Not to mention that large-scale grain farming is only economically possible because it is subsidized.

The paradigm of confinement comes from a foundational approach, often called factory farming. The "factory" metaphor is an apt one, because the goal is to convert labor and inputs into meat and dollars. The focus is on producing meat or milk without concern for the other parts of a well-functioning ecosystem. And since the animals and plants are still biological, and therefore dynamic, it doesn't often work out. The planting season for corn is too wet, or too dry. Animals in confinement develop diseases and deformities, which must be addressed with antibiotics. And the soil is continually robbed of nutrition, while the manure on the other end is overwhelming and becomes a destructive force on the land.

So what is an alternative? At the simplest level we might call it ecosystem-farmed meats—a system in which we take into consideration not only the animals, but the plants and soil, too. If we scrap the current dominant livestock production system and design one based on our ideals, it would include:

• Keeping soil in constant cover, with pasture, trees, and forages.

Figure 1.14. The majority of beef we buy in the United States comes from feedlots where grazing animals don't see grass or exercise, and are often subject to unsanitary conditions. Photo by Randy Heinitz via VisualHunt.com.

- Providing a diverse diet of foods animals are designed to eat.
- Producing foods (for the animals, and people) on-site or locally.
- Allowing animals space to move, exercise, and express their innate character.
- Creating a profit center based in on-farm resources.

As Gene Logsdon writes in his seminal book, *All Flesh Is Grass*:[41]

We can have all the meat, milk, eggs, animal fabrics, and horsepower needed to feed the world without cultivating all those nice seedbeds that for centuries have been considered the beginning of the agricultural process by which human civilization survives. With modern grazing methods, farm animals can get their food mostly from grazing forages, thus avoiding the current crippling expense of annual cultivation and grain harvest. . . . I don't mean that we should quit growing grains altogether. But if the mix of 90 percent annual grain and 10 percent pasture that now reigns on our better farms was turned around to 80 percent pasture and 20 percent grain, farming could again be profitable and ecologically sane.

Silvopasture is arguably one of the highest forms of ecosystem farming. It combines all the aspects above, and especially diversifies the plant contributions to the system, which offers all sorts of benefits to the whole. Rather than clearing forests to grow corn and soy to feed animals in confinement, we are restoring forests while growing good forages to feed animals, who live a natural existence on pasture.

Figure 1.15. Silvopasture is a beautiful scene, and the story can be used to market the products as both regenerative and healthy. Photo by Brett Chedzoy.

5. Improved human health through diverse diets.

We've discussed feeding a growing world, and doing it in a way that supports not just livestock production but also ecosystem health. The fifth leg of the wide view of the benefits of silvopasture is how such production affects human health. The research is not something specific necessarily to silvopasture, but to good pasture-based livestock farming. One research review that compiled work from three decades found that grassfed beef had a "more desirable lipid profile" (good cholesterol) as well as higher precursors for vitamins A and E, and more amounts of cancer-fighting antioxidants.[42] Pasture-finished beef is also notably higher in omega-3 fatty acids while having a lower fat content.[43]

Sometimes people used to high-fat beef say the taste of grassfed beef is chewier or "grassy," but many others seem to prefer it, especially when learning of the health benefits. Clearly consumers are showing preference for it, as grassfed beef has seen a growth in demand of 25 to 30 percent a year in the last 10 years.[44] This has important implications for silvopasture, too. If more consumers link the way meat is raised with their health, and ecosystem health, they can do a lot to support more growth in the practice. Perhaps someday a "silvopasture-raised" or "tree-friendly" label will exist alongside "grassfed."

6. Increased demand for local foods and storied foods.

All these aspects combined result in the reinvigoration of consumers and markets to a fundamental truth: People want to know where their food comes from, and they are willing to support that by paying a higher percentage of income for food. Ultimately a label is no substitute for knowing a farmer and being able to learn directly how he or she approaches raising livestock. As farming continues to shift from larger commodity crops to smaller, diversified farms across the United States, what is becoming more important than price is the individual story of a farm and its products.

The rise in demand for grassfed and pasture-raised meat indicates that, increasingly, our society wants to see animals grazing and foraging outdoors, on the land. The organic movement, too, is rapidly increasing in demand, growing nearly 20 percent a year. Yet challenges abound, as a market trend inevitably catches the eye of the industrial players, namely large food corporations that do not respect farmers, local communities, and ecological health as part of their business plan. They want their share of the pie, but only if it means being able to produce food on their terms. This can lead to watered-down standards and must be carefully monitored.

And while the growth in consumer interest is certainly promising, we have to recognize that the choice to spend more money and support local foods is one of privilege. Small farmers often seek out niche and specialty markets, which fetch higher prices but are both only financially and physically accessible to wealthier people. These problems also relate to institutionalized racism, which permeates the food system as much as every other sector of society.[45]

And while many look to government and institutions as the problem, it also hits home at the community level, where farmers markets are predominantly "white" spaces, and white people make up the vast majority of new farmers entering the business, due in large part to the need for land access and capital to start a farm. There are systemic barriers that prevent people of color and the poor from getting access to nutritious, high-quality food as consumers, and to opportunities to farm as well. These are issues that cannot be ignored at the food systems level, no matter what the scale of production. We are all involved.

Both farmers and food policy advocates need to be aware of the issues, and find ways to address them

from their role in the overall food system. At the end of the day, silvopasture is a wonderful system for the land, but the food needs to be also wonderful for the larger society. This means that, at least at this point in time, farming is as much a cultural and social act as an environmental one.

Challenges to Silvopasture

The preceding list of benefits silvopasture offers is quite remarkable. The question then is, given all the clear benefits to silvopasture, why aren't more farmers and homesteaders doing it? As with any situation, the reasons are many, but the matter boils down to two main barriers. We might call them paradigm and practice.

The first, paradigm, is the challenge of wrapping our heads around the complexity of combining forestry and farming, which have traditionally, deliberately been kept separate, especially in the United States. There is a stigma around the idea of livestock in the woods, and many foresters today will still scoff and protest, as they were often trained in forestry school that livestock in the woods was a no-no. To be fair, this legacy comes from a poor habit farmers developed, where the woods was a place to dump animals when pasture resources were low or summers were excessively hot. There was no management. Using the woods as a dumping ground inevitably created issues and was eventually seen as a bad idea. More on this in the next chapter.

This bias, combined with a farming culture that has been more about separating species into monocultures than mixing them, is a huge obstacle to silvopasture. The good news is that the trend is changing. More farmers, especially young and beginning farmers, are more interested in diverse and integrated systems. The understanding that we need to think and act like nature in the way we farm is quickly gaining ground. In many senses the time is ripe for silvopasture to really take hold. Enough people are ready for it.

The caveat is that we have to do it right, and proceed with caution. This is the second major challenge: practice. In part because of the perspectives already mentioned, research and development of silvopasture, especially in cool temperate climates, has been slow on the uptake. Very few universities or organizations are conducting ongoing research, which is uniquely challenging due to the systems taking up to 10 years to establish, while most grant cycles are 2 to 3 years in duration. We have a lot to learn about the right species combinations and practices that work. That said, we have enough confidence to get started. We just should not assume something works until we see it working. Luckily, farmers are well suited to this approach to learning.

A third barrier that relates to these first two is that silvopasture has not been widely embraced by the USDA and other institutions. Once government, universities, and extension services are on board, adoption can be quite rapid. And while the research basis is pretty solid for silvopasture, few people in these sectors know about it, or can advocate or provide educational support for it. Slowly, partnerships between individuals are changing this, but it already

falls short of the demands from farmers to know more. Hopefully this book is a stepping-stone to advance the practice, and the necessary conversations around it.

THE TIME IS NOW

Given that farming paradigms are shifting toward management strategies that integrate, along with the very real prospects of a changing climate, there has not been a better time to adopt silvopasture more widely. Along with these changes, a significant technology has emerged that has really become a game-changer for rotational grazing, and by extension for silvopasture. This is the invention and refinement of portable fencing.

The cost of fencing is arguably one of the biggest hurdles in raising animals, but the emergence of relatively cheap, effective, and easy-to-use portable fencing systems has literally changed the grazing game. It allows farmers the flexibility to adjust their paddocks as needed, and to fence out trees as they are getting

Figure 1.16. The relatively recent availability of cheap and accessible fencing is a real game-changer for rotational grazing, and silvopasture. Photo by Jim Gerrish.

Table 1.2. Appropriate land-use types for silvopasture (SP)

Appropriate for SP	Possible, with Caution	Not Appropriate
Hedgerows	Even-aged forests with a mix of older trees above browse height	Forests in fragile habitats (wetlands, rare species, for example)
Plantation forests	Orchards, vineyards, Christmas tree farms	Mixed-aged forests with good species regeneration
Overgrown land	Wet zones, seasonally	Old-growth forests
Sites producing poor-quality trees		

established, without breaking the bank or consuming too much of their time. We will discuss fencing options in chapter 3, but no doubt their wide availability is going to be one of the biggest boosts to silvopasture in decades.

Priorities in Silvopasture

The rest of this book will continue to expand on the preceding points. At this juncture it is worthwhile to pause and recognize that while silvopasture could technically be implemented anywhere and with any number of species, a reality check is necessary.

Our food and farming system is in a very vulnerable place right now. The need for economic rural development, sustainable food-producing systems, and responses to the continued variables our changing climate is putting forth all point in one direction: We need to implement and scale up other ways of farming, and fast.

In my opinion, we are beyond the point where we can discuss the merits of planting trees and waiting for the returns 40 to 60 years down the line. We need to focus in on the ways we can get systems established quickly and effectively. Others may disagree.

In regard to silvopasture, I see this as focusing on the following approaches:

1. Making use of **overgrown or underutilized portions of land,** including hedgerows, plantations, scrubby lands, and sites producing poor-quality trees.
2. Bringing **trees into pasture,** and/or connecting patches, hedgerows, and the like to offer more continuous shade and shelter for animals.

Figure 1.17. Utilizing fast-growing trees such as this red alder (*Alnus rubra*) is an effective strategy for launching into silvopasture. We don't have a lot of time to get started.

3. Favoring ruminants and grass-based systems, which can provide increased forage resources on-site with fewer outside inputs.
4. **Planting fast-growing, highly adaptable trees** with low disease and pest pressure and focusing on the benefits of shade, shelter, and fodder for animals.

While I would love to wax poetic about black walnut silvopasture as a retirement fund, I sense that we have less time to get things moving than that. This doesn't mean that long-term species should not be part of a silvopasture design; instead, the dominant species we should favor are those trees that are most robust and fast growing, at least for starters.

There are ample lands where these first two priorities can be implemented, and because we don't know as much about the practice as we would like, it is best to experiment a bit on parts of the land that already have a lower ecological value rather than working within healthier forest stands.

We do know that the carbon benefits of implementing silvopasture where trees are planted will be better than with extensive thinning of more developed woodlands. The focus on tree selection is to encourage more rapid conversion of open areas to silvopasture, and to keep the stakes lower due to the reality that many trees may die during establishment (more on this in chapter 4). Working with poplar, willow, or black locust carries a lower burden than trying to establish fruit and nut trees for production in combination with livestock, though some are up for the extra work, valuing the return of these high-value crops. If you just want to raise livestock, keep it simple.

The overall approach, then, could be summarized as:

Start with most marginal pieces of land, and convert these first, utilizing hardy, fast-growing trees as the basis of the work.

These are recommendations, but of course you can choose your own adventure. As with other ecological farming methods, it's important to remember as you read through these pages that your silvopasture will look different from all others. Use the case studies and template designs throughout the book as inspiration, but ultimately your landscape will dictate the specifics of your unique silvopasture. Consider throughout the process what aspects of your land are special and unique, and how you can work with them. Nature is subtle, and adapts to even the slightest change in elevation, aspect (the direction a slope faces), soil composition, available water, and so on.

Consider this book an invitation to explore a more interesting and dynamic way of farming. How wonderful that there isn't a given recipe for this, and that we must each find our own way forward! Farming has become far too simplified, and we all benefit from having a challenge to our work. That said, in the pages that follow we will explore many patterns, design strategies, and techniques for silvopasture that aim to help you better think about your particular situation. Hopefully this information, along with a number of real life examples and case studies, offers a compelling case that silvopasture is not only fun and interesting, but achievable by all who are willing to put themselves fully into its development.

EXPLORING SILVOPASTURE IN MICHIGAN
By Don Kilpela

We live on an old 40-acre farm in Michigan's Copper Country, a rocky finger of land that juts into Lake Superior. Known for its heavy snowfall, long winters, and scenic beauty, the region was developed by mining interests in the late 1800s and attracted large numbers of immigrant laborers, including my own Finnish forebears. Our farm is surrounded by plantations of red pines that were planted in the early 1950s, shortly before I was born. I grew up with them, seeing them change from small trees to the tall timber that has now been mostly harvested.

My father was not a farmer, but he kept his "clearing" open with a brush mower. This changed about 20 years ago when I began planting pines, inspired by Aldo Leopold's *Sand County Almanac*. This became a regular spring ritual for my family and myself for many years. Since I wanted a diverse forest, not a simple monoculture, we intentionally planted seedlings in widely spaced rows and included some white pine, Douglas fir, spruce, and larch for variety. I also figured that maples and oaks would colonize the open areas through natural seeding from the fencerows.

My experiment has produced some interesting and challenging results. As any forester can tell you, a pine growing in the open grows long, stout branches—the bane of anyone looking for good sawlogs. Pines are normally planted in a tight stand so they are self-pruning; the shaded lower branches die off as the trees grow. My trees, sadly, looked like Christmas trees.

Figure 1.18. Planted pines on the land. Photo by Don Kilpela.

Once the old farm was mostly filled with pines, I embarked on a new venture. Not content with making silvicultural mistakes, I branched out into animal husbandry and acquired a small flock of Icelandic sheep. As my flock grew, I soon came to question the wisdom of planting all those pines. Much of my remaining open pasture is on low ground that is too wet for grazing for much of the summer, but the high and dry ground was full of pines. Could I graze sheep in the pines?

This ran counter to all I had heard and read about growing trees, where "Keep livestock out of the forest!" was the cardinal rule. But if deer can graze in the pines, why not sheep? I had seen that grass grew thick in places in a well-thinned pine plantation, and that deer bedded there regularly. The partial shade seemed to encourage lush growth.

I decided to follow my instincts. I bought a pole saw and began trimming the lower branches of my taller pines about 8 to 10 feet from the ground, letting more sunshine reach the grass. I now have several acres of open woods where the grass grows well and have begun pasturing sheep in it using electric netting. I move the netting and sheep to new areas every two weeks or so. The sheep seem to relish the variety of grasses, forbs, and leaves. They are good browsers and will even stand on their hind legs like deer to reach sugar maple leaves—a preferred delicacy.

I have also pastured the sheep in semi-wooded areas that are filling with naturally seeded conifers, hardwoods, and brush. These areas contain some toxic plants—bracken fern, pin cherry, and milkweed—but so far I have not observed any problems with the sheep. I have plucked up the milkweed where I find it, but my gut feeling is that as long as there is a variety of other food available and the area is not overgrazed, the sheep will not ingest harmful amounts of toxic plants. I have never heard of deer poisoning themselves in our area. I also suspect that in small quantities toxins might even

Figure 1.19. Icelandic sheep. Photo by Don Kilpela.

have medicinal value to the sheep, and that providing a smorgasbord of grasses, forbs, and leaves allows the animals to select what they want and need. It is also rewarding to observe my big ram, Gunnar, comfortably lying under the low branches of an old spruce, which provide shade and a degree of relief from biting insects.

In the interest of full disclosure, I should confess that I also suspect that biting insects confer some benefits to all of us that science will eventually discover, so take my advice with a grain of salt.

I am both fascinated and puzzled by the feeding habits of sheep. In some places they will clip the grass right down to the ground while leaving a nearby clump of seemingly lush grass untouched. I have wondered if it has something to do with the mineral composition of our glacial soil, which likely varies from place to place depending on what rocks are found in a particular spot. Interestingly, they graze the grass on an old railroad bed that runs through our land right down to the soil.

To help locate my sheep in thick brush, I have put bells on a number of them. When I want to gather them to count heads, all I need to do is rattle corn in a pail and they come running, bells jingling.

Sheep adapt to new environments. One day I was inspecting an area full of maple seedlings and found that the sheep had eaten all the leaves within reach—with the striking exception of those on two saplings that stood close together. Why hadn't the sheep touched those? Then I saw it—in the smaller sapling, hanging close to the ground, was a large gray hornets' nest. Who said sheep are dumb?

In Europe, sheep are widely used to maintain scenic landscapes, and it is possible that the game of golf would never have been invented were it not for these four-legged lawn mowers. My own long-range goal is to use the sheep to create open, parklike areas on our farm, spaces that my children and grandchildren can enjoy in the future. I also want to grow quality timber. This is a work in progress, but so far my flock seems to do well on my wild, unplanted, and diverse silvopastures. They haven't had any major health issues, and they produce a good crop of tasty lambs every year.

Don Kilpela lives in Michigan and maintains a blog about his adventures at donwkilpela.blogspot.com.

Figure 1.20. Sheep grazing among pruned pine trees. The experiment has been largely successful and pleasurable to manage. Photo by Don Kilpela.

Figure 1.21. Fencing through a previously planted field. The sheep and trees have proven to work well together.

2 Perspectives from Ecology and History

To understand silvopasture, we must first recognize that we are both giving a name to something that has happened before, and also engaged in a practice that is underdeveloped and emerging as a novel way to farm. Competent practitioners, then, must verse themselves in the aspects of ecology and history that we do know, and exhibit caution and curiosity as they approach the parts we don't.

In this chapter we examine some of the patterns in the ecology of the temperate forest, grassland, and savanna. With these tools we can then examine historical examples of silvopasture systems, and the lessons we can learn from them as we move forward. This portion of the book is not intended to offer a complete picture of forest, grassland, and savanna ecology, but rather to take a look at a few of the major principles from each that help inform silvopasture as a practice.

It is important to remember at this point that much of the geography this book focuses on is the cool temperate climate, where relatively even rainfall throughout the year has most often resulted in forests as the dominant ecotype. Grassland and savanna ecosystems are more typically found in drier climates, where precipitation distribution is much more of an

Figure 2.1. Examining the ecology and patterns in forest, savanna, and grassland ecosystems offers some excellent directives for approaches. From left: deciduous forest in New York; slash pine savanna on the Mississippi/Alabama state line (photo from Wikimedia); tallgrass prairie grassland in eastern Illinois (photo from Wikimedia).

ebb and flow. Practitioners should weight their own local climate and microclimate conditions in relation to the points made in this chapter, and always use their original ecosystem type as a model for the way to design a silvopasture.

Readers are encouraged to seek out the additional resources cited within, and to continue the inevitable lifelong journey of better learning the finer details of ecology, something that is best done as a combination of reading, learning, and on-the-ground experience and observation. What we don't know about relationships in the natural world is much less than what we do know. We must always check ourselves, and remain humble and open to new information.

Lessons from Forest Ecology

While an extensive view of forest ecology can be found in my previous work with Ken Mudge, *Farming the Woods*, as well as other reputable forestry resources (see the sidebar), in this book we will focus on some of the specific dynamics and properties of forests that will better inform the way we design and implement silvopasture. Many of these key principles come in response to myths that many people believe in regard to the ways forests develop.

FORESTS ARE *NOT* WILD PLACES, AND HAVE A STORY OF PAST USE AND ABUSE.

When we're approaching a set of acreage destined for silvopasture, it's important to investigate and consider the history of the land, because almost all land has a legacy of past impact and human intervention. Most often, the land was cleared at some point. Acres that are flat to gently sloped were likely tilled, while steeper slopes, rolling hills, and varied terrain would suggest pasture and/or management for hay. Barbed fencing growing through trees and rock walls meandering through the woods are telltale giveaways for the location of old paddocks and property lines.

It's important to recognize that historical grazing patterns are in most cases a scenario where the farmers merely fenced in the perimeter of their land and left animals to graze at will. Only relatively recently,

---

A FEW MUST-READS ON FOREST ECOLOGY

Deciduous Forests of Eastern North America by E. Lucy Braun (1950). Widely considered the best categorization of forest types in the temperate forests of the East Coast. The best resource for templates for the types of forests that have historically occupied different regions.

Changes in the Land by William Cronon (1983). A historical look at the patterns and attitudes of native peoples and colonizers and how these views of land and woods shaped their management (and exploitation) of the forest, the remnants of which we see today.

The Redesigned Forest by Chris Maser (1988). A personal and philosophical journey comparing patterns of forest development in nature and contrasting them with the shortsighted and misguided approaches modern forestry has imposed on the woods.

Forest Ecosystems by David Perry (1994). A more academic read on the patterns and succession of forests in all climates around the world. An essential text on the fundamentals of forest ecology.

---

with the advent of more reasonable portable fencing methods, has the idea of rotating animals and resting pasture come into serious consideration and practice by farmers. Some old farms did have multiple paddocks, but rotational grazing was certainly not practiced at the pace we find today.

Another important aspect of history is the stories of the native peoples who inhabited lands in North America for thousands of years, only to be displaced and exploited by settlers who sought to claim land for their own benefit. If we are farming on land, we are a part of this story, whether we know it or not. And while history books often talk of native peoples in the past tense, these cultures remain intact, with sovereign lands and the right to govern themselves and determine their future. As landowners and farmers, we have a duty and responsibility to learn the particular stories and support indigenous rights. If you don't know the stories of the places you inhabit now, learn them.

Learning history means reconciling with the fact that many of the previous land-use practices were damaging to both the ecology and the original people who lived on these lands. Farming has been largely destructive since the colonization of North America, and only recently are more ecological approaches being widely adopted.

Silvopasture practitioners have the opportunity to change the story of their lands, where they can play a positive role in supporting healthy forests while they farm. In different contexts this will mean introducing elements of pasture into a healthy woodlot, or slowing the natural transition nature performs from open field into forest.

FORESTS UNDERGO A TREMENDOUS PROCESS OF SELECTION.

One of the most important factors to consider, whether we're working in existing forests or planting new ones, is that while trees in a forest shed a lot of seed, most of the resulting trees will not survive to become mature, canopy-dominant species with the potential to live out their entire life cycle. This is also known as the process of forest stand development, and is critical to the healthy evolution of a forest in a place. As farmers and land stewards, we rarely consider this as we examine and plan for our woods.

Another way to say this is that for each acre of woods, in one generation at least 100,000 seeds might fall to the ground. Of those, 10,000 may germinate and begin to grow to ankle height, while only 1,000 make it to the sapling stage, or trees more than 2 inches in diameter and above head height. At the climax of the forest, as few as 100 to 200 trees might occupy the canopy. This represents just 0.1 percent of the original gene pool of the woods.

The logical pattern of succession dictates that the number of trees per acre will decline as the remaining trees get larger. Those trees that grow most rapidly will shade out other trees; some will succumb to disease, some to animal browse, and others to weather-related

Figure 2.2. The crowding of young saplings encourages upward, straight tree growth. At this stage it's often hard to tell which will be the dominant, successful trees in the long term. Photo by Jen Gabriel.

damage. In this way nature sets forth a grand experiment where the strongest and most vigorous specimens ultimately make up the forest. As Cornell extension forester Peter Smallidge notes, "About 20 percent of the stems in a fully stocked stand must die to allow for a one-inch increase in average stand dbh (diameter at breast height)."

Through this evolution of a forest stand development, many ecologists recognize four distinct stages of a woods:[1]

1. **Stand initiation**, where a larger disturbance (or farmland abandonment) offers ample sunlight and trees begin to establish.

Figure 2.3. The position in which a tree finds itself in the canopy is a reflection of genetics, site conditions, and sometimes plain old luck. In this case the white oak (*right*) has overtaken the hickory (*left*) and will eventually shade it out, at which point it will likely die prematurely. Photo by Jen Gabriel.

2. **Stem exclusion**, the point at which sunlight begins to be limited, and additional plants are excluded.
3. **Understory reinitiation.** As the canopy grows higher, more space emerges and seed is produced. As other trees die, new species and trees fill in spaces.
4. **Steady state**, where a small number of the original trees occupy the canopy.

These stages offer a good starting point for woodland owners to assess their woods and determine which stages different patches of forest are in. (More on this in chapter 4.)

The concept of disturbance is an unknown variable in this process, and can be anything from a single tree falling in the forest, to a thinning undertaken by the farmer, to an ice storm or hurricane. Each of these disturbances has a different scale of impact, and good management helps support the forest's ability to withstand devastation from these unknowns.

This process of development is a pattern all forests exhibit—that is, if we allow it to. The pattern of most forestry in the temperate United States over the last several centuries has been to enter the woods and harvest the prime specimens, which fetch the best price at the mill. The less successful stems are left and never quite fill in the space left behind by the removed trees. Oftentimes invasive brush moves in next, and arrests the successional path of the forest, slowing it and changing its ecological value.

Ideal management, then, is to understand and support and concentrate growth on the healthiest stems, while removing and utilizing the less desirable ones. Much of this work can be summed up as "thin as Mother Nature would thin"—in other words, observe and notice the trees that are already in the process of decline, and remove them first.

In the pasture we can also support this process by changing the way we plant trees. Many people approach tree planting by researching the mature spacing a tree would need and planting based on this measurement. So black walnut seedlings, for example, might be planted 40 to 60 feet apart. But this isn't how nature plants trees, and it's really betting severely against the odds. Instead, farmers would benefit from seedling trees at close spacing—even as close as 2 to 3 feet apart. This method provides multiple benefits, as trees growing this close together will help keep each other straight. Over time the manager can select the trees with the best qualities, and thin the others, essentially mimicking the process found in the forest. More on tree spacing considerations and patterns in chapter 5.

TREE SIZE DOES NOT EQUAL TREE AGE.

This relates directly to the previous principle. One of the myths many of us were taught at a young age was that older trees are bigger, and smaller trees younger.

Figure 2.4. The stump and "tree cookie" pictured are from two sugar maple trees that were growing right next to each other, with all the same site conditions at play. Both are 90 years old, illustrating the challenge of associating tree diameter with age in a given forest.

Walking into a typical forest, people assume they see two ages, with trees maybe 14 to 20 inches in diameter occupying one generation, and those that are maybe 4 to 8 inches a second—we assume they are the kin of the larger trees. Next time you cut trees in your woods, however, count the rings.

Figure 2.4 pictures two maples that are both approaching 90 years old. It's hard to believe until you count the rings; it's a facet of the forest that many of us have overlooked. The stark differences in the sizes of these trees, given that they are the same age, are a remnant of the stem exclusion phase of forest development.

So what causes one tree species to do so well, while another does so poorly? With trees, the soil type, aspect (direction a slope faces), and microclimate are all major factors. Within a single species we can interpret what

we see as a genetic expression—and also, in some cases, the luck of the draw.

All this prompts a question: What are the implications for management in a silvopasture setting? The most dangerous error is to assume that cutting a large tree will open the space for that smaller tree to grow into a large one. It just won't happen. So when you're starting with an existing woods, it is important to identify the best trees and retain them whenever possible. If you're converting a woods to silvopasture, a crucial step is a relatively aggressive thinning, so that the canopy cover is around 40 to 60 percent. This creates the light conditions for grasses, legumes, and forbs to grow.

Likewise, in a pasture setting, this pattern will occur, but in reverse. Trees are best planted at a high density, and then thinned as necessary to maximize growth. In pasture, the focus is on optimizing growth to get above the browse height of the grazing animals, so one can integrate the trees and animals into a single paddock. Trees are planted densely so that thinning of the less vigorous can occur over time, so that the best trees are ultimately favored.

IN TEMPERATE FORESTS, TWO-THIRDS OF BIOMASS CREATED GOES DIRECTLY TO DECOMPOSITION.

There are many idyllic pictures of silvopasture floating around that depict a clean image of animals, trees, and grasses. While these images are beautiful in their simplicity, the woods are not that groomed in their natural state. Healthy forests rely on a varied assortment of downed trunks, limbs, branches, and twigs—known officially as coarse woody debris—to support a variety of beneficial animals, insects, bacteria, and fungi, all of which together support a healthy soil food web.

Biomass (the plant material that results from trees, shrubs, and herbaceous plants capturing and transforming sunlight) should be maximized in a silvopasture system. As producers, we value biomass when it is living, as leaves on trees and lush, vibrant pastures. Perhaps this is because of the direct correlation of such grasses with pounds of meat or board feet of wood. But when biomass dies, we call it "waste," and many woodland owners seek a clean woods, with the debris removed from the forest floor.

Ultimately, in temperate forests the true wealth is in the soil. The health of the soil determines how well

Figure 2.5. This recently thinned silvopasture has a lot of debris remaining on the ground, which has multiple benefits over cleaning it all out. The material supports regeneration, improves soil biology, and even serves as a scratching post for livestock.

your trees and forage will grow. And research from forest ecologists, including David Perry, has demonstrated that while we were taught that plants make food for herbivores, which are then eaten by carnivores, and then decomposed by all manner of microscopic critters, the process isn't actually so linear. In fact, two-thirds of the biomass in a forest skips this chain and goes directly to decomposition.[2]

Healthy forests equal healthy soil, and biomass is necessary for this process. It might seem messy or inconvenient, but animals don't really mind grazing amid downed treetops and branches. Diversity in the size of biomass is also important. Leaves, twigs, branches, and logs all decompose at different rates and offer different types of value to the forest. Standing dead trees, known as *snags*, also have important value. All told, a mix of these elements into silvopasture is a good thing; it should be encouraged and balanced with other needs, such as the ability to mow, place fencing, or move animals.

REGENERATION IS AN EVENT IN THE LIFE OF A FOREST.

The regeneration of a forest is the period when favorable conditions allow a new generation of seed to germinate and grow into the next canopy. In modern woodlots, regeneration is challenged by the impact of human management, as well as overbrowsing by deer. One of the most common questions silvopasture practitioners ask is, "What about forest regeneration?" There is often concern expressed that grazing animals won't allow for trees to regenerate from seed, due to trampling and grazing.

Thinking back to the stages of forest development on pages 35 to 36, regeneration occurs at the end of stem exclusion, though it can realistically happen anytime light is opened up or the forest is disturbed in some way. In the context of developing a silvopasture then, at some point parts of the forest will need to be excluded from animals so that trees can regenerate. This might be an entire patch of the woods spanning several acres, or smaller portions that are fenced off. New seedlings may come from the existing seed stock in the woods, or with the addition of plantings.

Regeneration is an important part of the planning process, but depending on the life stage of the forest, it may not even occur in your lifetime. A good forester can assist you in determining the stage of your woods and when to plan management around the need to think about a forest for the next generation. More on this in chapter 4.

Lessons from Grassland Ecology

Grasslands and landscapes dominated by grasses are a significant portion of earth's land; it is estimated that 30 to 40 percent of total terrestrial area encompasses these communities, which is more than any other type of biome.[3] There is a long history between grasslands and grazing animals, but also between grasslands and humans, as these ecosystems have developed some of the richest soils known, which often have been valued by agricultural humans for their fertility. By extension, the cultivation and breeding of certain grass species, such as corn, rice, wheat, and sorghum, formed the foundation of many civilizations.

Grasslands on the global scale are disturbance-rich, primarily from the forces of climate (dramatic weather, long extended periods of drought) as well as from grazing and fire. These three major forces are often acting not alone but in relation to one another, at least in a historical context. For instance, periods of drought contributed to the accumulation of flammable plant materials, which often led to fire from lightning strikes or other sources, including native peoples.[4]

The major difference between grasslands and forest is, of course, the dominant vegetation. While forests hold a substantial amount of biomass aboveground, grasslands have most of their biomass underground, resulting in vegetation with a large root-to-shoot ratio. This allocation of energy into the ground, coupled with relatively slow rates of decomposition (depending on soil type) and weather, often results in the production of very deep and fertile soils.

It follows then that most of the highest-producing agricultural soils in the world were originally grasslands. In parts of the Midwest, topsoil was measured in feet. Now, due largely to intensive farming and tillage over the past 100 years, it's measured in inches. And no doubt the fertility of these landscapes was due not only to the vegetation but also to the thousands upon thousands of grazing herbivores that passed over it.

Grasses: The Main Course

Grasses send shoots, known as *tillers*, upward from their base. Tillers are sometimes vegetative and other times reproductive (flowers and seeds). Their growth is initiated from plant tissues that exist just below the surface of the soil in a growth pattern very different from those of trees and shrubs, whose newest growth occurs at the tips. Instead, the oldest portion of a leaf of grass is at the tip, while the youngest resides down near the soil. This quality allows grasses to quickly replace lost vegetation when grazed. The root structures of grasses are highly variable, but can be generally described as branched, fibrous, and concentrated in upper layers.

Good grasslands have a mix of annual and perennial grasses. Annual grasses are fast growing, reproduce through seed, and are more prevalent after large-scale disturbances. There are two general types of tiller development in perennial grasses: bunch-forming and sod-forming (rhizomatous). Sod-forming grasses utilize *stolons*, which are aboveground stems that run along the

soil surface, or *rhizomes*, which run belowground. In either case the grasses spread laterally and produce new tillers. Bunch-forming grasses focus their production around a central stem.

The leaf structure of grasses is narrow and composed of thick cell wells that provide strength to the plant, allowing them to remain upright or recover from environmental and animal impacts. Many grasses also have what are called *bulliform cells* that allow for leaf rolling in response to water deficits or excessive light stress. One indicator of the incredible strength in grasses is that science now widely agrees that the evolution of abrasion-resistant teeth in modern grazing animals came in response to the wearing of teeth over generations of grazing predominantly grasses, a history that has been traced back to the time of the dinosaurs.[5]

In the context of grazing systems, of maybe 10,000 species of grasses globally and 2,000 in North America, about 40 are used commonly as forage, making up generally 60 to 75 percent of the forage diet.[6] Grasses are further divided by the ways in which they synthesize sunlight, through what is called either a C3 (cool-season) or a C4 (warm-season) pathway. The anatomy and use of enzymes of these two groups evolved in response to the climate conditions in which

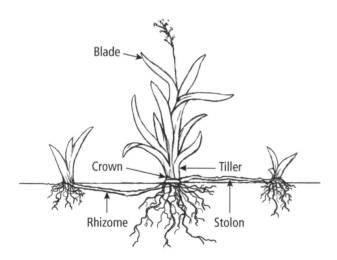

Figure 2.6. Grasses concentrate new growth at the base, with some remaining in clumps, while others send out runners aboveground (*stolons*) and others belowground (*rhizomes*). Illustration by Camilo Nascimento.

Figure 2.7. This map shows preferred grass types based on seasonal temperatures, with the transitional zone being most tolerant of both types in pastures.[7]

they evolved, which offers important options for those managing grazing systems.

Cool-season grasses like to grow in temperatures ranging from 65 to 75 degrees F (18–24 degrees C), and can initiate growth when soil temps get into the 40s F (4–10 degrees C). Thus they do best in temperate grasslands during spring and fall, and actually go into a subtle form of dormancy during hot summer months and during short-term droughts. Annual cool-season grasses include wheat, rye, and oats, while common grazing perennials include fescues, orchardgrass, and perennial ryegrass.

C3 grasses produce a higher percentage of crude protein than C4s. They utilize a strategy of photosynthesis that is highly efficient; less CO_2 and water are lost during the process than among C4 grasses. These grasses initiate growth when soil temps reach 60 degrees F (16 degrees C), and thrive in the 90 to 95 degrees F (32–35 degrees C) range. Annual grasses in this group include corn, Sudan grass, and millet, with perennials including Bermuda grass, bluestem varieties, tall fescue, and switchgrass.

Depending on your climate, your pasture may favor cool- or warm-season grasses, or you can introduce these to support a more diverse and resilient pasture. Naturally, as you go farther north, cooler grasses prevail, while the opposite is true for points farther south, which offer conditions favorable to warm-season grasses, as shown in figure 2.7. This map shows the dominant types that should make up the majority of pasture, but doesn't mean that experimentation isn't warranted, especially in northern climes where warmer summers are becoming more common with a changing climate.[8] The mixture of grasses, along with legumes, can potentially pair well so that the warm-season grasses fill the gap, as shown in figure 2.8.

FORBS AND LEGUMES: SIDE DISHES

Forbs and legumes are also important members of the ecosystem, but usually make up a smaller proportion of pasture plants. A *forb* is technically any flowering plant other than a grass, which would include legumes. The reason legumes are differentiated is their ability to harness nitrogen-fixing bacteria on their root structures,

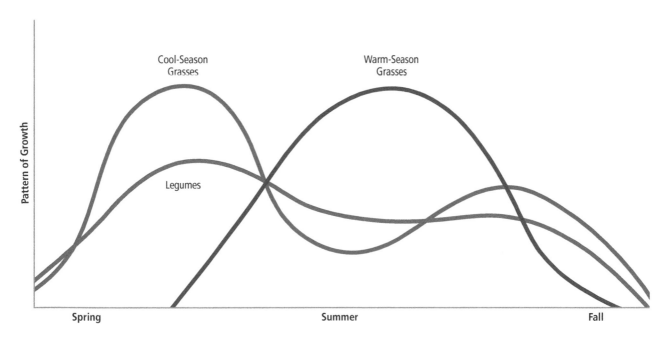

Figure 2.8. The combination of cool- and warm-season grasses can potentially help balance seasonal growth curves, depending on your local climate.

which synthesize nitrogen from the atmosphere and bring it into the plant tissue, which is then made available to grazing animals. And while there are some 12,000 species of legumes worldwide, only about 40 are commonly used in managed grazing, including the clovers, trefoils, and medics (such as alfalfa and lupines). These plants might make up 15 to 20 percent of an ideal pasture distribution.

Forbs are, in some senses, everything but grasses. Many of these plants have taproots, and they often accumulate or concentrate nutrients from the soil into their plant tissue. Common plants in cool temperate pastures might include dandelion, comfrey, chicory, lamb's-quarters, yarrow, sorrels, docks, and brassicas such as mustards and radish, which are often planted. Forbs bring structural diversity to the soil roots, attract pollinators, and provide nutrient-dense foods and even medicinal compounds to grazing animals.

All in all, the species diversity of grasses, legumes, and forbs underfoot is often much greater than in most forests. But like forests, a patch of pasture or grassland is a mosaic of species and patterns rather than uniform across the landscape. Like forests, there can be "gaps" in the "canopy" of the grasses, where smaller disturbances such as animals bedding down or pooling water can lead to different species compositions.

GRASSLAND COMPOSITION AND GRAZING

Almost all grasslands have experienced grazing as a selective force in their history, sometimes consistent, and at other times periodic. The actions of grazing animals are one of the primary factors determining both the composition of species and the patterns of biomass production in grasslands. Historically, we see that many of the large grazing herbivores such as the African buffalo or North American bison primarily consumed grasses, opening up niches for less abundant forb species.[9] This pattern has been applied in many different places worldwide as a method of conserving and enhancing biodiversity in grassland ecosystems.[10]

The choices animals make in their natural ways of seeking out food have inevitable effects on the plant composition in a grassland. Since most grazing animals are highly selective, they spend time seeking out their preferred foods first, working down a chain of preferences from most desired to least. This relationship is one that has developed over millions of years: Grazing animals impact grasslands in selectively consuming some species and ignoring others, all the while trampling vegetation, disturbing soil, and recycling and redistributing nutrients.[11]

This process can lead to problems, especially in continuous grazing situations, where the undesirable (and even problematic) plants, which are ignored, are given an advantage over those that are desirable. Anyone who has ever explored an unmanaged continuous pasture sees this: a patchwork of grasses among towering thistles, mullein, and horse nettle.

For this reason, as you engage with more active management in new pasture, you'll often need to follow grazing animals with clippers, a scythe, or a mower to knock those undesirable plants back. This form of human disturbance might, in effect, be substituting for the effects of fire or other environmental disturbance, though perhaps more crudely. Often, over time, rotational grazing can shift the species composition and favor native and perennial plants, which have adapted to grazing and are among the more desirable from the animals' perspective.

One of the major reasons grazing animals are so critical to functional grassland is that they offer a major disturbance force to unlock and cycle plant nutrients. Nitrogen, for example, can remain bound up in plant tissue, stems, and roots for years or even decades, slowly decomposed by soil microbes over long pulses of time. Grazing animals rapidly consume plant tissue, process it, and put it back on the landscape in a form that is readily available for uptake by plants (poop).

This means that the addition of animals can not only increase the volume of available nutrients in a pasture but also potentially redistribute those nutrients, since animals eat in one place and poop in another. Research finds that ruminant animals retain only 5 to 25 percent of nitrogen, 25 to 35 percent of phosphorus, and 2 to 12 percent of potassium in their bodies. The rest is given back to the pasture.[12]

Figure 2.9. At Wellspring Farm we mimic grassland disturbance by targeting areas of poor pasture with our winter feeding, leaving the sheep in one place for four to six weeks. During that time they are fed high-quality hay with seed, and any waste is actually seed for the soil. The two images show the "before" of an impacted area in early spring (*left*), and the "after," about a month into the grazing season (*right*), where seed has taken hold and greatly improved the pasture.

Animal activity also causes gaps in vegetation, often around areas where watering, loafing, and other impactful activities occur. These activities are the result of many combined forces, including grazing, trampling by livestock, rodents burrowing the soil, plant competition, and root die-back after grazing.[13] This concept is applied in managed grazing as *winter bale grazing*, where hay is placed in areas that need renovation and species diversification, which the animals perform as they tear through a bale during the winter, thereby trampling existing vegetation and spreading seed through missed hay and their manure deposits.[14]

While globally distributed, grassland ecosystems are hyperlocalized in terms of their species composition, distribution, and dynamics. An important lesson for any farmer working to develop silvopasture is to consider the interactive forces of climate, the type of grazing animal, and other potential disturbance factors. Details are important.

For instance, climate factors can be highly variable, even within the span of just a few short miles. In the Finger Lakes region of New York, the beginning of the growing season can come several weeks earlier for farms along the Finger Lakes than for those just a few miles inland, due to the temperature-moderating

effect of the lakes. Also important to consider is not just the type of animal (cow versus goat versus pig) but also the breed, as there are many observed and a few research studies[15] that show great variation in breed habits, preferences, and patterns in grazing behavior. More on this in chapter 3.

Lessons from Savanna Ecology

We could say that the savanna ecosystem is the middle ground between the forest and grassland, where there is a balanced coexistence of trees and grasses.[16] John Curtis of the University of Wisconsin defines a savanna as a place where trees are a component, but where density is "so low that it allows grasses and other herbaceous vegetation to become the actual dominants of the community."[17]

In effect then, a savanna is a grassland with scattered trees; often in ecology texts they are simply described alongside each other.[18] In some cases conservationists express concerns with trees showing up, using the favorite word *invasive* to represent concerns over some species threatening the integrity of what they see as native grasslands.[19] This fits nicely with all the presumptions that one species belongs and another

Figure 2.10. Historical maps of the estimated temperate oak savanna of the midwestern United States and longleaf pine savannas of the Southeast. Image courtesy of GR McPherson.[20]

does not—a notion that is being challenged by many today as we reconsider what *native* versus *invasive* really means.[21]

Savanna ecosystems are generally defined in terms of how open the canopy is, with 50 percent being perhaps the upper limit of cover, though arguably 10 to 20 percent cover would still qualify. What then are the key points that a savanna has to offer?

For one, the trees grow very differently than in the forest. In open conditions trees don't face competition for the canopy, so they retain their lower limbs and often grow in a round, full form. This could have a variety of implications, depending on farmer goals. A larger crown on the tree, for instance, would yield more sap for tapping, but poorer-quality timber wood. The scattered pattern of tree dispersal also creates a unique and diverse light dynamic, where the entire continuum of full sun to full shade can be expressed throughout a pasture area. This allows for the greatest range of possible species to find a niche in which to grow well—if, that is, those species can survive in the delicate balance a savanna maintains.

From a biomass standpoint, savannas offer the best of both worlds, maintaining a delicate balance of various forces.[22] The balance of trees and grasses is maintained because of several factors: Less precipitation and soil moisture favors the grasses, and the presence of fire severely limits the tree species that can exist. Grasses also recover better from grazing, limiting the amount of successful tree regeneration.

Water use of grasses and trees is very different: Grasses have a strategy to capture water mostly from surface layers of the soil, while woody plants and trees usually use considerably more water from the deeper layers.[23] Trees can also help draw nutrients and water from lower portions of the soil in a process known as *hydraulic lift*, through which trees and some plants are able to move water upward in the soil, where the demand is higher.[24] Research has found in particular that grasses thrive as a result of this process, and that the presence of trees in this way helps support stable savanna ecosystems.[25] This strategy is likely more important in drier climates, or when humid climates experience abnormally dry years or periods of drought.

While savanna ecosystems tend to show up in more arid regions worldwide, one excellent example of a cool temperate savanna persists in the Upper Midwest, where a transition from tallgrass prairie to deciduous forest offers a template for modern agroforestry farming, a concept popularized by permaculture farmer Mark Shepard in his book *Restoration Agriculture*.[26] The historical oak savanna[27] of the Midwest in fact stretched from Wisconsin to Texas, covering a vast amount of land, and mixing and bordering many grassland and forest ecosystems along the way.[28]

Historically, the openness of the oak savanna was maintained by fire, and of the major tree species in the Midwest the oaks are uniquely fire-resistant (the reason they are the dominant species). Fire maintains an open pattern on the savanna as well as consuming debris and releasing available nutrients in duff material. Low-intensity fires "reset" the grasses and other ground vegetation, allowing for species composition to shift and reorganize.

While the managed use of fire may have some potential use for silvopasture practitioners,[29] to a degree intensive rotational grazing and mowing can replace it, at least by knocking back woody plants that attempt to establish themselves, as well as by redistributing nutrients through animal waste. This type of management, however, doesn't replace fire with its ability to restore and reshape soil fertility.

The ecological concept of savanna, then, is perhaps our best image of what we want to emulate as we design silvopasture systems. The widespread adaptability of these ecosystems, coupled with their tolerance of fluctuations in precipitation worldwide, offers a promising model. Our major challenge is to mimic the pattern of the savanna yet consider the unique combination of species that best fits local bioregions. Also, we need to recognize that intensive rotational grazing is in many ways different from the types of ecosystems we look to as examples.

Take-Home Lessons from Ecology

The goal of examining the ecologies of forest, grassland, and savanna ecosystems is to consider our current understanding of these ecotypes, and how

Figure 2.11. This longleaf pine forest at the USDA/US Forest Service Southern Research Station is a prime example of a managed planting maintained with fire (and grazing) to both sustain a native ecosystem and provide economic yields to farmers.[30] This pattern is less appropriate for northern climates, though pine plantations were planted in many places in the 1950s and might benefit from similar management. Photo courtesy of USDA/US Forest Service.

their patterns can translate and support a robust silvopasture practice. To this end, let's name a few key lessons to take along with us as we think about applying them to silvopasture.

1. Silvopasture is a restorative practice.

Since no forest is wild, we have some confidence that our work can add value and diversity to the forested landscape. Focus first on the most marginal lands before considering how to integrate silvopasture into more mature woodlots.

2. Plant densely, and thin to support the best trees.

Based on the patterning of seed dispersal in trees, silvopasture implementation means we need to plant *a lot* of trees densely in pasture and assume many will perish as they attempt to establish. In existing woodlots we can support the process of thinning that is already under way, leaving the best and removing the inferior trees.

3. Plan for succession and for regeneration events.

By engaging in tree-based systems, we are managing on a much longer timescale—on the order of a human generation or longer. This is both challenging and perhaps a relief. We have time! Recognizing which stage our forest is at successionally is an important starting point. Many silvopasture projects will be either in the stand establishment or in the stem exclusion phase.

4. See our work, and the work of animals, as disturbance events.

Every activity in silvopasture is an act of disturbance, and clearly is a critical factor in shaping healthy ecosystems. Disturbance can have positive or negative consequences, and by naming and acknowledging it, we reframe the ways we contemplate our actions.

5. Support pasture diversity with grasses, legumes, forbs, and woody plants.

As we will discuss in more detail in the following chapters, there is only benefit from increasing the variety of vegetation found on our grazing lands for animals. These benefits extend to the ecosystem as well, supporting soil health, carbon sequestration, and moisture distribution. Within a paddock, we should strive to develop a diverse salad bar of options for our animals.

6. Abhor a monoculture; allow for many types of pasture and structure.

At the same time that we want lots of diverse plants in our silvopastures, we should not pattern for uniformity. In other words, we don't want each paddock to look the same. Trees can be clustered in some, and in rows in others. Some paddocks in our system may not have trees at all, which offers its own benefits. A proposal to consider: Think about designing for one-third open pasture, one-third pasture with trees (savanna), and one-third woodland pasture. These proportions are relative, not precise.

7. Aim for 40 to 60 percent canopy.

Regardless of whether we are adding trees to pasture or thinning woods, in silvopasture we want to aim for a balance of trees and forages. Grazing, mowing, and potentially fire are all forces that can support maintaining this dynamic.

8. Mimic the native ecosystem.

Study up and learn the native ecotypes that have adapted to local conditions for a long time. The historical assembly of trees, forage patterns, and disturbance events can be the basis of your silvopasture design.

As we've seen from our exploration of ecology, these recommendations are guideposts, not boundaries. Practitioners are encouraged to trial different ecological patterns on their landscape, with the general concept that diversity is good. There is no one perfect pattern to emulate, but rather some concepts that can be applied broadly. Many farming activities fight against the direction of nature, but the beauty in silvopasture is the potential to engage in a more ecological type of farming, one that to many outsiders may not look much like farming at all.

Traditional/Historical Silvopasture

In addition to understanding some of the key ecological patterns that inform good silvopasture, there are a number of traditional systems globally that offer lessons learned that we can apply to our planning for silvopasture moving forward. In several cases these examples provide words of caution or concern rather than clear directives for what works well. They certainly don't offer the whole picture, as the documentation of these systems is not very thorough, nor does it seem complete.

Also important to acknowledge in any exploration of history is that none of these systems called themselves silvopasture. Many stem from native and indigenous peoples' methods of land use, and in learning from them, we want to avoid appropriating them by using the terms we might assign in this modern time. These systems stand on their own merits, and we appreciate the lessons that they can offer us. Unfortunately, history does not usually offer an accurate or fair representation of all the people involved. We must learn what we can, while recognizing there is likely more that is unknown than known.

Spanish Dehesa and Portuguese Montado

The dehesa, or *montado* as it is called in Portugal, is widespread in southern and central Spain and southern Portugal. The focus is not on timber production but on promoting the maximum crown size per tree to produce acorns as browse for animals, along with fuel wood and, in the case of the montado, cork production.

These systems are often cited and used as justification both for silvopasture as a concept and for the use of pigs in such systems. Unfortunately, too often some of the more subtle details are missed, which we will examine here, to consider how this long-standing system might relate to grazing practices today.

The dehesa is traditionally seen primarily as a method to maintain pasture and grasslands, preventing the encroachment of shrubs and providing food for animals; the harvesting of any other products is secondary. In other words, the primary benefits are enough to justify the system. The two main ways shrubs have been excluded over time are through manual removal and by leasing lands to landless peasants, who grow cereal grains and make charcoal from the brush.

These systems are ancient; the first documented record dates back to 924 CE, though evidence exists to suggest the systems date even further back.[31] They arose in the context of the Mediterranean climate, with dry summers and moderately cold winters, along with low soil fertility. The low productivity made tillage and crop agriculture less successful, and so the ability of these approaches to capitalize on marginal lands was a major reason for their continued existence. Currently, the two systems occupy around 15 million acres in the west and southwest of Spain and southern Portugal.[32]

Figure 2.12. The dehesa can be thought of as large rangeland ecosystems at a scale similar to large open lands in the western United States, comprising millions of acres. Photo by Pravdaverita/Wikimedia.

Trees are usually in a wide spacing, closely mimicking the appearance of a savanna. Oaks are the most common species, the type depending on microclimate and elevation. Species of oak used include mainly the holm (*Quercus ilex*) and cork (*Q. suber*), and sometimes melojo (*Q. pyrenaica*) and quejigo (*Q. faginea*). Historically the trees have been intensively pruned, with different techniques for acorn versus cork production.

There are three major factors that have influenced how these systems have developed and changed over time:

1. Water scarcity, experienced in seasonal and prolonged droughts.
2. Nutrient availability in low-fertility soils.
3. The presence and activities of humans.

The extreme nature of these systems can provide some valuable insight into our own practice of silvopasture, if not additionally some appreciation of how good we have it in cooler, wetter climates.

Dehesa/montado are mostly found in regions that receive low and sporadic amounts of rainfall, coupled with long-term droughts. In the region of Sevilla, Spain, for instance, while the average yearly rainfall is around 22 inches, it can range from 15.7 to 39.3 inches. Several occurrences of 3- and 4-year droughts can be found in the data from the years 1865 through 1993, and summer droughts of 150 days have been calculated at a return rate of every 4.25 years.[33] This means that trees must survive a wide range of extreme conditions.

Since water and nutrients are the critical building blocks of any ecosystem, these two dynamics interplay in the dehesa/montado. The two main components, of course, are the areas located outside the tree canopy, mainly composed of grasses and herbaceous plants, and then the oak trees and various plants that make up the understory around them. Research has shown that the trees improve the bulk density of the soils around them, as well as create conditions richer in nutrients and organic matter.[34]

These improvements come from various mechanisms, including the actions of trees to pull nutrients from deeper soil horizons,[35] as well as the connection

Figure 2.13. The classic image of a dehesa is one of large, happy Iberian pigs rolling among the oak trees, happily feasting on acorns. The reality is that this event was only periodic, when there was enough mast and other conditions allowed for it. Photo by Guanbirra/Wikimedia.

between the animals' desire for shade and the inevitable deposit of nutrients under trees, from their dung. These actions result in conditions where soil water content is significantly higher under tree cover when compared with open areas.[36] It's important to note that the authors of these studies indicated that these qualities are emergent from a long-evolved and managed system, and that the same effect would not necessarily be present in younger systems. Tree age appears to be an important factor in these cycles of nutrient improvement.[37]

On the economic side of the equation, often the most attention is paid to the production of Iberian pigs, which are largely finished on masting acorn crops and said to produce the finest-tasting ham in the world, *jamón ibérico*. There is a deep cultural relationship with the processing and production of the meat into various products, known affectionately in some regions as the *mantaza*.[38] It is important to note that, in these systems, pigs are given a lot of land for foraging (up to 5 acres per pig), and that this is only a small part of the production cycle, when acorns are masting. Though pigs often get the spotlight in these systems, merino sheep production for high-quality wool has been of equal or greater importance, given that the sheep can graze more often on the available vegetation.

In regions where cork oak trees thrive, the sustainable harvest on a nine-year cycle provides a significant income to farmers. Portugal is responsible for over 50 percent of natural cork production globally. Cork is harvested during the dry period of spring/summer when the bark of maturing trees begins to naturally separate. This harvest is a significant income source for many rural communities, with sustainable management passed down as a legacy over centuries.[39] Utilizing the pasture space for grazing sheep and finishing pigs supports the longer rotational harvest of the cork trees,[40] a key concept to apply to silvopasture everywhere.

In addition to the ways these systems offer benefits to people and agriculture in rural communities, the ecosystems that have been maintained are important places for wildlife and, as an extension, hunting. Several endangered animals, such as the Iberian lynx and Spanish imperial eagle, have long found homes in the dehesa. The National Wildlife Federation has argued that the large-scale switch from cork to plastic wine stoppers ultimately has had a negative effect on wildlife habitat, a connection few people would ever imagine.[41]

Since the 1950s several significant changes have come to dehesa. These include a decrease in pasture management, an increase in deforestation, and a large decrease in human population, with issues of inequitable landownership leaving much of the land unmanaged. Because of a shortage of shepherds, a switch from sheep and pigs to cattle has occurred in some regions. And regeneration of oak trees remains an issue for this system, amplified by a fungal disease and other factors, which can severely affect the oaks and lead to rapid mortality.[42]

These systems, which have a relatively long history in practice, offer an example of not only how environmental conservation and rural economic development can work together, but also that arguably they are necessary factors to achieve the goals of both. In addition, the patterns of these systems suggest some powerful benefits to soil and ecosystem health, and also to wildlife.

We should be careful, though, because climate conditions, along with the extremely long time line of development of these systems, mean that some of the specifics may not apply to other situations. Many use the concept of the dehesa and the montado to justify woodland grazing of pigs, but miss the important factors of scale, timing, and animal density, which are necessary to avoid inflicting harm.

Kyushu Hill Farming (Japan)

Another traditional system offers insight into how silvopasture can be beneficial in a very different climate. Kyushu is one of the southernmost islands of Japan, where there is evidence that grazing in forests has occurred since the thirteenth century.[43] This region is characterized by wide seasonal variation in temperature and rainfall that totals well over 80 inches per year, and by its volcanic and mountainous terrain. These conditions set the stage for a set of agricultural practices that need to consider the dangers of soil erosion and take measures for soil conservation. In this context a production system combining farming, animal husbandry, and forestry has long persisted.

In a region known as Oita, 80 percent of the land is in forest. Some farmers there have introduced a species of oak called *Kunugi*, using the logs from the trees for shiitake mushroom production. Sometimes, the forest is also used as woodland grazing land for cattle. In one report farmers said they derived 40 percent of their income from shiitake and 30 percent from cattle, making the integrated system a significant portion of their income.[44]

To establish trees, sawtooth oaks (*Quercus acutissima*) are planted and cut when they reach an initial height of 40 feet, or every 10 to 15 years. These are done in small clear-cuts, rather than over large swaths of land, so as to provide an annual harvest of mushroom logs to sustain production, as well as to reopen the canopy for grasses and browse, all while retaining the overall conservation functions of the larger forest. The stumps are left and allowed to resprout in the practice widely known as *coppicing*.

As trees in these oak plantations grow, they are inhibited by wild grasses such as Japanese plume-grass (*Miscanthus sinensis* Andress) and sasa (*Bambusoideae* and *Pleioblastus*) growing on the forest floor. Farmers find it hard to find time to scythe or mow, and so have used low-intensity fire, as well as introduced cattle, to keep the understory managed. The cattle graze in

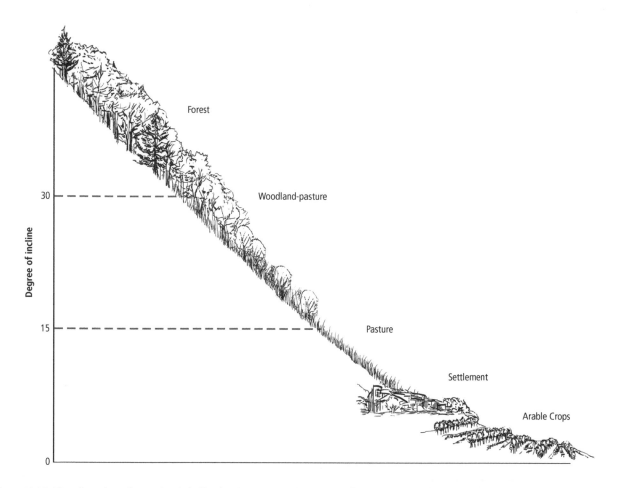

Figure 2.14. The allocation of steep lands in Kyushu, Japan, to various types of management. It's a good idea to always leave steep upper slopes in forest and not graze them, because of excessive erosion potential. Illustration by Camilo Nascimento, inspired by documents from the Hokkaido National Agricultural Experiment Station.

groups of 10 to 100 in very large paddocks, sometimes 50 acres or more, and are moved every 40 to 70 days, based on the observed condition of the vegetation and available feed for livestock.

Researchers from the Hokkaido National Agriculture Experiment Station documented farmer experience and noted in a summary paper that animal nutrition was not sufficient in pastures with only wild grass, recommending that pastures be planted with improved forage grasses to increase productivity. The anecdotal reports from farmers were that the manure from the cattle appeared to speed up growth of the Kunugi oak trees, shortening the interval from one harvest to the next.

A more thorough research project examining the dynamics of the grasses, cattle, and trees occurred in a study in the Minamioguni region.[45] Researchers examined tree growth, vegetation dynamics, and effects of grazing density over eight years and with thorough analysis. Their findings provide some valuable considerations for managing an ever-changing system.

The first area of concern is how stems are able to regrow after initial cutting and the introduction of grazing animals. Researchers found that, even though stems that regenerated from stumps were (not surprisingly) browsed and broken by cattle, as long as the density of cattle was low enough, regeneration was not negatively affected.

Figure 2.15. Coppice often looks like small clear-cuts with residual "standard" trees left behind, for seed or ecological benefits. This system is short-rotation and can produce a variety of beneficial wood products, including firewood and mushroom logs, which are the most common uses in Kyushu grazing systems. Photo by Johnaapw/123RF.com.

Figure 2.16. The long-term evolution of the Kyushu forest suggests that shiitake log production could mesh well with silvopasture systems.

The study also revealed that damage to sprouts occurred until the fourth year, with the most intense impact of browsing in the fall of years one and two, where damage was observed on 70 to 80 percent of trees. Still, the percentage of this damage considered fatal (from stem breaking and bark removal) was in the 4 to 13 percent range, which led to the stabilizing of stems and, eventually, forest.

Researchers also measured diameter at breast height (DBH) of trees over time, to monitor growth, and discovered that, in the short term, the no-grazing plot did better, because cows were prone to browse leaves and sprouts. However, after eight years average DBH per tree was actually higher in grazed than in non-grazed areas, likely in response to added fertility from the cattle manure.

As far as ground vegetation goes, the project found that grazing suppressed wild grasses over time, and also tended to favor improved forages. As stems grew to trees, the amount of forage was of course reduced each year. A clear pattern of management emerged over the decade-long rotation: In early years sprouts were abundant and were beneficial to manage, whether manually or by browsing. During years four through six, a sort of balance occurred among the cattle, emerging trees, and browse.

Overall, the density of animals, along with the timing of moving them, was clearly one of the more critical management factors. The study found that 60 to 80 cow-days/acre was the maximum on areas with wild grasses; up to as much as 150 cow-days/acre for paddocks with improved forages available. (A cow-day is the amount of feed one cow will eat in one day.)

The other critical management factor for growing good trees is stem density. As with all animal grazing and browsing, proper management was found to stabilize good stem density in managed oak resprout, with additional human pruning helping to increase the growth rate. The research concluded by discussing an important relationship: Cattle could be useful in managing resprout, but daily observation was critical to avoid too much of a good thing.

Again, as with the dehesa and montado, this system suggests a beneficial stratification of both products and management, where animals provide annual yields and help maintain the desired character of the woods. In turn this helps manage for the longer-term productivity, whether it be for cork, acorn production (which ultimately feeds pigs), or shiitake log production. While cork is particular to just one species of oak, shiitake is suitable on a wide range of species that would work well in a silvopasture, including oak, alder, birch, and beech.

TRADITIONAL GRAZING PATTERNS OF EUROPE

In many parts of Europe, there are centuries-old examples of livestock being integrated with tree systems. Only because livestock and other production systems have become more segregated in recent times are these systems really novel, because for a longer history they were known simply as "farming." Unfortunately, in many cases the record exists to describe that the practice did occur, but with less detail than is necessary to understand the finer points of management. Still, enough tidbits remain for some takeaway lessons that we can apply today.

One of the earliest mentions of mast feeding in literature occurs in Homer's epic work *Odyssey*, where the goddess Circe turns the hero Odysseus's men into pigs and feeds them acorns while they are captive. The casual mention of this relationship is suggestive, along with other references from Greco-Roman authors, including Strabo and Cato, who frequently mentioned pigs, acorns, and woodlands in their writing. In England this practice became known as denbera by the Anglo-Saxons, and later as pannage by the conquering Normans. While the former was a word relating to the practice of feeding acorns to pigs, pannage is derived from an old French word, *pasnage*, which refers to a payment given to a landholder for the right to bring swine in for feeding, during the season generally from late summer through midwinter. Pannage has a well-documented and varied history in many parts of Europe, cited again and again as a critical part of many economies and cultures over time.[46]

While pannage was important, it was seasonal, and cyclical. Oaks generally mast (put out copious amounts of seed) around once every five to seven years, a phenomenon only somewhat understood even today.[47] This means that, realistically, pannage formed

Figure 2.17. This historical painting from 1416 depicts a peasant knocking acorns down with a throwing stick to feed his pigs. Image courtesy of Réunion des Musées Nationaux.

only a portion of the swine diet. Much of the well-documented nature of the system had less to do with the practice and more with the economics, since rental payments were recorded as a matter of necessity as landowners gave access to their land. This has perhaps led to an unbalanced historical focus on its prevalence, which, while important, does not tell the full story of the farming system.[48]

In England there has long been a distinction made between woodland and wood-pasture, the latter, of course, being woods where animal activity occurred. In other words, not all woods were grazed, intentionally, so that other forest products could be procured.[49] Within the grazed wood-pastures, there were two main types:

Compartmentalized wood-pastures were stands of trees that were routinely *coppiced*, or clear-cut, so that the trees would resprout from the stumps. After felling, a coppice stand was often fenced for

six to nine years, until regrowth was resilient enough to withstand browse, at which point animals were given access, until the woods were cut again.

Uncompartmental wood-pasture was managed to include trees plus grasses and forage, where the trees were often *pollarded*, or cut high above browse height and allowed to resprout without the danger of damage from grazing. Or the trees were left to grow normally, in wider spacing, while forages developed underneath. This type of land use was developed both to graze domestic animals and to create "deer parks" that offered favorable conditions for wild game animals.

In practice it's most useful to see that land use was varied, with a mosaic of woodland types and intensities of management. Animals were given access for periods of time, depending on the type of management being practiced as well as the stock of available foods. Over time these landscapes have generally become more

Figure 2.18. A pasture with pollarded trees two years after cutting in Sluis, in the Netherlands. The remnant of this functional practice is mostly used as management for street trees in urban settings today. Photo by Charles01/Wikimedia.

homogenized, in part because of changes in the legal structure of landownership (from commons to private ownership) as well as the decrease in people managing the land as industrialization took hold.

OTHER EXAMPLES WORLDWIDE

There are many other places in the world where humans have long managed tree and animal production systems together. Unfortunately, documentation and details are often scarce, leaving us with just a few notions of what these systems might have looked like.[50] Given the lack of details, it's hard to distinguish when we might refer to these practices as silvopasture versus simply woodland or forest grazing.

In our ideal, silvopasture is something more deliberately and intensively managed, not a product merely of circumstance. It is important to note that in the instances where grazing in woodlands occurs, careful observation and the controlling of grazing access seem critical across the globe.[51] Due to a lack of good documentation, only a whisper of some systems remains, which doesn't give us a lot of good information as to how sustainable the practices might have been, including:

- Systems similar to the dehesa and montado discussed on page 47 in Italy and Sardinia (known as *seminativo* or *pascolo arborato*) as well as in South America (known as *espinal*), Slovakia, and Spain.[52]
- A global phenomenon of what is known as *transhumance*, or the practice of moving animals from lowland to highland pastures during the summertime, to preserve the home pasture for haymaking. Some transhumance events took animals above the tree line, and likely they grazed woodlands along the way. One example is the Swedish *fäbod*, where cultural practices are embedded in woodland grazing in the northern taiga forest during summer months when herds are moved to high-elevation pastures. This practice, which has been going on for hundreds

Figure 2.19. Cows wander into the taiga to graze in the Swedish Fäbod system during the day before returning to be milked and enjoy the safety of more protective paddocks in the evening. Photo by Costa Boutsikaris from the short film series www.Woodlanders.com.

of years, even developed specific songs and customs around the seasonal movement of the animals. Studies have found over 100 species that would not occur without the centuries-old grazing activity.[53]

- Traditional orchard grazing techniques in Germany, known as *Streuobstwiesen*, where animals were sometimes incorporated into management practices.[54] Some modern examples of orchard grazing are explored further in chapter 5.
- Remnants of pollard systems all over Europe, from Norway to Greece, dating back to the Iron Age.[55] These systems were developed as a way to harvest "tree hay" for animal feed during the dormant seasons.

None of these examples provide enough detail to inform our desire to learn from the past, so we can do better moving forward. What can be said is that the interaction of trees and grazing animals is definitely widespread around the world; in some cases the effects are positive, and in other cases likely destructive. When looking at North America, what we do know about are the perspectives central to colonization by Europeans of both land and people in the development of what is now called the United States and Canada. This story is one that must be remembered today, as it is one that makes the promise of silvopasture difficult, to say the least.

"Turning Livestock into the Woods": A Disappointing Legacy of American Forest Grazing

A few scattered references indicate that, historically, a variation on the Plains bison lived in many parts of what now constitutes the midwestern and eastern United States, perhaps traveling as far north as New York,[56] though some historians are skeptical that what is known as the eastern woodland bison made it very far.[57] Estimates indicate that at least some parts of the eastern forest had as many as 30 to 60 million of these beasts roaming the woods. This subspecies, known as *Bison bison pennsylvanicus* (though some dispute the distinct classification), was hunted aggressively by settlers, with total extinction having occurred by 1825.

This, and the killing off of the passenger pigeon both had radical effects on the ecology of the eastern forest.

These species tell a story of a forest very different from the one we see today. What we inhabit is a legacy of land that was stewarded by native peoples for thousands of years, then rapidly degraded and cleared by European settlers over just the past few centuries. This destruction included the mass extermination of species such as those mentioned previously, the wholesale clearing of the forest (mostly for farming), and, perhaps most damaging, the infusion of the European mind-set that draws a stark boundary between the forest and the field, labeling one "wasteland" and the other "productive."

This separation and categorization is rooted in the word *colonization*, which is defined as "the action of appropriating a place or domain for one's own use."[58] In order for this to work, colonizers must discount and declare both the land and any people inhabiting it as inferior, and declare their own worldview and approach superior. Much of the wholesale destruction of native communities was based in a belief that it was an act ordained by God for a people destined to achieve greatness. And in relation to the woods, if we label the woods as lacking value, it's easier to engage in its destruction.

The relation of colonization and farming is something that persists even today, in our attitudes and approaches to how we see and value land. Our societal notions of land ownership, dominion, and the right to do what we want on land we own all stem from this attitude. It's a perspective that has resulted in the devaluing of forest lands, and the viewpoint that they can be managed as sacrifice zones when it comes to livestock.

This has, in part, led to the idea of "turning livestock into the woods," where the need for additional forage, or to rest pastures, or to offer animals respite from hot weather has led farmers to fence a perimeter around a woods and just let the animals in. And while the USDA sugarcoats this as a "practice" with "less structured management goals," there is no doubt in the minds of many silvopasture practitioners that this type of unintentional management is nothing short of exploitative.

Table 2.1. Grazing and woodland grazing data for selected US states, 2012 USDA Ag Census.

State	Pasture Acres	Woodland Acres Total	Woodland Pasture	Percentage of Total Pasture That Is Woodland	Number of Farms with Woodland Pasture	Number of Farms with Alley Crop or Silvopasture	Percentage of Farms Practicing SP (at most)	Number of Farms Practicing Rotational or MIG Grazing
Alabama	2,269,315	3,333,046	496,015	21.86	11,999	119	0.99	5,548
Illinois	1,169,013	1,449,212	207,875	17.78	6,381	35	0.55	4,366
Iowa	2,478,116	1,165,549	347,743	14.03	7,565	13	0.17	6,642
Kentucky	4,214,208	2,745,655	665,010	15.78	21,224	96	0.45	14,652
Maine	118,980	773,652	27,105	22.78	1,103	58	5.26	1,372
Massachusetts	85,760	209,111	21,853	25.48	1,093	59	5.40	1,005
Michigan	619,986	1,175,893	110,067	17.75	4,641	38	0.82	5,065
Minnesota	1,877,600	1,641,521	439,332	23.40	10,172	62	0.61	5,604
Mississippi	2,382,767	3,469,315	470,724	19.76	9,216	65	0.71	3,998
Missouri	9,372,783	4,551,644	1,741,089	18.58	29,793	141	0.47	16,882
New York	985,494	1,613,045	146,995	14.92	5,286	186	3.52	5,878
North Carolina	1,416,886	2,145,710	270,242	19.07	12,675	119	0.94	7,207
Ohio	1,426,694	1,511,638	263,800	18.49	11,144	54	0.48	8,905
Oregon	10,824,500	1,764,937	1,167,078	10.78	5,346	87	1.63	6,705
Pennsylvania	814,210	1,804,157	134,964	16.58	8,420	141	1.67	9,280
South Carolina	872,080	2,036,260	181,880	20.86	5,711	51	0.89	3,274
Tennessee	4,059,581	2,303,156	737,308	18.16	22,524	51	0.23	11,766
Vermont	195,000	536,075	37,100	19.03	1,184	68	5.74	1,801
Virginia	3,047,595	2,465,061	464,186	15.23	12,531	74	0.59	9,315
Washington	5,785,508	2,139,141	1,141,696	19.73	4,624	82	1.77	5,798
Wisconsin	1,668,912	2,526,754	472,079	28.29	11,586	109	0.94	7,569
Average				19.92	204,218	1,708	1.69	142,632
United States	456,111,132	77,012,907	27,999,006	6.14	370,297	2,725	0.74	288,719

The number of acres in this type of land use is remarkable: Farms use around 15 to 20 percent of their forestland for grazing, relying on this for around 20 percent of their total pasture acreage. (See table 2.1.) All this, while on average just 1.61 percent of these same farms report practicing either silvopasture or alley cropping, a specific data point the USDA Ag Census asks for. This widespread use of woods in an unmanaged fashion has inevitably created a belief among many foresters, extension educators, and government ag officials that animals don't belong in the woods, and can only do harm. This perspective is arguably justified, given so much of what is currently being practiced on many farms, where root compaction, tree girdling, and soil degradation all lead to the long-term destruction of the woods, and often to tree die-off.

These issues are inherent not to silvopasture, but rather to farm mismanagement. You could argue, as Joe Orefice does so well in his article "Silvopasture—It's Not a Load of Manure: Differentiating Between Silvopasture and Wooded Livestock Paddocks in the Northeastern United States,"[59] that these problems accompany any continuous, unmanaged form of grazing. Poor management is poor management, regardless of the land cover.

Further confusing the dialogue is when farmers see themselves as doing silvopasture, or outwardly claiming it, when in fact they are doing woodland grazing. This is analogous to the current rage over "grassfed" animal products, which defines one variable in a wide range of management approaches. Grassfed meat can come from farms that give their animals little pasture and feed them only hay standing in one place, as well as from some of the most well-managed pasture farms in existence. Details matter.

Fortunately we're now seeing a renewed respect for and understanding of what good grazing practices really look like, and we can extend this education to the woods and silvopasture. The good news is that while nationally 370,297 farms report utilizing woodland pasture, 288,719 farms say they practice rotational grazing or management-intensive grazing. (These numbers likely overlap, but not necessarily.) In other words, while there are a lot of folks grazing their woods, there are also many farmers interested and likely convinced that rotational grazing has its merits, too.

No doubt some farmers reading this book might be among those who graze their woods, and we are not writing and fleshing out these details to make you feel bad, but rather to shine a light on a large, systemic error in American agriculture, where the historical context—rooted in colonization, which held the forest in low esteem—paved the way for misuse. The vast majority of folks may not understand the difference between woodland grazing and silvopasture, something that indeed has not been well articulated by the USDA, institutions, or educators. What is needed is not only education to distinguish the two, but also a shift in perspective so that we value and honor the forested landscape as important in its own right, and as a key component of a farm ecosystem. If that belief isn't changed, the system won't change, either.

Farmers are certainly open-minded, adaptable, and willing to learn, and these are the perfect traits for learning silvopasture. The practice must be grounded in both a respect for nature and a commitment to protecting the welfare of the soil, trees, forages, and animals. This is at the heart of silvopasture: not only a set of techniques, but a worldview holding humans and farming are a part of the ecosystem, not an overwhelming force that oppresses it.

LEARNING FROM HISTORY

Even though we might not know all the details of historical silvopasture systems, a few notable lessons can emerge from looking at the patterns they offer collectively. They provide both guidance and a few cautionary tales as we venture into developing our own systems.

Developed Over Long Periods of Time

Each of these systems has a long story of evolution and change. It is important to recognize that in many ways the silvopasture we are engaging with is the new introduction of an old idea. The lands we implement it on are perhaps at the start of a very long evolutionary journey. And it's going to look a lot different from many of the centuries-old practices. We are, at best, just getting started.

Large Ranges of Land Used, Seasonally and Cyclically

In many of the examples provided, the acreage in silvopasture was massive and animal stocking relatively low. Often the patterns of grazing were seasonal, more closely mimicking the patterns of migratory animals.

Mosaics of Woodland Patterns

While it might be easiest to picture a forest with uniformly spaced trees, it's clear from reading about these systems that the types of tree patterning, spacing, and forest cover were dynamic and different. This diversity was a valuable and important aspect of having many habitats available to respond to forever-changing conditions.

Human-Labor-Intensive

In all of these systems, people play a critical part as herders. Much of the decline of traditional systems can be correlated with the loss of people working the land. And observation and timing of when animals were moved were key to avoiding destructive behavior. It's clear that silvopasture demanded, and will continue to demand, significant human interaction to be successful.

Cultural

Another important aspect history can teach is how agriculture and the seasonal cycles of management and harvest relate to cultural attitudes, norms, and foods. The notable reverence for oak masting, and its translation into high-quality meat production, was and continues to be, a celebrated event. We can build culture, as we build silvopasture.

WHAT WE DON'T KNOW

The stories presented here are likely just the thinnest slice of the pie, as there is evidence on almost every continent of farming systems that included trees and livestock together. Sometimes the lack of information is because these were simply traditional farming practices that weren't distinguished with a name. In some cases there is mention of but little expansion on the types of species, interactions, and management undertaken by farmers. What is clear, above all, is that the practice is rich and woven into the complex story of agriculture around the globe.

It's only recently that farming meant single crops, grown in isolation, as well as the reduction of trees and forests valued mostly as a timber resource. At some point in history, agriculture largely shifted to mono-culture, tillage, and field crops as its main outputs, with animals being taken from ecosystems and fed in confinement, and the only productive trees being arranged in orchards, in straight rows. This is the story of modern agriculture, but we must remember that this is but a small blip in time. For a much longer period, things operated very differently.

There is a great opportunity now to shift the way agriculture works back toward the image of an ecosystem, in which a more complex plant community structure and animals interact in different ways, at different times of the season, or differently from year to year. In order to get there, however, we have to challenge one of the most potent attitudes in farming today: that animals don't belong in the woods, and trees don't belong in the pasture. The only way we can do this is by understanding the ways to steward animals, woods, and pasture together, to which the remainder of this book is dedicated.

KATAHDIN SHEEP AND LOBLOLLY PINES AT BRIERY CREEK FOREST FARM
By Chris Fields-Johnson, PhD

One of my great-uncles was a sweet potato farmer in Louisiana, and another was a dairy farmer in Virginia, but I grew up in the city and suburbs of Richmond, watching the forests I played in as a child being removed to make room for more houses and shopping centers. Growing up playing and exploring on these family farms, as well as hiking and camping in the Blue Ridge Mountains with my family, gave me a foundational connection to both agriculture and wilderness that has shaped my entire life. When I settled on a major in college, I chose to study forestry.

Studying at Virginia Tech opened my eyes to how much more there was to learn in pursuit of my interests. In addition to studying the core classes of traditional forestry, like timber cruising and silviculture, I took classes on animal agriculture, urban forestry, fire management, land reclamation, soil science, and eventually agroforestry, which brought together so many other disciplines. Not satisfied to just study these topics at a distance, I spent all of my free time during my college years working on a leased 300-acre private forest, composed primarily of a loblolly pine plantation, developing my skills and making improvements to the land. This never-ending project evolved organically into Briery Creek Forest Farm.

During a field tour in forestry school in 2007, we visited a loblolly pine forest that was being managed with intensive thinning and controlled burning to

Figure 2.20. Sheep grazing in loblolly pine at Briery Creek Forest Farm. Photo by Chris Fields-Johnson.

make habitat for the red-cockaded woodpecker. It was becoming similar to the longleaf pine savanna ecosystems of the Deep South, and it clicked in my mind that there was no reason I could not do the same thing with the forest I was working on. With the beauty of those large trees, the diverse ground covers, and the abundance of wildlife in that forest imprinted on my mind, I took a semester-long class on agroforestry that fall. There I realized that I could use silvopasturing of livestock as a tool to achieve the same effect while increasing the productivity of the land.

My problem from that point forward was figuring out how, as a broke college student with no land and no farming resources, I was going to transform the forest I was leasing into a viable silvopasture operation. I began graduate school in 2008, working on reforestation of Appalachian mined lands, and studying crop and soil environmental sciences. I literally rented a closet to live in at a house where some friends lived, made do without a car, biked everywhere, and went hungry often so I could save money and pay off my undergraduate student loans. There was no way I was going to be able to break into farming with the burden of student loans.

By 2010 my loans were paid off and I moved to a very small farm plot, renting it with some friends and family. Finally, I could start to buy animals and learn to care for them and breed them before bringing them out to the silvopasture project. I got East Friesian sheep, Katahdin sheep, Dorper sheep, Nubian goats, Boer goats, and Pygmy goats as I searched for the ideal breed for my operation. Buying animals from all over the place and bringing them to one small farm, I quickly learned about anthelminthic resistance and parasite management, realizing that an ideal system would move animals off a plot before parasite eggs grew to become infective, and would not return to that plot until infective parasite larvae were dead.

In February 2013 I began moving the animals out to Briery Creek, and by the end of March they were all in place. I had built a small barn with a greenhouse extension on the south side as a central shelter, but had no permanent fencing beyond the small barnyard, no electricity, and no running water on the site. I collected rainwater from the barn roof for the

Figure 2.21. Guard dogs have been integrated to protect the grazing animals. Their puppies also provide an additional income stream for the enterprise. Photo by Chris Fields-Johnson.

animals to drink. Electric netting was powered with solar chargers. My system was to move the electric fence and then the animals at least once every week, and to not return to a given spot for at least a year, to break the parasite life cycles.

I let the animals eat everything they could reach, then I would cut whatever else I wanted them to eat with a chain saw or brush cutter, leaving just the pines, and some high-value hardwoods standing here and there. Once they had stripped the felled trees of green leaves, they were ready to move again. At the peak under the original system, I had about 50 sheep and goats of different breeds.

The main problem with the original system was the ease with which the goats could escape. They would get out frequently, often knocking over the electric net fence and letting the whole herd loose with them. I sold off the goats, then found that the sheep were still

Figure 2.22. Thinning out the understory of things the animals won't eat. A lot of sweat equity has gone into this project. Photo by Chris Fields-Johnson.

eating all of the same plants that the goats were consuming. The next apparent problem was that the East Friesian sheep were not very hardy when consuming a large proportion of browse in their diet, and they were really bad at raising their own lambs under such extensive management, so I sold them off next. The Katahdin hair sheep, by contrast, were thriving at all life stages under the system, so I decided to continue to breed and purchase that variety exclusively.

Currently I have 24 Katahdin hair sheep rotating on about 30 acres of the 300-acre forest. Between new births and purchases of new breeding stock, I expect to almost double the flock every year until I find the optimal size, which I anticipate to be about 200 ewes. The flock is largely parasite-free, and is accustomed to eating nearly every understory plant among the pines. The forest is gradually transforming to more of the desired savanna landscape with every annual pass of the flock.

Figure 2.23. Ultimately, for this system, the Katahdin breed of sheep has worked out best. Photo by Chris Fields-Johnson.

3

Taking Care of Grazing Animals

Without the animals, there is no silvopasture. Animals bring the element of movement to the system, cycling nutrients and organic materials back into the soil and around the pasture. Livestock also offers the shorter-term yields that enable the finances to work for the farmer in a silvopasture system. In some cases livestock products may be the only yield, with trees and forage working to support optimal health and growth of the animals. In others, animals provide the foundation of a system that evolves to produce wood products, fruit and nuts, or timber as the long-term payoff of investments.

The inclusion of animals in a farming system is not just practical or functional but also, as we will discuss

Figure 3.1. Our sheep at Wellspring found their way onto the land first as a way for us to manage and maintain pasture and build soil, but we quickly fell for their charm and character, and relish the cycles of managing them throughout the seasons. Photo by Jen Gabriel.

THE GOOD GRAZIER'S LIBRARY

Managing good pasture is a lifelong learning pursuit. Several good books offer a variety of great perspectives and insight critical to good grazing:

- *Grass Productivity* by Andre Voison (1959). A true classic text for grass farmers worldwide, this book outlines many of the principles and patterns for sound grass-based livestock management.
- *All Flesh Is Grass* by Gene Logsdon (2004). A treatise on the true virtues of grazing, told in the poetic voice of an experienced farmer-philosopher.
- *Management-Intensive Grazing: The Grassroots of Grass Farming* by Jim Gerrish (2004). An incredibly practical, straightforward, yet inspiring outline and approach to grazing that maximizes productive use of the landscape.
- *The Art and Science of Grazing* by Sarah Flack (2016). A more recent addition to the "must-read" category, where the latest research on and knowledge of grazing are offered in a user-friendly format.

In *Farming the Woods* the concept of animals was explored briefly in the context of those animals that could forage food from the woods, in short pulses of exposure. The concept of silvopasture was deliberately left to a minimal discussion, as we knew that the topic deserved a separate and specific book unto itself.

Here we dive into the basic tenets of good grazing, noting along the way the relevance and relationship specifically to silvopasture. This is not a comprehensive text on good grazing practices, and readers should consult and learn from the range of excellent resources already available on the subject, noted in the sidebar. Where we do dive deeper is into the fascinating aspect of behavior-based management, a concept that honors an animal's own intelligence to seek out her or his own diet. This approach is critical to silvopasture, both in understanding the ways we design good grazing systems, as well as in highlighting the true offerings of silvopasture to provide a diverse diet of grasses, forbs, legumes, and woody plants, which taken together can provide the healthiest animals possible.

What Is "Good" Silvopasture Grazing?

While it can be argued that across the board animals will benefit from trees in the areas they graze, it's harder to prove that the presence of animals won't do damage to the trees, soil, and forages. Certainly there are plenty of examples of cows overrunning saplings, goats and sheep stripping tree bark, and pigs rooting deep holes around tree roots. And in some circumstances these behaviors can be used as a tool, while in other situations they need to be avoided, because they can inflict serious harm on the landscape.

In any given area of land, there is only a finite amount of food. The type of animal dictates the amount of food that can be utilized. The role of the farmer, then, is to assess the available food in a paddock, and ensure that animals are removed not only before that supply is diminished, but also before it is degraded so much that it cannot recover in a reasonable period of time.

Overgrazing is overgrazing. This isn't a question of how much the land can take, but of managing things in

in this chapter, a living and evolving relationship. Those who choose to raise animals also choose to distance themselves from them, or become intimately connected to them. The latter choice results in a dynamic where the farmers must think constantly about their animals' well-being as well as grappling with the ethical dimensions of choosing to kill some and keep others. Animals offer an intimacy that is unlike any other, and bring their caretakers into a complex world where the aim is to optimize the grazing conditions each day of the growing season.

Animals must be carefully selected, both in the general species and in the breed, in order to work best in a silvopasture system. In many cases farmers may already have the animals, and to some degree any animal can be trained to work within a silvopasture, though some will prove easier than others. You must take extra care if your goal is to have additional yields from the trees, whereas if shade and shelter are your main objectives, you have more leeway to experiment.

Figure 3.2. Pigs grazing a woodland at Joel Salatin's Polyface Farm. Grazing a paddock means ensuring that an animal has enough food, and supplementing if not, as noted here by the grain feeder in back. Once animals run out of food, damage is being done as they look for more. Photo by Jessica Reeder/ Wikimedia.

a way to ensure there is no harm done to the landscape. Overgrazing is simply giving animals access to a parcel of land for too long, which depletes both the existing vegetative resource and potentially the soil resource, too. Overgrazing also means that the recovery of the grazing area is very slow, or impossible. In silvopasture, this is further complicated by the fact that trees don't display signs of stress or decline as immediately as other plants do. While the evidence of damage to grasses and forages often appears in one season or less, trees may not exhibit the consequences of poor management until 5 or 10 years down the line.

All animals can overgraze, but there are additional challenges with larger animals (cows and pigs) and their potential impact due to soil compaction and damage to both surface roots and trees; there is even further concern with pigs rooting and destroying soil structure completely. Responsible farmers need to recognize this real threat and act accordingly. Management must be active and continual, and assumptions about the potential impacts need to be questioned continually.

Rather than attempt to define any absolute terms to say if one practice is right and another wrong, it's more important to keep the following questions in the forefront of your mind as you design and implement silvopasture:

- Is there bare ground as a result of grazing or foraging activity?
- Is there a loss or decline of forage quantity and density as a result of grazing?
- Has the percentage of organic matter increased, or decreased, over time?
- Is the soil more or less compacted over time?
- Is there any evidence of abnormal decline in the trees?

A key factor in answering these questions comes back to animal behavior, which is discussed in more detail starting on page 70. Animals that have access to ample food and are removed from an area before food becomes scarce and they become bored are less likely to do damage to plants, soil, and trees. A basic indicator we can name is: *Bare soil equals degradation*. But this alone isn't sufficient, because bare ground means you've gone far beyond the threshold, and any remedial action will be too little, too late.

Timing is of the essence, and in all cases with all types of animals, it's better to err on the side of too soon, rather than too late. A farmer's flexibility with this relates directly to the amount of available pastureland. All too often farmers push this capacity to the maximum, rather than allowing themselves ample room to be flexible and adaptable. In practice this means always being ahead of the game, having the infrastructure in place so that animals can move when they need to. At our farm we have learned over time to have enough fencing on hand to build two to three paddocks at a time, which allows us to move the sheep when they *should* move, not when we can scrape together the time in our busy schedules.

Assumptions are what get us all in the end—assuming that no damage is being done from overgrazing, for instance, or that the season will have optimal rainfall, temperature, and weather. The best farmers expected the unexpected and already have a Plan B in their pocket to fall back on.

On the flip side, one of the wonderful things about pasturing animals is the potential to always improve the on-site food source in both quantity and quality. This is the big advantage of utilizing farming systems with grazing ruminants (sheep, goats, cows), who can digest forages as the majority or entirety of their diet, versus grain-fed animals like pigs and poultry, which benefit only marginally from pasture improvement in comparison. Because of this factor, you will note that this book favors promoting the use of ruminants in silvopasture over monogastric (single-stomach) animals, though the options for regenerative use of several common species will be considered.

The Mandate: Rotational Grazing

The way we avoid overgrazing is by moving animals from one space to the next, controlling when they no longer have access to grazed land (so it can rest) and when they can gain access to new pasture. If you aren't ready or willing to incorporate frequent and intensively managed rotational grazing into your management, you aren't ready for silvopasture. At the very least, for the sake of the rest of us, please don't call it that.

As discussed in chapter 1, while the broader climate benefits of rotational grazing are a challenging dynamic to pin down, the more direct benefits to the farmer's fields and to animal health are much clearer and more straightforward. By moving animals, we reduce their exposure to disease, increase the forage capacity of our land, and build healthier soils. The formula for rotational grazing is not fixed, but accounts for the moving target of how ecosystems dynamically change over the course of the growing season.

Continuous grazing, on the other hand, enables someone to do as little as possible and often squeak

Figure 3.3. Fencing is the critical element of a rotational grazing system and performs two main functions: (1) keeping animals out of areas that need rest and recovery time post-grazing, and (2) preventing them from grazing new spaces until the timing is right. For well-trained cows this can be as simple as a single strand of hot rope fencing.

by. Generally animals are fenced in and given free rein over the whole pasture. This results in a feedback loop, where preferred plants are grazed repeatedly without being able to rest and regrow adequately, which eventually kills them. At the same time, undesirable plants dominate, seed themselves, and take over the pasture. Rotational grazing is more work, but it's work that, along with good planning, can pay off.[1]

Some farmers treat rotational grazing as a more optional and open-ended prospect than it is. They figure, dividing a 10-acre pasture in half is better than not, and while this might be marginally true, it's ultimately only less bad, and damage is likely being done in the long term.[2] In fact, it's very easy to improve practices that damage pastures, and yet very hard not to do any damage through grazing. This almost certainly will happen in the beginning, and is an ongoing opportunity to learn, adapt, and optimize.

Adding to the learning curve of the farmer is the challenge of how grass growth and recovery change throughout the season, as well as how they vary from one season to the next. The animals, too, change their behavior, preferences, and attitudes over time. These factors all mean that grazing is a moving target, and good practices are based on keen observation, flexibility, and adaptation. On the one hand this might sound like a lot of work; on the other this is an invitation, and even a responsibility, to engage in the dynamic cycles nature offers us as stewards of the land.

This prospect can be incredibly empowering, because the farmer has a large degree of control over this process, and can harvest X amount of forage, turning it into Y pounds of meat (or milk, or whatever yield is desired). In this equation X can be very low or high, and the more forage the animals harvest, the more food and fiber they produce, all while their health and happiness are supported.

Understanding all the dynamics at play in grazing takes some time. A great starting point is to understand the main variables that determine the outcomes, which include:

1. The ways forages grow.
2. How forages recover from grazing.

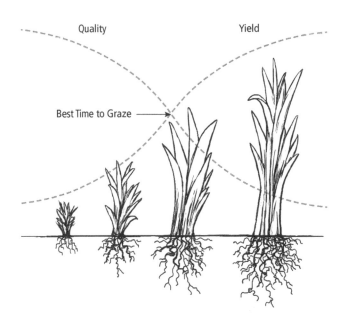

Figure 3.4. As grass grows, the volume increases, yet the quality for animals decreases. The sweet spot is somewhere in the middle, and always a bit of a moving target a good grazier aims for. Illustration by Camilo Nascimeito.

3. The effect that seasonal variations have on growth and recovery.
4. How different animals display preference for forages.

Then you can moderate the effect by managing:

1. How *many* animals are in a given paddock.
2. How *long* animals are in a paddock, and how long a paddock rests post-grazing.
3. The *size and number* of paddocks in the grazing system.

In the next few chapters, we will dive into a more organized approach to determining these factors, but for now we will examine the basic principles and patterns, deliberately keeping them separate from the calculations because, at the end of the day, good graziers rely more on experience, observation, and adaptability. An ecological system is not precise, but fluid and dynamic.

GETTING STARTED

As a starting point, the function of rotational grazing is to get the animals on a given area of pasture at the point when forages are optimal in both quality and quantity. These factors are actually opposing forces, and bigger isn't better. The highest quality occurs at the beginning of the plant growth, while volume obviously increases over time. The sweet spot is somewhere in the middle of this cycle, as shown in figure 3.4. The goal, then, is to put animals into a paddock at this point, keeping them there until they have harvested the bulk of this material, then moving them onto another area, allowing the impacted space adequate time to rest and recover before it is grazed again.

According to Sarah Flack, a grazing consultant who has worked with many farmers and the author of *The Art and Science of Grazing* (a highly recommended read), while there are many variables, the basic tenets of good grazing are:

1. **Pasture needs to rest and recover.** The amount of time is variable and based on the farmer observing how plants are growing, but is generally between 15 and 40 days depending on your climate and the time of year.

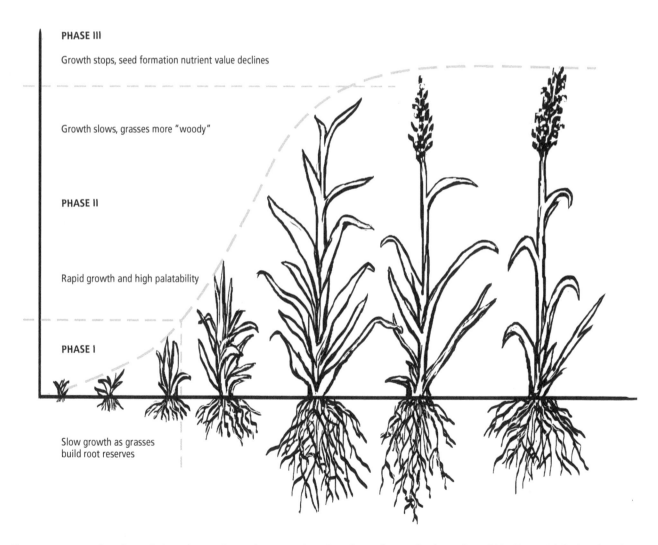

PHASE III

Growth stops, seed formation nutrient value declines

Growth slows, grasses more "woody"

PHASE II

Rapid growth and high palatability

PHASE I

Slow growth as grasses build root reserves

Figure 3.5. Forage has three distinct phases of growth. Targeted grazing aims to keep animals grazing within Phase II, bringing them in at the top of the curve and taking them out before they graze down into Phase I. Illustration by Camilo Nascimento.

Coupled with:

2. **Short period of animal presence in a paddock to prevent regrazing of a plant that is trying to regrow,** ideally around two to three days and no more than five. Many advanced practitioners move animals once a day, or more frequently.

Another lens to more accurately view this is through a forage growth curve, which has three distinct phases, as shown in figure 3.5.[3]

Phase I initiates after grasses are grazed or cut low, when the small leaf area of the remaining material can only photosynthesize at a low level and thereby grows slower. This is a time of rebuilding. Grazing during this time can be detrimental to the reserves of the plant. If forages are grazed down into this region, regrowth is slowed.

Phase II is when growth is rapid and extensive, and nutritional value is very high. This is the rapid upward growth of the curve.

Phase III is still exhibiting good growth, but slowing down. Fiber is notably high and so digestibility for animals decreases. Lower leaves on the plant are often dying off. Allowing forages to reach this stage results in uneven grazing and lots of residual forage.

Watching the patterns of grazing and animal preferences is key to identifying these phases accurately. For starters, going by the general height of the forages can be a useful measure, as shown in table 3.1.[4] An oversimplified starting point is to bring animals into a paddock when forages are 6 to 10 inches tall, and rest it when they are grazed down to 3 to 4 inches from ground level.

A further complexity that many new graziers quickly discover is the high variability in regrowth depending on the time of year. In many temperate regions, recovery from one grazing event to the next is as little as 12 to 15 days in the spring, as many as 30 to 36 days in the later summer months, and often over 40 days in the fall.[5] This effectively means that, at optimal grazing, the farmer must either double the number of paddocks in rotation or decrease the number of animals she or he is managing by 50 percent as the season progresses. Many resort to mowing their pastures in the spring, targeting animals on their highest-quality paddocks, and then grazing more and more of the pasture as the growth balances out. Most important is to target grazing to the slower parts of year, to ensure you have enough paddocks to cover the longer rest periods. We go through the calculations for paddock sizes and rest periods in chapter 6. They are left until the end, since we first want to think more about the animals, trees, and forage considerations of silvopasture before pinning down any numbers.

RISING TO THE CHALLENGE, AKA THE LEARNING CURVE

The main reason farmers avoid rotational grazing is because, in all truth, it can be time consuming and infrastructure-intense. Each paddock needs fencing, water, and shelter for the animals. It's certainly easier to

Table 3.1. Approximate heights for various forages at which to initiate grazing, and to remove animals from a paddock

Species	Plant Height at Start of Grazing	Plant Height at End of Grazing
Tall cool-season grasses (orchardgrass, canary grass, tall fescue, timothy)	8–10″	4″
Tall legumes (alfala, red clover, trefoil)	8–10″	4″
Ryegrasses	6–8″	2″
Short cool-season grasses (Kentucky bluegrass, white clover)	4–6″	2″
Warm-season grasses (big bluestem, sorghum / Sudan grass, switchgrass)	12–14″	4–6″

just have one place for these things, which often is the barn. Yet the benefits of some additional planning and work will translate to a better end in the longer term. Animals belong on the pasture, and their health, and your enterprise, will ultimately benefit from establishing a rotational system.

It is easier to teach rotational grazing to someone new to livestock management, since they don't have preconceived notions or established infrastructure that needs editing. That said, more experienced livestock handlers have the advantage of knowing the animals; they just need to retrain their brains to the tasks of observing forage growth, moving fence (in most cases), and a bit more planning. To be fair to those who've been at this a long time, rotational grazing was incredibly cumbersome until very recently, before the advent of various fencing systems that are inexpensive, and flexible to the needs of rotating animals. Portable fencing might arguably be the most innovative development of farming in recent times, if only we could learn how to use it well.

At first, the prospect of moving fence seems daunting and is often full of frustration, especially if you are learning how to use portable net fencing and it all becomes a tangled mess. Over time, though, the system becomes very manageable. Know that it will get easier, especially if you pay attention and cultivate patience around it. More on fencing systems and strategies later in this chapter.

With the assumption that we are ready to move on with a rotational grazing system, the next step in the equation is to understand the ways in which animal behavior forms the foundation of a good grazing system.

The Role of Animal (and Human) Behavior

Animal behavior is complex and currently understated in its role in grazing management. There can be behaviors of individual animals in our systems, and also collective behaviors of the mob. There are inherent behaviors for a given breed's genetics, evolutionary history, and age of animals, as well as many learned behaviors that animals exhibit. Individual animals, too, have personalities, and express these in a variety of ways. Darrell Emmick, longtime extension researcher and author of the wonderful guide "Managing Pasture as a Crop,"[6] notes that:

While the behavior of animals is often quite complex, understanding the mechanisms underlying their behavior is actually fairly easy. Behavior is a function of its consequences. Unfortunately, very few ever take the time to look past the immediacy of what an animal is doing at any given moment to try and understand what has motivated the animal to behave in a particular manner in the first place. Thus, many opportunities to take advantage of what is known about the principles of behavior to make our lives easier, the lives of our animals less stressful, and the overall operation of our livestock enterprises more efficient and profitable, have simply gone unrealized.

The invitation here is to see behavior as its own complex system of cause and effect, and seek to actively learn why as part of our work in managing animals. This means especially not assuming that an observed behavior is a result of whatever we first think it is. Questioning the possibilities and developing a long view of the herd are key—as is recognizing that animal intelligence and sentience exists, though not necessarily in the ways we think of these elements of behavior in humans.

BEHAVIOR IS RELATIONSHIPS

Livestock science often focuses overwhelmingly on animal nutrition and attempts to formulate some sense of an ideal diet for all animals at all times. A smaller but compelling body of research in a program known as BEHAVE (Behavioral Education for Human and Animal Vegetation and Ecosystem Management) based at Utah State University focuses instead on the ways animals select their own habitat and diet.[7] This effort has been led largely by Fred Provenza, whose background in wildlife biology paved the way to explore livestock management from the perspective of

Figure 3.6. Each time we visit animals in a paddock, we learn something new about their behavior. The times they are given new pasture are some of the most interesting, as they then take time to explore the whole space, discover the boundaries, and sample the buffet in a full circle before settling into grazing different areas.

the animals' behavior. This approach, which supports the needs of the animals, rather than predetermining them, leads not only to better systems but also to more profitable ones.

In practice, good grazing is a mixture of good understanding of the preceding concepts, a relationship to the animals, and an understanding of your own behavior, too. In reducing a complex system such as silvopasture to aspects of right and wrong, we do a disservice to perhaps the most important part of grazing: building relationships among the humans, animals, pasture, and, in the case of silvopasture, trees. Suffice it to say, from my experience witnessing many different grazing operations, those farmers who speak of their animals in kind ways and know them intimately always have the best operations.

Behavior also combines the genetic nature of an animal with the ways it learns from its environment (nurture). We can think of the evolution and specifics of the type of animal as the container, while the learned behavior from Mom, and interactions both with the social group and with the farmer, all dictate how the animal behaves within that container.

BEHAVIOR AND FEEDBACK

At the core of the concept Provenza promotes is that animal behavior is in relationship to the fact that the natural ecosystems and their participants are constantly adapting to change. Since change is the only constant, the way we manage livestock also needs to be adaptive. This isn't how most animals are kept in the modern world: fed in confinement, with a

PRINCIPLES OF LIVESTOCK BEHAVIOR

From www.behave.net.

1. Behavior depends on consequences.
2. Mother knows best.
3. Early experiences matter most.
4. Animals must learn how to forage.
5. Animals avoid unfamiliar foods.
6. Palatability depends on feedback from nutrients and toxins in food.

   - Nutrients increase palatability.
   - Toxins decrease palatability.
   - Changes in food preferences are automatic.
   - Toxins set a limit on intake.

7. Variety is the spice of life.
8. Everybody is an individual.

predetermined ration of food and a predictable, predetermined routine. Provenza notes, "Understanding how animals learn may enable us to train animals to fit our landscapes, rather than needing to manipulate our landscapes to fit our animals."[8]

This concept is particularly important for silvopasture, where we are looking at both potentially utilizing less common forages for food (like invasive species) as well as possibly wanting animals *not* to eat crops in the silvopasture (such as fruit trees, Christmas trees, and so on). The silvopasture landscape is inevitably one where change will be an important constant. And in the process of change and adaptation, there is a tension between what an animal knows and what it does not know.

There are several ways an animal builds learning and understanding that all work in tandem as a given animal interacts with pasture:

1. **Flavor feedback** from the exploration of forage in the landscape.
2. **Social group learning**, most notably in the mother-kin relationship.

3. **Satiety**, from the opportunity given for animals to choose variety in their diet and explore their surroundings.

FLAVOR FEEDBACK

At the core of understanding behavior is the composition of plants, which are each unique and complex in their available nutrition, and in how desirable they are to an individual. When we think about feed in the pasture, often the word *palatability* is thrown around, with the thinking that it relates to the taste or food preference of animals. The assumption is that animals like certain things but not others, and that little can be done to change this relationship.

Yet research shows otherwise, and if we can better understand the feedback loops between animals and forages, we can potentially shift the dynamic and train our animals to eat (or not eat) certain things. Palatability is not just taste, but also feedback in response to chemical characteristics of foods ingested, and the feedback animals receive in the gut as information. When looking at the composition of plants as a food, often compounds are identified as primary and secondary.

Primary compounds are those critical to plant growth and reproduction, and are those most often thought to be the foundational nutrition of a plant: proteins, carbohydrates, lipids, and acids.

Secondary compounds appear to play no role in plant growth and reproduction, and are often seen as repulsive to grazing animals, including alkaloids, tannins, and terpenes. These compounds often have evolved to support a plant's defense against environmental stressors.[9]

Much of the common thought can be summarized as: Primary compounds are good, and secondary compounds are bad. In fact, much research and focus within the world of secondary compounds have concentrated on toxins in plants. Lists are generated that strike fear in the minds and hearts of farmers not wanting to lose their animals.

Our farm encountered this early on in our grazing experience. We learned that black cherry was toxic to animals. We freaked out, discovering that our new pasture, which had been neglected and was dominated by woody vegetation, contained an overwhelming number of cherry sprouts. We spent hours clipping every last one, only to learn that the toxicity really was only possible in older developed leaves that wilted after being separated from the main trunk. With more research, we learned about the compound prussic acid, which is a cyanide toxin that forms when two components are combined with enzymes. Normally these are kept separate in the plant tissue, but during the process of wilting they combine and concentrate, thus presenting a potential threat.

This isn't to say that farmers should not worry about cherry or other potentially toxic plants; certainly we should be aware of where trees exist, and be sure to monitor paddocks, especially after storms when blowdowns are a possibility. But we should also recognize the potential benefits of our animals consuming secondary compounds and maintaining a diverse diet, one that includes woody vegetation. Our sheep, in fact, love consuming young black cherry leaves, with no ill effect whatsoever. The same can be extended to many "toxic" plants, including milkweed, black locust, and oak. Anything in excess can be dangerous, yet animals adjusted to a diverse buffet in the landscape are able to handle it. Their evolution demanded it.

Tannins are some of the most promising secondary compounds.[10] They have been shown to essentially slow down the digestion of ruminants without compromising their ability to access the food value in foraging. Research has shown that the moderate consumption of tannins offers several positive effects for ruminants, including increased milk production, better growth of fiber, increased lambing percentage, and most notably a reduction in risk of bloat and various problems associated with internal parasites.[11]

Equally compelling is research indicating that diets rich in tannins reduce methane emissions in grazing ruminants, which is a major critique of grazing systems' having a net positive or negative effect on climate change. Trees and shrubs with tannins essentially slow down the rumen, which decreases the emissions of methane (aka farting and belching).[12]

Figure 3.7. Willow is a phenomenal silvopasture tree, as it is inexpensive to establish, fast growing, productive, and high in tannins, which are good for digestion and health.

So where can tannins be found? In the pasture, birdsfoot trefoil is notably high in tannins; trees include birch, beech, willow, oak, and even the "invasive" buckthorn. And while overconsumption of tannins can have negative effects, much of the body of research fails to account for a paradigm underlying the overall conversation, assuming instead that animals lack the intelligence to regulate their own intake of tannins to a stable level. Research, again by Fred Provenza and colleagues, indicates that animals are able to self-regulate intake to a beneficial level.[13]

All this is to say: There is some compelling evidence that secondary compounds are beneficial to animals, in moderation, which they are able to assess and manage as a function of the feedback they receive. This further makes the case for getting tree fodder into animal diets as an important strategy for good grazing.

CHOOSING WHAT TO EAT

From an outsider's perspective, grazing can appear to be aimless and random. A closer look, however, reveals this is very far from true. Selecting what to eat has always been a dangerous prospect from an evolutionary perspective, and so animals have developed a complex strategy to evaluate and choose foods. The actions of grazing include searching for forage, selecting and grasping it, and taking it into the mouth. When entering a paddock for the first time, animals engage all their senses and work systematically to locate food, working down the chain from most to least desirable. They use their sight to survey the area, and smell to get a preliminary "taste" of the various options. Most remarkable is their sense of touch; whiskers on the snout are incredibly sensitive, being neurologically connected to the

Figure 3.8. Good pasture is a multistoried, diverse mix of grasses, legumes, and forbs, which offer a buffet for grazing animals to choose from. No two animals will make the same choices, as each has its own dietary needs.

brain. These sensory aspects activate memory and base choice on past experiences, while the consumption of forages and the resulting post-ingestive evaluation offer information that feeds into the next set of choices, all in a constantly revolving feedback cycle.

Choice includes not only *what* to eat, but also *where*, as the landscape can have a profound effect on eating. The boundaries of a paddock, how plant communities are distributed, and elements such as slope and sun exposure all factor into the experience an animal is having. At the foundation, animals work to satisfy their physiological needs, in order of necessity, which some suggest as:

1. Thirst.
2. Heat/cold.
3. Hunger.
4. Orientation and predator avoidance.
5. Rest.

Depending on the context, these five factors will change in their relative importance, affecting behavior and dictating how animals make their way through their food. As they feed, they seek the most desirable stuff, and when that resource is exhausted they move to mature green leaves, then green stems, and then, often from desperation or boredom, dry materials.[14] In this way they work from best to worst, and as managing farmers we move them to new spaces before they have to resort to the bad choices in a given paddock.

All this happens in the context of time of the day. While animals graze at any given point in the day, they tend to prefer sunrise, sunset, and the afternoon, and their rumen has evolved as such to largely function during off-times and overnight.[15] These patterns in time are particularly challenging to those who are milking: Optimal grazing times often coincide with the regular milking schedule.

MOTHERING WISDOM AND GROUPTHINK

While feedback and choice are behaviors an individual expresses, the process begins with young animals learning from Mom and, in some cases, from the group as a whole. It's remarkable, but perhaps not surprising, that

Figure 3.9. From even before her birth, this calf has received biofeedback about food preferences from her mother. She continues to learn from Mom as she milks for several months, eventually learning to graze herself as her rumen becomes fully developed. Photo by Uberprutser/Wikimedia.

this learning begins in utero, before birth. Preferences begin to evolve in the womb, and continue as animals observe and interact with their nursing moms, receiving the tastes of preferred foods in the milk they consume. Research also shows that lambs learn to graze sooner and spend more time grazing when they learn alongside Mom.[16] Where mothers offer stability and experience, the young bring creativity to the picture.

In addition to this intimate relationship, the group mentality of animals greatly influences outcomes. Animals instinctively understand that safety lies in numbers, and this can have a great effect on grazing patterns.

For instance, research out of Missouri found that if water was too far away (more than 800 feet), individual animals would only travel in the whole group, whereas with distances less than 800 feet, individual animals would visit the trough and more regularly return to grazing.[17]

As with any group, leadership of sound individuals, along with community mindfulness of the greater good, proves to be an important aspect of grazing management. The main way these dynamics are changed is through culling, where less cooperative members of the herd, especially those who cause trouble, may need to be removed for the greater good over the long term.

LESSONS FROM FRENCH SHEEP HERDERS
AND THE INTERSECTION OF ART AND SCIENCE

Among all the resources available about animal behavior, the one that is recommended most highly for all graziers is titled "When Art and Science Meet: Integrating Knowledge of French Herders with Science of Foraging Behavior."[18] This journal paper is a good summary, while the authors Meuret and Provenza also ended up authoring a book on the subject, titled *The Art and Science of Shepherding*, published in 2014.

The paper and book both describe a unique partnership between herders and scientists, acknowl-edging that the hands-on knowledge of sheep herders in France stands on its own as a valuable set of insights, but that coupling this with the experi-mental knowledge of scientists might give a more robust picture of how animals, people, and land-scape work in cooperation. After all, those managing animals really just need to know what works, while researchers are often after questions with implica-tions on a larger scale.

The process of herding animals over vast land-scapes might at first seem irrelevant to conversations

AS: Appetite Stimulator
FC: First Course
B: Booster
SC: Second Course

Figure 3.10. An experienced herder has intimate knowledge of the landscape and plants a grazing circuit much the way a chef plans a fine meal, mixing the types of forages available in just the right way to get maximum performance from his or her animals. Photo from Meuret and Provenza, 2015.

about livestock management on farms, but in fact many important lessons are embedded within this study, with useful application to any grazing scenario. Rising to the top is this idea: As useful as fences may be, they can't replace a herder's knowledge and constant daily attention to orient grazing on rangelands.

This certainly applies to any landscape. As researchers explored the system, they discovered something remarkable: Animals in these systems routinely consumed up to twice the forage of animals in confinement. Why?

One factor is clearly the choices presented to the animals in both scenarios. In grazing situations animals may sample 50 or more plants per day, while maybe 3 to 5 of those plants make up the majority of their diet. Most research in confinement focuses on one or sometimes two species offered in combinations that simply don't stimulate interest in the same fashion.

But perhaps the most interesting discovery of this study is the importance of timing and sequencing in the animals' foraging behavior. Through both conversations with herders and data collection, it became clear that the location and mix of plants, along with what the animals just ate, and even expected to eat next, shaped their consumption habits.

From the paper:

For instance, if a herder makes sheep graze for days on similar swards—grasses and legumes—without also taking them to other patches of different plants, the sheep get "bored," especially sheep that have experienced in previous years the array of forages in the area; they know other forages might be available and they come to dislike the herder. This leads to lower daily take, as meal durations become ever shorter because sheep are not "boosted" by some diversity when they reach partial satiety for a specific type of forage mix. Day after day, they know the forages and locations will probably be the same, and they rapidly satiate on both the forages and locations.

This knowledge has led herders to effectively become chefs, designing meals for their animals through the grazing circuit they choose, and when and where they decide to move them. Research backs up this concept. For instance, cattle will decline endophyte-infected tall fescue (very common in the South and West) if they first graze the fescue, then eat tannin-rich birdsfoot trefoil and alfalfa. If the combination is reversed, their forage activity remains higher, and thus their overall intake.[19]

The researchers and herders further articulated a complicated methodology of assembling meals, which is analogous to offering the animals an appetizer followed by the main course (the desired area to target), then a diversion to a booster and then back to the main course for a second go, once interest is stimulated. The circuit might end with "dessert": the highest-quality, most desirable forage available (see figure 3.10).

This idea of planning a menu is intriguing, and warrants a lot of further study, though it has some important take-home lessons we can begin implementing today. And while many of us might not be managing our herds or flocks as intensively as a herder, we can apply many of the same concepts for the benefit of our animals.

The first and most important is that we don't want monocultures or uniform goals for our forages in pasture. To this end:

1. Design a mixture of experiences and value the diversity of patches within our paddocks.
2. Arrange paddock boundaries to provide a mixture of vegetation, versus paddocks that are more uniform.
3. Consider rotation circuits flexible, moving animals in different patterns at different times.
4. Move animals when they are least interested in grazing (for instance, late afternoon) to boost appetite and encourage grazing for more of the day.
5. Design and plant mixtures of trees that provide a diversity of forages for the animals to sample.

We'll have more on these strategies in subsequent chapters, when we dig into the design of tree plantings and paddocks. It's highly recommended that anyone interested in this concept further read the materials available on this subject, which offer many more insights than are captured here.

TRAINING ANIMALS TO EAT, OR NOT

Beyond the fundamentals of behavior lies a question: If behavior is learned, can it be changed? In the context of silvopasture, the question is especially potent, because if behavior can change, the potential exists to both train animals to eat the vegetation we want them to, as well train them *not* to eat vegetation we want preserved. And while those who've spent time with animals might be skeptical that it is possible, both goals can be achieved, provided there is a period of concerted effort.

Training animals to eat something they normally would not is the easier of the two prospects. Kathy Voth spent the better part of 10 years refining a program, and she can cite plenty of examples that work. She outlined her method in a book titled *Cows Eat Weeds*, where the main components are to know the plant, train the right animals, and establish a routine where new becomes normal. By starting with the most willing animals (usually the young ones), and using the inclination animals have toward novel foods, it is possible to shift the culture of a group.[20] Kathy has found this method to be most effective when integrated in the pasture alongside the normal grazing routine. More on this practice later in the book.

On the flip side, animals can also be engaged in avoidance training, which usually takes place in a more controlled setting. Basically, the process is to associate consumption of a plant with nausea. You serve the animal the item in question, then follow it promptly with a dose of lithium chloride, which creates a temporary upset to the stomach. Experience dictates that this only works when the animal has not been previously exposed to the given plant, and generally it's been easier to train older animals than younger ones. Using a method outlined in detail by Utah State University, farmers have trained animals to avoid everything from grape leaves to olive trees and even ryegrass seed plantings.[21]

THE BOTTOM LINE: OUR BEHAVIOR

Animal behavior is a relatively novel topic in the livestock world, but it isn't new; perhaps it's simply been forgotten as we've been pursuing farming methods to maximize output. There is ample research to support the notion that animals exhibit conscious and intelligent forms of decision making and are in many ways able to care for

---

CHECKLIST: BEHAVIOR-BASED MANAGEMENT

❏ Provide a diverse diet, and trust the animals' ability to balance it.
❏ Start animals on pasture young and let them learn from Mom.
❏ Observe and learn the unique character of your flock or herd and its individuals.
❏ Maintain diverse pasture species mixed within and between different paddocks.

---

themselves and their young. For too long the general attitude has been one of dominion: We humans control our herds or flocks, and we think we make all the decisions. Instead, a closer look offers a clear picture—that animals are able to largely maintain a social community and to thrive, if only we provide them the space to do so.

Our role as managers, then, is not to fully determine the future of our animals, but rather to observe and support an existence that meets their needs for satisfaction, and ours for production (though many of us keep animals also for the pleasure of relating to them).

In the context of silvopasture, we can clearly offer a life to animals that is more comfortable, more diverse in food choices, and, frankly, more interesting. Fred Provenza, widely hailed in the United States as someone who has devoted a career to understanding the dynamics of animal behavior, has said it best: "Choice, and ability to choose, might be the most important aspect to production; increase performance and decrease cost."

The beauty of behavior-based management is that the costs are low; it mainly takes an adjustment of our attitude and paradigm. This will be an easier task for new graziers, who are not yet biased in their approach. For more experienced animal stewards, a new learning curve may need to take place, alongside all the other aspects of silvopasture. In the end, though, adding trees and woody plants to grazing offers clear benefits to the animals, if only we are up to the challenge.

Let's Get Specific: Choosing an Animal

How do we choose what types of animals to integrate into our silvopasture system? To start, we can separate pigs from poultry, and both of these groups from ruminants and other minor animals, including horses, oxen, and others. This text will focus on the major livestock animals commonly raised in the United States today, mostly because of the available information on their relationship to silvopasture systems.

Modern notions of silvopasture largely revolve around ruminants. This is because managing the grasses and forbs ruminants need alongside trees is a simpler equation than managing non-ruminant needs, with fewer questions and red flags. This book, and many who actively advocate for silvopasture, admittedly favor ruminants in these systems, and the research is there to support what is clearly a beneficial mutualism. Other animals, especially pigs, need a considerable amount of time-consuming and cautious experimentation.

Farmers often select animals based on either emotional or financial considerations—perhaps a certain character appeals to them, or they think a particular animal can turn the best profit. While these are important factors, ultimately species selection should center on the appropriate fit of animal to landscape. You want to raise your animals without any danger of degrading the landscape in the long term. The size of your property, stage of development, available local markets, and farming goals all play into the process of determining the best animal or animals for your silvopasture.

There are two potential ways animals can interact with a given piece of land: on an ongoing basis, or as a pulse disturbance. As an example of the latter, many people have utilized pigs at the beginning of their endeavors, to help renovate pasture and scrubland for a few years, before transitioning to ruminants. Another strategy is to bring steers onto the land every so often, using their trampling and weight as a short-term impact to renovate pasture before moving on to maintain it with smaller goats and sheep.

FUNCTIONS OF ANIMALS IN THE LANDSCAPE, AND SOCIETY

There are many roles animals can play in the landscape while enjoying a healthy and happy existence and contributing to the bottom line. These services could include tilling the soil, removing brush, removing pest and disease pressure, mowing, and cleaning drops at the end of a harvest. For the right animal these tasks are not work, but merely a result of their way of being in the world. Indeed, the highest value of animals in the farm landscape should be to participate in ways that help

Table 3.2. Potential functions of various farm animals

Species	Mow/Graze	Clear Brush	Till	Eat Slugs and Bugs	Weed Grass Only	Clean Drops	Site Preparation/ Renovation
Chickens, Turkeys	No	No	Yes, light	Yes	No	Yes	Yes
Ducks	No	No	No	Yes	Somewhat	No	No
Geese	Yes	No	No	Yes	Yes	No	No
Pigs	Some breeds; Kune Kune and Berkshire best	Root around brush	Yes, can go deep	Yes	No	Yes, excellent	Yes, excellent
Sheep	Yes	Some breeds strip vegetation and bark	No	No	Can be trained	Somewhat	No
Goats	As supplement; prefer browse	Yes	No	No	Can be trained	Somewhat	No
Cattle	Yes	Some breeds strip vegetation and bark	No	No	Can be trained	No	No

Figure 3.11. The choice of animals for silvopasture depends on your land, goals, and personal preferences. It's easy to get attached and not think critically before setting your sights on a species. *Clockwise from top left*, Highland cow (photo from Jen Gabriel), Kunekune pig (photo from Brian Gratwicke/Wikimedia), Pekin ducks, Katahdin sheep, and Boer goat.

improve its health and resiliency. Meat, milk, and other products can be produced in tandem with or as a bonus of this activity. In this way we might bring animals back into the landscape from confinement, and find ways to mutually support each other's coexistence.

Indeed, the role animals play in our society today is at once hidden and right in front of our noses. Those living in the United States eat an average of 200 pounds of meat a year per person, and all that puts pressure on farmers, and most especially the meat industry, to produce lots of meat, cheaply. This, plus the surplus production of corn, has evolved into a twisted system in which animals are removed from habitats and fed in confinement with only one purpose in mind: Grow big and grow fast. Beyond the ethical issues of raising animals in this way, it is impossible to raise this much meat in pasture-based systems. In order to work with livestock in ecosystems, people need to eat less meat, period.

The remarkable role of animals on lands is their potential to be agents of restoration in addition to providing products. Making use of this gift means a shift on the part of the farmer, which can be largely driven by consumers, who see the value in supporting livestock farms that have animals involved in systems that sequester carbon, reduce soil erosion, and increase wildlife habitat, in addition to providing a high quality of life to the animals themselves.

Part of the work in silvopasture, alongside orchestrating the interactions of animals, trees, and forage, is to build beneficial relationships. With this type of approach, it's important to balance productivity with investment in the long-term biology of the farm ecosystem. For instance, we might need to raise animals that weigh less at slaughter time, because many of the heritage breeds that are well adapted to silvopasture and grazing systems aren't the largest creatures out there.

As shown in table 3.2, different types of animals each have their own niche and can be placed into the types of land and farm production systems you have available. Bringing animals into an inappropriate scenario, or not being able to put them to good use, usually results in a negative outcome. Selection of animal species is a balance of their ecosystem function, local market potential, and your own preferences.

BREEDS MATTER!

In addition to matching the type of animal to your landscape, you must pay careful attention to the breed, the breeder, and the breeding line your animals come from. Some breeds are simply better suited to the conditions silvopasture offers, and do well in more wild or natural scenarios. Table 3.3 offers some suggestions for the different animal types, by no means a complete list. The breeds are those mentioned in texts relating

Table 3.3. Good possible breeds for silvopasture

Species	Desired Characteristics for Silvopasture	Some Recommended Breeds
Chickens, Turkeys	Follow grazing animals and break up manure patties; ground disturbance, light tillage	Welsummer, Plymouth Rock, Rhode Island Red
Ducks	Pest, slug, snail control; foraging for pests without disturbing ground cover or mulch	Cayuga, Khaki Campbell, Rouen
Geese	Weeding grasses, protection of smaller birds	African, Cottonpatch, Shetland
Pigs	Rooting, deeper tillage, brush clearing	Kune Kune, Guinea, Berkshire are said to be least in need of grain supplements
Sheep	Mixture of grasses, forbs, woody plants; smaller acreage	Katahdin, Shropshire, Dorper, Icelandic, Jacobs
Goats	Wood-dominated landscapes, hilly landscapes	Alpine, Lamancha, Nubian, Saanen, Boer, Pygmy; most breeds appear to be good browsers, when trained
Cattle	Larger acreage, more open woods, less woody vegetation	Highland, Hereford, Belted Galloway, Longhorn, some Angus all do well with brush/tree fodder

to silvopasture, but likely over time more can be added through farmer experience.

The history and philosophy of the breeder are also important. Ask lots of questions of the folks from whom you purchase animals, and try to find those familiar with the scenario you present to them. Of course, behavior can be changed over time, but the more you can start out ahead, the better. If the person you are purchasing animals from doesn't have strong convictions, or doesn't seem organized in their thinking, it's often best to search further afield and settle on stock that is the best fit possible.

In the next several sections, we offer some considerations for various types of animals in a silvopasture system. There are many resources that offer the basic care needs and specifics for each animal, a level of detail outside the scope of this book. Readers are encouraged to seek out additional resources as they build their knowledge and understanding.

The Problem with Pigs

People who have extensive esperience with silvopasture often raise red flags at the idea of sustaining pigs in the woods,[22] at least on a regular basis.[23] Historically, pigs might be one of the better-known animals to graze woodlands, but the intention in the past was remarkably different, as discussed in chapter 2. In the context of silvopasture, pigs might offer benefits for short periods of time, when new systems are getting established. The ability to sustain them, and forests or trees, in the long term, is much more questionable.

Many cite the dehesa/montado and other historical examples of pigs foraging among trees as the base argument for why they are good candidates, though they

Figure 3.12. Saddleback pigs are employed at an arboretum in England to prepare woodland ground for tree planting. These pigs come on loan from a local farm and were stocked at a very low density, and in pulses. Photo by Rob Young/Wikimedia.

often miss the details critical to their sustainable persistence in farming landscapes. Adding fuel to the fire are popular farming articles that sell the concept, with few words of caution or concern.[24] If any progress is to be made, a serious and critical look is necessary, along with an effort to establish clear indicators to determine if pigs can be good or can only cause harm. To date only a few resources have examined this question in depth, citing the need for a lot more research.[25]

One important factor is the widely varying opinions about the number of pigs that can be sustained per acre. While some claim as many as 10 to 25 per acre, a research paper from the dehesa suggested that even a system this old is most sustainable at the rate of 0.5 to 1 pig to the acre.[26] As mentioned in chapter 2, history offers examples of systems where pigs were only *part* of the equation, passing through periodically when fruit or nuts were plentiful on the ground. This phenomenon,

pannage, is very different from the rooting that most pastured pigs practice today.

The critical difference between these two approaches to management is that pannage involves allowing pigs in only when there's enough mast to ensure they focus on it; rooting involves leaving pigs in one place for a continuous duration, or in short rotations, without necessarily considering the season or the volume of potentially available food. In this scenario pigs quickly graze surface vegetation, then begin exploring the soil for insects and roots. Rooting can very quickly destroy soil and damage trees. In general, bare soil means damage to the ecosystem, and it is really challenging to avoid this with pigs in the woods.

Some have observed that farmer behavior in this instance is an extension of the woodland grazing phenomenon discussed in chapter 2. Farmers know that pigs damage their pasture, too, and so sometimes

Figure 3.13. This silvopasture is reaching its pig capacity, where the ground is bare but still fluffy, with grasses still present in much of the paddock. On rotation, this stage of growth can recover with an adequate rest period, likely longer than with ruminant grazing. Photo by Brett Chedzoy.

default to the woods. What is odd is that sending pigs into the woods (or onto pasture) without mast is like sending sheep or cows into a paddock that doesn't have

GOOD PIGGY PLACES

There are few examples of farmers who raise pigs in a sustainable and creative manner, in clear benefit to the ecosystem. Some inspirational examples include:

- D'Acres Pig-Powered Potatoes in New Hampshire. In *Farming the Woods*, we profiled this innovative farm that uses pigs in a sequence to convert forest to farmland, as the original farm was all dense forest. Pigs are followed with potatoes, paving the way for annual production and forest gardens. The pigs are thus a pulse of activity supporting the establishment of new systems on the farm.
- **Sugar Mountain Farm** in Vermont. Members of the Jeffries family raise a wide range of animals and produce but are best known for raising pigs without purchased grain. The secret? Pasture supplemented with local waste whey and spent barley grain, along with a planted rotation of rape, kale, sugar beets, turnips, pumpkins, and sunflowers to meet the needs of the pigs on-site. Hay is the primary source of feed in the winter, fed in "winter paddocks" packed full of pumpkins for the pigs. Walter Jeffries reports that they have raised pigs on pasture only, and it can work, though it takes longer to finish the animals and the meat is leaner. http://sugarmtnfarm.com/

This is certainly not all the examples out there, but in all honesty there aren't many. Neither was developed entirely in a silvopasture context.

We need smart and clever people to cautiously work on more methods and understanding for how pigs can work in silvopasture, as the potential is good. This is a call to action! Are you doing good work with pigs, or do you know someone who is? Please get in touch, share your experience, and document your work.

grass. Without adequate food, only harm can be done, regardless of the animal.

This pattern is in fact the current overall trajectory of pig management, but it doesn't have to be this way. Possible strategies to improve pig management in relation to forests include:

1. The use of pigs only as a *restoration or renovation* tool, meaning that rooting is done intentionally, as a way to clear vegetation when the ecosystem can potentially benefit from a change. Examples may include land overrun with undesirable plants, or areas in woods where a patch or gap opening is beneficial. Pigs can also renovate pastures that are low in diversity, plowing the soil and preparing the seedbed for planting. In these contexts pigs can be incredibly effective, but not as a permanent fixture in the landscape—they must remain temporary or occasional. Rather than cycling pigs through annually, pigs might show up every three to five years, for pulses of impact.

2. The establishment of forage and the use of pigs that can utilize more grass forage as a significant portion of their diet. Some breeds, generally smaller in size, are touted for their interest in grass, including Guinea hogs and Kunekune pigs. It's important to note that these breeds are often significantly smaller than the common commercial hog breeds.

3. The dedicated redirection of local waste streams as the bulk of food for the animals. In his book *Meat: A Benign Extravagance*, Simon Fairlie lays out a convincing argument that given the massive amount of food waste cycling through our food system, pigs might be able to do good for society while cycling nutrients back to the farmscape.

It's important, even in these scenarios, to exercise extreme caution and vigilance, remembering that pigs can much more easily harm woods or trees than other grazing or foraging animals. Rotations will need to be frequent, rapidly responding to changing needs. Rest times for paddocks will need to be longer. Soil compaction and organic matter residue should be carefully monitored, and data collected. And farmers would

Figure 3.14. These pigs have good pasture to graze, though they should be moved very soon, with the beginnings of bare soil showing. Photo by Brett Chedzoy.

do best by limiting their scale of production (that is, their number of pigs) and keep them only in the more marginal parts of the land.

While it's easy to see the abusive impact of pigs on surface duff or pasture, what's harder to know is how pig activity affects a site in the long term. Often trees don't show signs of stress or decline from this type of disturbance for many years—sometimes a decade or more. This is true for other grazing animals as well. Overgrazing is overgrazing, regardless of the system.

As suggested earlier, if you are going to go with pigs, then start with very low stocking densities and plan on moving them frequently. Know that supplemental feed is a must, and seek waste resources in your local community, especially those that are high in protein (whey from cheesemaking, spent brewing grains, and so on).

At the end of the day, pigs are a real challenge, and leave more questions than answers. They are not really the best animal to start learning with, but rather one to possibly work toward over time, and only to work with if you have ample energy to monitor, experiment, and remain vigilant. This is a challenging proposition given the generally busy lives of farmers. If you really care about stewarding the land, you may want to question whether to have pigs at all.

Chickens, Turkeys, Ducks, and Geese: Low-Risk Silvopasture

In some ways birds are the least likely animals to inflict long-term damage in silvopasture systems, purely because of their size. This doesn't mean you can't do harm with birds, but the stakes are almost certainly lower, especially when you start with smaller numbers. Some purists might not consider poultry in the woods or among trees to be true silvopasture, as the forage layer may be less managed, more a result of what's already there. Nonetheless integrating birds into a silvopasture system offers potential benefits for the birds, as well as for the woods, particularly in relationship to other, larger grazing animals.

Also notable with poultry is that although there is a rather large interest in feeding birds solely off the land, very few are actually doing it, and no one is really looking deeply into the question of doing it sustainably. While the Internet is littered with anecdotes of both success and failure, ultimately some sort of controlled trials will be needed to learn exact thresholds. Just as with the pigs, importation of feed from off-site will be a necessary part of the equation, at least at the outset. When I first began raising meat, it was with chickens. I was supremely excited to see them scratch, till, and fertilize our land and start to bring a dynamic, active presence to restoring ecosystems. One thing I didn't anticipate in my utopian vision was how many trips to the feed store I would take. We experienced this in an even more extreme way when we raised ducks for meat. Each duck needed 0.3 to 0.4 pound of grain per day to gain enough weight for market in a timely manner.

Our solution was ultimately to settle on birds for egg production. This drastically reduces their need for grain and allows them to rely more on foraged feeds. We are exploring ways to produce more feeds on-site, including producing black soldier fly larvae, and using spent mushroom spawn as a feed source. We also limited our flock size to 30 to 50 birds, which means that even feeding some grain doesn't break the bank or impact the larger ecological picture as much. We can justify this economically, but it's not a major income source for the farm. It is basically break-even, but we

> ### GOOD READS: POULTRY
>
> Essential and well-organized information has made the following manuals essential for all types of livestock:
>
> - *Raising the Home Duck Flock* by Dave Holderread (1978) is one of our favorite duck-specific books. Lots of candid and hard-to-find tips you won't see in other books.
> - *The Small-Scale Poultry Flock* by Harvey Ussery (2011) takes the conversation to a more systems level of thinking. Great information for a range of scales and producer goals.
> - *Storey's Guide to Raising Poultry* by Glenn Drowns (2012) is a good place to start, and then move into specific books for chickens, turkeys, and ducks.

value the animals largely for their pest control benefits to other systems on the farm.

Within the context of silvopasture, getting poultry into the woods offers them shade and shelter. They absolutely thrive in more diverse, shaded landscapes, especially ducks and geese, who easily overheat during hot summer days. Certainly a plan for silvopasture can include plantings of forages, such as legume cover crops. In many instances we can also leverage poultry to address pest issues on the landscape, a true win-win: We're reducing the impact of the pest and feeding the animals at the same time.

Additionally, there is abundant opportunity to capture feed from local waste streams, especially byproducts of grain and flour processing, as well as spent grain from breweries. For chickens and turkeys, food scraps can be a great resource, especially when they're composted—a process that inevitably attracts and breeds all sorts of insects and bugs to eat. This importation of food wastes, along with the easiest input of all (from a time perspective), store-bought feed, can be seen by the savvy land manager as valuable nutrient inputs on the landscape. In other words, outside inputs bring in outside fertility, which can then be captured and cycled through the system. This ultimately helps the soil, forages, and trees thrive.

Birds are in many senses the lowest-risk animals to incorporate into forest- and tree-based systems, mainly because they are small and their collective weight is low. It would be hard to set back the woods in a serious way with birds, but this doesn't mean you can't do damage. Rotational grazing is still a must, especially in existing woodlots. One of the most notable effects of poultry presence in the woods is the rapid degradation of leaf cover on the forest floor.

Considering that the leaf fall in a forest generally covers and mulches the forest floor for about one year, it's not surprising that a bunch of birds would degrade it much faster. What this means for silvopasture is that supplemental organic matter is going to be necessary, even when rotating animals for a shorter stay in a paddock. Our farm always has hay, straw, or leaves on hand to cover any bare soil after our ducks have moved on. Bare soil means damaged soil. (Are we getting repetitive yet?)

Each type of poultry brings a different set of advantages and disadvantages to silvopastures. For this reason you should select birds for the needs of your system, rather than what is often done, which is selecting the product first, then figuring out next how to work that into the landscape. Let's distinguish scratch birds from waterfowl, and get into more detail about the advantages and disadvantages of each. We won't elaborate on all the management needs for each type, instead commenting on the considerations specific to silvopasture. Consult the sidebar for recommended reads if you choose to go down the poultry path.

CHICKENS

Scratch-and-till chickens seek a diet heavy in seeds, insects, and worms. Their sharp talons help them seek food, stirring up the soil and, when not moved frequently, degrading the surface organic matter rapidly. If you use them to follow grazing animals, they will break apart manure clusters, accelerating its decomposition while also consuming any remaining insects and larvae.

They can also be rotated through with proper monitoring to till up areas and establish new seedbeds to

Figure 3.15. Chickens seem at home in the woods, though they can gain only so much from the setting. They should be frequently rotated within the forest, and in pasture could be used to clear ground for tree planting. Photo by GaylaLin/Wikimedia.

improve forage diversity and quality. While we don't have an exact figure, some report that chickens can utilize forages for up to around 5 to 20 percent of their diet.[27] The rest usually comes from grain. Egg layers may be able to survive on little to no grain, but meat birds especially will almost certainly need it, and are generally considered to be poorer at foraging, especially as they get bigger. Those interested in chicken silvopasture will likely need to focus on egg production and find breeds with more experience in active foraging.

Chickens tend to stick close to home and, while they live in groups, have less of a herd mentality than other poultry. While their activities can have some positive effects on pest populations in cropping systems, they must be closely monitored in any existing cropping or around newly planted trees. In relation to plants, they are best used in the establishment phase. Rotate them frequently if you want to keep them from stirring up the soil. Or you could concentrate them in an area you're prepping for tree planting, where their activity will remove the grass layer and help prep the site.

In the context of silvopasture, trees and brush clearly provide benefit to chickens, sheltering them from heat and high winds, and protecting them from predators, especially those lurking overhead. They can do good in orchards, where their curious pecking can help with post-harvest cleanup. As chickens are small, they are less amenable to overgrown pastures than are larger birds, but also able to make better headway through thick brush, though they won't do much in the way of cleaning it up.

Chickens can be immediately integrated into existing woods, though prolonged access to a confined area may prematurely wear down the delicate duff layer that protects the forest soil. Gaps and patches in the forest will provide a more diverse layer of vegetation, though mostly the birds will seek out the critters in downed logs and in the soil. Choose breeds that are proven as excellent foragers—a title usually reserved for the egg-laying and mixed breeds, and those with heritage bloodlines.

All in all, pasturing chickens in silvopasture has some benefits, but is not as compelling as some of the other options out there. Their biggest benefit is arguably in a leader-follower rotation, where they can compound the benefits of grazing ruminants by cleaning up after them.[28] In this scenario they can help break apart manure pads and integrate them into the soil faster.

TURKEYS

Like chickens, turkeys scratch and till, though generally with less intensity. In the wild, turkeys are long-distance travelers, moving in tightly knit packs. Some literature suggests that this pattern of movement situates them well to be herded like sheep to areas that would benefit from their form of scratch and peck, which ideally tends to focus on bigger seeds, nuts, and insects. They can also utilize a wider range of forage legumes and fallen fruits than chickens, though supplemental grain is almost definitely going to be a necessity if weight gain is a production goal.

Figure 3.16. Turkeys can use a wider range of forage than chickens, and also offer pest control benefits to the farm. They don't have an extensive track record in integrated farming practices, but as woodland native animals they have great potential. Photo by Curt Gibbs/Wikimedia.

One of the best possible contributions of turkeys is their ability to address pest problems around the farm, especially around orchards with fruit and nut crops. They hunt with incredible precision and are able to utilize a wider range of forages than chickens. These could include acorns, beech nuts, pine seeds, roots, wild fruits, and clovers and alfalfa.[29]

The biggest challenge with turkeys is their vulnerability to disease, which is most dramatic during their first months. Some report that they also can be difficult to corral into a roost at night, and so can be vulnerable to predation. Careful breed selection and some extra care early on can reduce some of these issues.

The role of turkeys in silvopasture systems is similar to that of chickens, where they're mostly on the receiving end of the benefits, except where a specific crop pest can be mitigated by including them. Turkeys would be excellent in silvopasture systems with hazelnuts or chestnuts, where they could harvest the excess and break potentially damaging pest cycles.

Guinea Hens

Mostly foraging birds, guineas are best known for their odd appearance and behavior—which depending on whom you ask ranges from charming to annoying. These ground-nesting birds originate on the African continent and are from savanna and open pasture systems. They are most often free ranging, as they can fly and evade some attempts to contain them. Their biggest advantage is their expert ability to forage without engaging in the intense tilling and scratching associated with chickens and turkeys.[30]

Their wild nature can be challenging, though. It leads to lower maintenance, as guinea fowl often avoid returning to a coop and prefer to roost in trees around the property (readily available in a silvopasture!). While free ranging, they eat all sorts of beetles, fleas, grasshoppers, and even ticks, which is what they are most famous for. Yet this can leave them especially vulnerable to predation.

Guineas make a lot of noise—which can in some cases deter predators and plant pests. It's been reported that farmers have used them in the past mixed with chickens and turkeys as predator deterrents, as well as in orchards to drive off pest birds.[31] Still, many people report that the noise drives them crazy; that the birds' wild habits lead to them wandering off; and that they are very hard to contain. Few raise them commercially; in most cases they are free-ranging birds valued more for their ecosystem services.

Ducks

Contrary to what many believe, ducks can survive without constant access to open water. Such access would be ideal, perhaps, although in many cases too much of it can lead to issues. These waterfowl need water both to clean themselves and to clean out their nostrils, which get muddied up as they root and forage in the soil. All this leads to dirty water, and as such duck access to

Figure 3.17. Guinea fowl are interesting creatures, suited best for a free-ranging life. They can be great in areas with high populations of ticks, and they enjoy the opportunity to roost in trees.

ponds and creeks should be occasional and monitored. For the rest of the time, smaller 5- to 10-gallon tubs that are filled and emptied daily suffice for their needs.

Places where water accumulates during heavy rain events but dissipates quickly, sometimes known as vernal pools, are also good options. In our experience, ducks have enjoyed and benefited from the range of environments our system offers, from the forest garden to tree-planted swales to the woods surrounding our mushroom operation. Depending on the level of heat intensity and precipitation, we can rotate them to an appropriate environment to make the most of their attributes.

Most ducks are descendants of the mallard breed, known as "dabblers" because they forage by digging in the soil or muck with their beaks—hence the need to clean themselves frequently. With resident ducks, you can't keep water clean for more than a day, but that's just something that you accept as a reality of keeping them.

In their foraging, ducks can take on a lot of pests, most notably slugs and snails. This makes them ideal candidates to integrate with mushroom production (see more

Figure 3.18. When in the woods, ducks should be moved frequently; they will deplete the leaf duff layer with prolonged residency. We use them very effectively as part of our slug control in the mushroom yard, and they in turn enjoy the cool shade the forest canopy offers. Photo by Jen Gabriel.

Figure 3.19. Ducks and geese are particularly excellent in mulched orchards, where they can perform pest control and fertilizer duties without completely messing up the mulch layer.

on this in *Farming the Woods*), as well as other cropping systems susceptible to slug damage. They will avoid doing harm to all but the most tender of plants while seeking out these pests, including mature adults, young, and eggs. Ducks are incredibly thorough, and your plantings and crops will thank you for incorporating them.

Like chickens, some breeds are better for meat, others for eggs, while some are touted as "mixed" and can provide a little of both. Our experience dictates that it's best to choose one yield or another, and focus on that. For meat, Muscovy, Rouen, and Pekin are the better breeds, while Khaki Campbell and Cayuga ducks are our favorites for eggs (though only Khaki lay prolifically).

Ducks are pleasant to have and easy to care for, provided your landscape can handle their near-constant quacking and chatter. We use a multilayered approach to deterring potential threats: a few male guard geese, a hot fence, and our dogs.

Geese

Of all the poultry options, geese may be the best one if your goals are mainly weeding, some pest control, and avoiding damage to crops, especially those occupying the herbaceous layer. Studies have found geese to be more effective than chickens in this capacity.[32] Of all the poultry, geese can sustain themselves the best off pasture alone, and this in combination with their preference for grass situates them to be an excellent potential addition to orchards, fruit and nut plantings, and even Christmas tree farms.

Figure 3.20. Geese are perhaps the ideal choice when weeding is a desired outcome. They are the smartest and most trainable of all the poultry breeds, as well as the species best able to utilize forage for the widest range of their diet. We use male geese as very effective guard animals for our duck flock. Photo by Jen Gabriel.

That said, geese can be a pain, especially during breeding season. Some breeds are more aggressive than others, which can help reduce predation, and even protect other birds such as ducks and chickens from threats. The concept of "weeder" geese suggests that they have long been valued for this activity.

Geese are perhaps the most intelligent among the poultry, and the most likely to imprint on people. This can be problematic, especially if you live in an area with predator pressure and would find some aggression desirable. It's a fine balance, to train geese to be reasonable with people, but otherwise on guard. Careful choice of breed, along with some training, will reduce the chance of a minor disaster on the farm. At the same time, aggressive birds can come after humans and children alike.

The general care of geese is very similar to that of ducks, where they need access to a significant amount of water. They often do well integrated with other poultry, and even with small ruminants on some farms. Markets for geese products are highly localized and unproven, with meat birds the most likely outlet.

POTENTIAL APPLICATIONS FOR POULTRY—AND MANY UNKNOWNS

A good strategy for integrating poultry into silvopasture follows the logic that a particular species can be matched to the situation and need of a given system, all while yielding marketable products in meat or eggs. The opportunity presented by silvopasture is to enhance their environment and comfort, and provide some beneficial services to productive systems. While they can exist on their own, poultry seem best positioned to integrate and support other elements on the farm, such as:

Pairing with other animals. Chickens or turkeys can follow large animals like cows to help break down manure and integrate it into the soil; ducks and geese can reduce snails and slugs that transmit parasites to sheep and goats.

Pairing with other tree and woody crop systems for pest control. You can use geese very successfully to weed berry and orchard plantings, or ducks

to prevent slug damage to woodland mushroom production.

Pairing with one another. Geese or guinea fowl provide protection for more vulnerable birds, while a mix of chickens and ducks will provide a mix of rooting, digging, scratching, and tilling.

In the end farmers tend to settle on the animals that work practically, provide a positive emotional experience from interaction, and interact well with their system. There is no right or perfect option for all situations, but rather one or more for a particular scenario.

With poultry, what we don't know in regard to silvopasture certainly exceeds what we do. The interactions mentioned above are, at best, good concepts. Of all the animals, birds are the most easily integrated into trees and forests, with the least potential for inflicting long-term harm. Practitioners are advised to proceed with an exploratory mind, ideally seeing poultry as a support member of a larger ecological farm system.

A Rant on Ruminants

Not all animals are created equal, at least not in *their ability to improve the on-site ecosystem while solely or mostly using on-site resources.*

Enter the ruminant. A true force and phenomenon of nature. Cows, sheep, and goats have rumens, which are multichambered stomachs where a community of bacteria and fungi are cultivated, creating a fermentation site that rivals the most advanced technology out there. The rumen allows the animal to consume green vegetation and process it, turning it into bones, flesh, skin, hair, and milk. If we want to talk about clear positive impact on acres, ruminants are no contest. They the most ready for large-scale silvopasture application, as there are ample research studies and clear parameters for managing them well in these systems.

If it's not already obvious, this book is wholesale advocating that we raise and eat more grassfed animals, or ruminants. We cannot continue to ignore the glaring difference in animals who are adapted to grasses and forbs and their tremendous potential for sustainable livestock production. Especially in the United States,

Figure 3.21. Cows, sheep, and goats form the basis of a clearly regenerative grazing silvopasture system. With these animals every square inch of pasture is valuable feed, with almost limitless potential to increase in food value over time.

where we overwhelmingly prefer chicken and pork to other meats, a shift to pastured ruminants would be a huge step in the right direction.

Unfortunately, in the current livestock paradigm, animals have largely been removed as players in a grazing ecosystem, as discussed in previous chapters. Part of the work, then, is to get them back on grass. In silvopasture we might best seek the edges of a species, where heritage breeds reside, relatively unaffected by modern industrial farming practices.

In this section the text is unabashed in its promotion of ruminants as the core of regenerative silvopasture practices. This is because of:

- The nature of the rumen, and its ability to utilize on-site resources (grasses, forbs and fodder).
- A significant body of research that enhances our understanding of best practices.
- The potential impact of grazing systems for environmental and climate health.

Pigs and poultry have a place, but from many viewpoints they can, at best, supplement their feed needs from a silvopasture. Perhaps this will prove wrong in the longer term, but at the moment more research and systems development is needed. Ruminants, on the other hand, are largely ready for prime time, so long as systems are approached with intention and fore-thought, especially if the animals aren't used to forage.

THE RUMEN

Where humans and other animals use mainly enzymes and mechanical means to digest food, the rumen is essentially a fermentation chamber that maintains an anaerobic environment, along with several chambers necessary to help break down and process forages.[33] The animal first chews the material and then swallows, providing a rather large quantity of saliva to wash it down. The saliva helps balance the roughage and also assists in buffering the acid-rich rumen pH. Within the rumen, resources are partitioned, with gases rising to the upper layer, fluid-saturated materials (aka yesterday's feed) sinking to the bottom, and new material floating in the middle. Bacteria cultivated in the rumen are the main digesters of fiber, fatty acids, and proteins.

After this stage the material heads to the reticulum, which resembles a honeycomb and separates any large remaining particles from the digested bits. These particles are what head back for rumination, after being regurgitated and chewed a second time in the mouth. This "cud" is mixed with even more saliva before heading back to the rumen for round two.

Digested solids head to the omasum, which helps dry out the material, then to the abomasum, which is lined with glands and enzymes much like a human stomach, for more digestion. The material finally exits the stomach after being processed for up to 48 hours, heading to the small and large intestines, and eventually out the other end.

All this is to say that ruminant animals' mix of mechanical, chemical, and biological processing means they are able to digest an incredible array of plant foods, along with providing the critical feedback that perpetually affects the feeding choices they make the next time.

Choosing a Ruminant

While ruminant stomachs are mostly similar, there is quite a difference in the effects one ruminant has over another on both your landscape and your grazing strategy. Careful selection of the type of animal, and then further attention to a good breed, is critical.

Characteristics to consider include:

- Size of individuals and group.
- Preference for forage type.
- Herd mentality habits.
- Methods of eating.
- Height of forages.

It can be said generally that cows, sheep, and goats live on a spectrum, with cows wanting the largest amount of grasses and pasture forbs, goats more in favor of woody shrubs, and sheep somewhere in between (see figure 3.22). Thus, selection is in relationship both to the existing forage and to what may be established down the road.

Since ruminants are herd animals, the acreage you have access to will help determine the best fit. A good starting point is about one cow, or four to six sheep or goats, per acre. A minimal herd size of eight to ten animals is also recommended to maintain a competent group, though of course homesteaders may choose to just have a couple of animals at a time. These should be considered very rough guides, as the carrying capacity is largely determined by the quality of the forage. In addition, with grazing animals, everything is always in flux.

We keep around 15 to 20 ewes, which after lambing means around 40 to 50 total animals; we can sustain this number comfortably on the roughly 30 acres we graze. Over time we should be able to increase this rate, but stocking at perhaps 50 percent of what the literature suggests means we aren't ever stressed about having enough food, especially during drought conditions. It should be said that we started with just four pregnant ewes the first year and grew into the present flock size, getting feedback on our maximum based both on the capacity of the land and on advice from other farmers.

COWS

Among ruminants, bringing cows into silvopasture systems literally has the most impact: A mature cow weighs 900 to 1,200 pounds. This means grazing herds

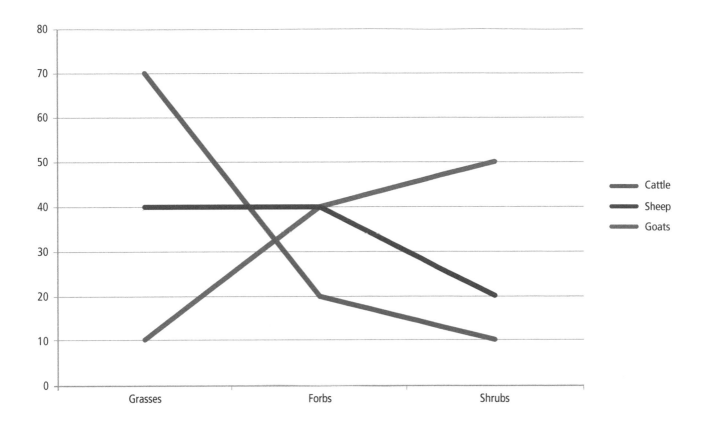

Figure 3.22. The rough proportion of preferred forages for cows, sheep, and goats. Adapted from ATTRA.[34]

can have a dramatic effect on vegetation, as whatever they don't eat, they trample. This can be a great benefit to a pasture that needs renovation. Within the pasture, cows also tend to be the most willing to consume grasses and forbs, even those that might be past prime. Sheep and goats tend to be more picky, leaving behind more residual forage in the pasture.

The heavy mass of a herd of cows can be beneficial, but it can also be too much of a good thing. Especially in wet pastures and silvopastures, where trees have shallow root systems, extra care and caution are needed to ensure no damage is done. Depending on the season and even the year, you may need to skip paddocks in the rotation.

All told, this means that cows fit best into grazing systems with larger amounts of acreage, and where the farmer has the ability to invest in more substantial handling infrastructure for moving and treating animals over time. That said, the investment in a squeeze chute is perhaps offset by the ability to keep cows contained with minimal fencing (a single strand of electric is all it takes) as they are moved from one paddock to the next.

Because of their size, the biggest impact cows may have on young trees is in rubbing, breaking, bending, and trampling. This is not because the cows want to maim trees; rather, it's a byproduct of their activity. One common need of all grazing animals is places to scratch an itch, especially on the rump where they can't reach with their heads. Lacking a place to do this, they may try to use a vulnerable tree tube or fence as a scratching post. This is another reason to leave a few downed trees available in the silvopasture.

Cows have long and flexible tongues, which they use to wrap around grasses and forbs and pull into their mouths. Because of the way their lips, teeth, and jaws are structured, they aren't able to eat down as low as

Figure 3.23. Black Angus cows grazing at Brett Chedzoy's Angus Glen Farm in New York in late autumn. Cattle offer the best option for larger-acreage farms and are the most common animal used in traditional silvopasture systems. Photo by Brett Chedzoy.

sheep or goats, nor can they strip woody vegetation and trees as finely. Cows don't tend to go after bark as much as the small ruminants, at least not with their mouths.

While many breeds can likely be trained to better use brush, some appear to be inherently better than others. In browsing farmer forums online, and in discussions with those working on silvopasture systems, you'll find that Scottish Highland cattle seem to be a crowd favorite. Others prefer Belted Galloway, Angus, and even Jerseys for the task.

To be clear, these comments center on their ability to utilize brush, not to graze in silvopasture. All species can be grazers of grasses and forbs, which are going to be the primary source of food. It's just that if a farmer is looking to utilize woody shrubs, either to reduce their presence or to provide sustenance, cows may sometimes be less effective than the smaller ruminants.

Cows offer a stark advantage over smaller ruminants: They're much less susceptible to parasites, a huge consideration especially in warmer climate zones. They also don't require hoof trimming, shearing, and

Figure 3.24. Scottish Highland cattle are great candidates for silvopasture, being extremely cold-hardy as well as interested in utilizing a wider range of vegetation than some other breeds. Photo by Jen Gabriel.

other maintenance that goats and sheep often need. And across most breeds they tend to be tolerant of the weather, not requiring expansive shelter even during the coldest months of the year.

Operations usually choose one of two approaches. Some manage a cow/calf operation, where they keep animals year-round and breed the mothers, selling a mix of live animals to others while likely also finishing some for harvest. Others choose to simply purchase stocker cattle and graze them for the growing season, selling them off before winter sets in. Those interested in only grazing and not dealing with selling or slaughtering animals should also consider the possibility of grazing replacement heifers for neighboring dairy farmers, who sometimes pay a premium to have these cows taken care of by another farmer. In this scenario you only have to manage the grazing and not worry about transportation off-farm, or marketing the products.

The main yield outputs from grazing cows and cattle are, of course, milk and meat. While demand is high for these products, they are relatively abundant on the national and global scale, which can undercut the prices of a smaller operation. Producers will need to develop niche markets, and silvopasture provides a good selling point (this is discussed more in chapter 6).

SHEEP

There is considerable difference in the browsing interest among the sheep breeds. Some seem to take to it willingly, while others are less interested. This can be used in either direction; Katahdin sheep can utilize brush as a major food source, for instance, while in England, Shropshire sheep are intentionally used in Christmas tree plantings precisely because of their lack of interest in browsing the conifer trees. Among all the ruminants, the differences between sheep breeds in terms of forage preference, habits, parasite resistance, and tolerance of weather are most stark. Careful breed selection is critical.

Because of their tighter lips and teeth, which they use to strip branches and clip forages, sheep can trim grass lower as well as strip leaves and bark from shrubs and trees. Like goats, the browsing breeds also

Figure 3.25. Sheep devouring multiflora rose willingly as a preferred forage. These hair sheep are often mistaken for goats. They are agile climbers and can be rambunctious. Photo by Brett Chedzoy.

Figure 3.26. Sheep loafing underneath hybrid hickory trees at the Badgersett Research Farm in Minnesota. These animals help maintain a clear understory and make the nut harvest easier.

SMALL RUMINANTS AND PARASITES

A big advantage of cows over sheep and goats is the drastic reduction in worry over parasites. Anyone who gets into raising these animals will need to verse themselves in the basics of parasite management. Ongoing monitoring is essential.

For years the general recommendation was to dose animals with de-wormers, which only exacerbated the situation since the parasites proved to develop resistance. Currently, the thinking is actually that some low-level exposure to parasites is a good thing, as it can build up resistance in the animal to more serious outbreaks. Becoming familiar with the various species and management practices that increase exposure is important to getting ahead of any issues that may arise.[35] Generally, the recommended approaches include:

1. Choose and select for parasite-resistant breeds.
2. Keep animals on rotation.
3. Don't graze too low, and rest pastures.
4. Provide forages high in condensed tannins.
5. Clip pasture after animals to expose parasites to sun and heat.
6. Monitor for parasite impact using the FAMACHA system, which specifically targets the barber's pole worm.

Many report success from a strategy that focuses on these grazing practices not as stand-alone actions, but in combination with one another.[36]

have a passion for stripping the trunks of young trees. Inevitably, when introduced into a pasture, sheep will go right for the trees if they aren't adequately protected. Browse height for sheep is effectively the length they can stretch their necks upward, so trees need firm bark with lower branches around 5 to 6 feet above ground level if you wish to integrate sheep safely.

From an infrastructure standpoint, sheep make perhaps the fewest demands (besides fencing). They can be handled in a set of basic rigid panels for medical attention or hoof trimming. They need only a moderately hot fence and generally stay put even if the change is low. The largest management challenge is parasites. See the sidebar for more details on this aspect of supporting a healthy flock.

Sheep are raised for their wool, milk, and meat. Those uninterested in developing a use or a market for the wool should consider hair sheep, which don't require shearing and instead shed their hair each spring. This can reduce cost and stress for both the sheep and their handler. Hair sheep also seem to be more interested in browsing woody plants than the wool breeds. A further consideration is that the diets of both milking and wool sheep must be more closely monitored to maintain the quality of the product. Meat production can be more flexible, although of course it's still very important to ensure a diverse and healthy diet.

In the context of silvopasture, sheep offer a compromise between the more extreme diet preferences of cows (for grasses and forbs) and goats (for woody vegetation). For small and medium-sized farms, they can be easily integrated and managed with minimal infrastructure. Sheep are quite tolerant of inclement weather, though this is also very breed-dependent. Good breeds and breeding can result in a high level of forage utilization, potentially more than with either cows or goats. And markets and interest in lamb are growing, provided the focus is on niche products. In some ways, as we envision silvopasture systems, sheep offer a middle ground with much appeal.

Goats

When people first think of grazing among trees, they often think of goats. "Goats will eat anything!" is the common phrase. Well, almost anything. They certainly have a preference for woody plants, which is great if your land is covered with brush, but hard if the goal is not only clearing an area but also sustaining a herd on woody plants alone, something much trickier than getting good regrowth from grasses and forbs. And while goats will munch and graze on pasture, it certainly isn't their preferred meal.

Goats are curious and exploratory creatures, much more so than cows and sheep with their more

Figure 3.27. Goats have a unique mouth structure, which allows them to work many types of vegetation. Photo by Michel Meuret.

conservative attitudes. This leads to a lot of entertainment, but also means you must be more attentive to keeping fences secure and hot. Goats will happily find any means possible to extend their browse height—including standing on their hind legs or available machinery or structures. This means that their browse height is effectively even higher than with sheep.

Goats have a mouth structure similar to that of sheep, but use their tongues more and as such are able to more carefully strip vegetation while avoiding thorns. This can be a wonderful attribute for clearing woody brush, and a horrible one for protecting desirable trees.

With respect to shelter, goats are the neediest of the ruminants, preferring to remain protected from driving rain and snow. This is in part because many breeds come from warmer, drier climates, and also because they don't develop as thick a coat as sheep or some cow breeds. So good shelter is an important part of management.

Goats might be best suited for a traveling weed-eating road show. Their preference for brush means they

Figure 3.28. Goats graze on invasive weeds at Naval Base Kitsap in Washington State. Photo by Michael E. Wagoner/US Navy.

have a lot of potential to support landscape restoration efforts. Landowners are willing to pay a premium for this ecological service, which will likely fetch a herdsperson a better price than selling milk or meat. And because brush is slower to recover from grazing than grasses and forbs, it might be a more reasonable prospect than keeping them on only one piece of land. It's definitely harder to establish a grazing system for goats that utilize on-site resources entirely.

Overall, goats certainly haven't found a niche in North America at the rate that cows or sheep have. One challenge is that many consumers are less familiar with their primary products, meat and milk. This challenge can be reframed as an opportunity, as goat meat often finds good specialty markets, especially when closer to urban centers that have a higher population of ethnic diversity. Globally, goats and sheep are a much more common form of livestock, while much of the United States food system has preferred to favor cattle.

Infrastructure: Water, Shelter, and Fencing

Forage and species selection are not the only aspects of making a silvopasture system work. For optimal animal performance, adequate water and shelter from the heat

and weather are important, and fencing is what makes it possible to contain animals and give them access to certain areas of land, while also excluding them from others. The choices for each of these elements vary widely in cost as well as in the time and energy required of the farmer. We will finish up this chapter by considering the various components of infrastructure you'll need for the various livestock breeds.

Water

In silvopasture we've established that the animals will move, and so should their water. Chickens, turkeys, sheep, and goats are among the species that use lesser amounts of water, while ducks, geese, pigs, and cattle need significantly more. Across the board, any drastic increase or decrease in temperatures usually results in an increase in water needs.

Water is heavy, and takes energy to move. There is a distinct advantage to animals that require less water. Water systems should always be planned with multiple options that offer redundancy if one option fails. We invested in some tanks and a trailer that we can move anywhere on the farm, and fill easily from the well or a pump from the large pond during the warm months. It's a priceless investment that makes life incredibly easy (see figure 3.29). A trough and float valve mean peace of mind, knowing the animals always have what they

Table 3.4. Infrastructure needs for various animals

	Water (gal/animal/day)	Fencing	Shelter
Chickens & Turkeys	0.2–0.5; dehydration can lead to mortality	Poultry netting	Should be locked in at night to avoid predation; taller structure needed for roosting habit
Ducks & Geese	0.5–1.5, for drinking and cleaning beaks from rooting; need clean water daily	Poultry netting	Should be locked in at night to avoid predation; more floor space needed for nesting habit
Pigs	1–6	High-tensile wire, pig netting, or woven wire with openings less than 2″ × 2″	Three-sided shelter, low to ground
Sheep & Goats	0.5–2	Woven wire or netting, high-tensile for perimeter	Sheep need shelter only from inclement weather; goats are much pickier
Cows	15–20 for steers in hot weather; 20–40 for milking cows	Single- or multistrand high-tensile wire	Minimal depending on breed; shade is ideal

Figure 3.29. A sturdy cart and tank of water that can be easily hauled around the farm are among the best investments a livestock operation can make, allowing water access in any paddock or pasture.

need. The exception to this is ducks, who will splash and thrash around in the water, always demanding more, and eventually draining the reservoir.

Animals should never be given free access to live waterways, no matter how appealing it is simply to let them drink from ponds, streams, or creeks. It just isn't responsible, as it's hard to really know when ecological balance is being compromised. The water you provide should be as clean as possible, as poor-quality water can lower animal intake, leading to a whole suite of possible problems.

Winter presents a whole series of unique challenges to providing water to animals. It's best to design multiple winter paddocks into your overall grazing system, and make it easy to get water there without the danger of freezing. A frost-free watering system may prove well worth the investment, while tank and bucket heaters

can often do the trick with a lower price tag. For new grazers, the first few winters can often be a rude awakening, since the management needs are dramatically different than in the warmer months. Plan ahead, and have a backup option in place.

Shelter

As discussed in the first chapter, the desire for shade and shelter is one of the main reasons people are attracted to silvopasture in the first place. You can't really beat the quality of shelter from the heat, wind, and storm that trees can provide for any animal. Yet for most animals, tree shelter will improve their comfort but not totally meet their needs for protection from predation and the weather.

The mandate for silvopasture is rotational grazing, and shelter, too, should rotate. Some farms use a central

LIVING BARNS
By Brett Chedzoy

At Angus Glen Farms, we graze 100 cows on 300 acres of rolling pastures adjacent to the Watkins Glen State Park. This is the second most visited park in New York, after Niagara Falls. Its main attraction is the Glen Creek Gorge, which passes just a few hundred yards from both of our old dairy barnyards. Two miles downstream is the heart of the Finger Lakes region: Seneca Lake. Those who keep livestock year-round know how quickly 100 cows can generate mud at certain times of the year. In the early years of building our herd, we did what most other beef producers do in this area: park the herd at the barn and feed them hay—lots of hay.

At our current size, this would result in a clear and significant impact on the watershed. About 10 years ago I consulted with the local soil and water expert, who estimated that our barnyard needed at least $100,000 in "upgrades" to keep the nutrient-laden runoff from reaching Glen Creek. Building a new roofed shelter in a better location was estimated to cost $30,000 at that time. That's when I decided that we had to either stay small or do things differently.

One of the big shifts in our winter management was bale grazing and out-wintering the cattle. When the herd finishes the last of the stockpiled pastures in late December, we fence-wean for about a week and

Figure 3.30. Natural "living barns" are effective and economical shelters for livestock during the winter months. Photo by Brett Chedzoy.

then turn the main cow group back out to rotationally graze round bales until grass returns in late April. A one- to two-day supply of round bales is placed in a number of the 75 paddocks across the farm as time and weather permit. No bale feeders are used, and the sisal twine is left on the bales. We haven't had problems with animals ingesting too much twine to date, but it's a risk that every grazier should weigh individually.

We try to store most of our bales under cover so that the twine is mostly intact. This helps reduce the amount that the cows are likely to eat (long, continuous loops versus short strands). Access is left open to some of the previously bale-grazed paddocks so that the cows can reach water and can bed on waste hay. A combination of energy-free waterers, streams, and ponds provides winter water. As I write this, the streams, springs, and ponds are frozen solid here due to some good ol'-fashioned winter weather—so the lesson to be learned is not to be completely dependent on surface water sources. We watch the extended weather forecast and move the mob toward shelter when severe weather is imminent.

Green Shelters

Our shelters look very different from those on most farms. They are green, growing, and appreciating in value—unlike our two old dairy barns. Better yet is that the taxman can't touch them, and they require very little upkeep. We refer to these wooded shelter areas as living barns. They are managed primarily to protect the cows from extreme windchills and prolonged periods of driving snow, versus the silvopasture areas that are managed primarily to produce both quality timber and grazing. The main difference between the living barns

Figure 3.31. Use a diversity of species for living barns, in case a single species becomes afflicted by disease over time, such as the Scots pine in this photo. Photo by Brett Chedzoy.

and silvopastures is the composition and density of trees. Evergreen conifer species provide the greatest protection in living barns. Tree density (stocking) is maintained low enough to promote good growth and vigor of the best trees, but high enough to ameliorate nasty weather. By contrast, silvopastures are usually maintained at lower density to allow enough sunlight at ground level for optimal growth of both trees and forages.

Living barns require balancing density for both tree vigor and livestock protection through periodic thinnings. Consulting foresters can provide invaluable assistance when contemplating a commercial harvest. Herbicide injection can also be used effectively to remove cull trees, but requires caution due to potential root grafting in conifers. Girdling may work with some conifers, but trees are usually slow to die and could create a hazard.

Living barns are not sacrifice areas, and care should be taken not to repeatedly damage tree roots during soft ground conditions or smother roots with excessive waste hay and manure. Repeated damage and stress will result in the eventual decline or death of the living barn. Then it's back to investing tens of thousands of dollars in building a shelter.

If you're interested in creating living barns from scratch by planting trees, use a diverse mix of species to hedge against future pest issues. The Scots pines in figure 3.31 were healthy up until a few years ago, when they became infected by an invasive and lethal foliar fungus. The addition of other tree species may have limited the spread of the disease but—more important—would have ensured sufficient residual stocking of resistant species for continued utility of the living barn.

Figure 3.32. After many iterations we've settled on a duck house design we are mostly happy with. Adequate shelter from the elements, ease of moving, and strength and integrity of the structure are all important characteristics to balance in the design and construction of movable shelters.

laneway with access to paddocks, connecting animals to a central barn space, but more often than not this becomes a damaged, muddy mess. Land stewardship means spreading out impacts, and so, depending on

the needs of your animals, plan for your shelter to move along with them whenever possible.

Shelter should be designed to meet the particular needs of the site, animal, and breed. It's important to assess the climate conditions of your grazing sites, as well as microsites where conditions differ. Shelter may be necessary for:

- Shade from sunlight.
- Protection from high winds.
- Rain and snow.
- Predation (especially in the case of poultry).

Retrofitting a trailer or making a shelter that can be dragged or rolled along the ground is pretty easy and straightforward; open-floor shelters, however, are only feasible if the threat of digging predators like muskrats and weasels is low or nonexistent.

Across the board, poultry should be given a locked space for nighttime, due to their vulnerability to predation. Chickens and turkeys prefer to roost, so structures need to be taller to accommodate this. Ducks and geese nest, so their shelters can be shorter—as low as 3 to 4

Figure 3.33. This lightweight shelter is made from chain-link fence poles and fittings found online. Add some small cart wheels and it's easy to move by hand or with a tractor.

feet tall—but 2 to 3 square feet of floor space per bird is required.

Many breeds of pigs, goats, and some sheep need shelter from anything but the nicest weather conditions. They prefer to watch the rain or snow fall from under cover. Most cows and many of the sheep breeds happily graze during inclement weather, only seeking shelter during the most intense storms.

Different breeds can beat up on shelters in different ways. Goats love to climb, and if a shelter gives them a leg up they will get on top. With sheep, it's mostly the young lambs that love to play, but older ewes can inflict damage while seeking to satisfy their need to scratch and itch. Cows and pigs can topple a weakly constructed shelter outright. Especially in constructing movable shelters, you must strike a

balance between strength and how heavy the shelter ends up being.

Ideally, with silvopasture we can eliminate the need to drag a portable shelter around. Indeed, this is one of our stated goals of our farm: to phase out the need for a shelter at all. About one-third of our paddocks currently provide adequate shelter for the sheep, and we hope to continually increase this over time.

FENCING

As previously mentioned, the real game-changer that's put rotational grazing within the reach of many farms has been the advent of many options for portable fencing. These include net fencing, strands, and tape. The habits of your animals mostly determine what type of fencing is most effective, which ultimately has more to do with temperament than size. Of course, there are also more permanent fencing options, such as wire mesh, welded wire, and high-tensile fence. No one who invests the time and finances into permanent fencing regrets it, except for when they don't think ahead and end up putting posts in the wrong place.

Sheep and cows are mostly willing to stay put without challenging a reasonably hot fence, provided there is good forage available. Cows trained to single strand can be kept in with the most minimal fencing of all animals, though sometimes a second strand is needed to keep in young calves. Sheep need multiple strands, usually six to eight, though most producers seem to go with net fencing as the preferred approach. With proper training these animals learn to respect and avoid the fence.

Goats, pigs, and poultry, on the other hand, need fencing in the highest working order. Given their curiosity and intelligence, they will seek out weakness in the fencing and exploit it. Good fencing and a hot charge at all times are musts for these animals, as they will happily take advantage of any opportunity to explore beyond the fence should the opportunity arise.

Ideally, fencing consists of a permanent perimeter fence around a grazing property, with the addition of portable fences to subdivide space into paddocks. Unfortunately, many existing high-tensile and other permanent fences have been installed right at the edge of pasture and woods, effectively excluding and defining the stark change in vegetation, which poses a problem for silvopasture planning. In reality, both sides of the fence can be used in silvopasture, making even better use of it. It's just that your perimeter will change.

If there's no existing fence infrastructure, it's nice to start out with all portable fencing to get a feel for management and paddock placement before committing to expensive installation and material costs. Plenty of grazing operations choose not to invest in permanent fencing, getting by with all portable. This prospect is less feasible if escaped animals pose a particular risk— for instance if your farm is adjacent to a major highway or a neighbor who doesn't tolerate loose animals.

Choosing a fencing system is always one of the most significant investments in a grazing operation.

Figure 3.34. For birds of all sizes, netting is really the only option for portability. The fencing specifically for poultry has very small openings and is much heavier than other types; often supplemental poles are needed in a few places to keep it from sagging. Photo by Jen Gabriel.

The choice between portable and permanent often comes down to the cost of installing fence versus the time spent in the long run moving fencing again and again.[37] But portable fencing also offers one more great advantage: flexibility. Paddock sizes and shapes, as well as your needs to include or exclude animals, can change over time. The lack of permanent fencing also offers an aesthetic advantage.

Types of Fencing

High-tensile fencing is great for perimeter fences and consists of a system of permanent posts and wire that can be placed under high tension. It's often the preferred choice for perimeter fences, with single-strand electric used to subdivide paddocks. The cost of installation ranges from $0.89 to $1.50 per linear

Figure 3.35. This high-tensile fence was constructed from black locust posts harvested from a silvopasture. It's critical that corners and bracing are done properly, since the fence will be put under high tension.

foot. There are many technical details involved in installing the fence properly, though the payoff is worth it—the fence often lasts longer than the farmer.

Woven wire / welded wire is another option for more permanent fencing, though it's often one of the more expensive methods, ranging from $1.50 to $3.00 per linear foot. The process of installation involves sinking posts (T-posts are often used) and then unrolling the fence, stretching it as you go.

Portable net fencing has quickly gained popularity as an affordable and flexible method, with costs between $0.75 and $1.25 per linear foot. There is a steep learning curve in the beginning to properly fold and roll fencing to avoid it getting tangled, but once you get the hang of it, it's a breeze. With silvopasture, dragging the net through woody areas can be a real pain. It's important to maintain wide alleys,

and clip stumps and brush stubs completely down to the ground, to avoid tremendous frustration.

Single strand / rope / tape is by far the cheapest and least durable option. It's most practical for larger livestock that are well trained to the fence (cows). The cost can be as low as $0.20 to $0.50 per linear foot. Installation involves moving temporary posts and rolling out the strands to create the desired fence line.

Keeping a Clean Fence

Whether you choose permanent fencing, a portable option, or a mixture of the two, keeping fences hot with electricity is critical. What seems to many beginners to be a simple task quickly becomes complicated with the many environmental variables and demands of a fence. The basic needs to maintain a hot fence are:

Figure 3.36. One of the benefits of working with cows is that a single strand of wire is enough to exclude them from a paddock (once they are trained).

Figure 3.37. This three-dimensional fence at a Virginia Tech research farm is multifunctional, as it keeps deer out, preventing them from nibbling on seedling trees, and it can keep livestock in.

1. Fencing that is "clean," meaning that the electrified parts are not in contact with too much grass or brush.
2. An appropriately sized energizer that has a good grounding system.
3. Reliable power to support the energizer.

The way electric fences work is that energy originates at the energizer, which sends a pulse of power through the fencing, whether it be single strand, tape, rope, or net. This wave of power completes a circuit provided the system is grounded, which means it utilizes the soil to provide a reliable, low-resistance return of the power sent out to the energizer. When animals (humans, too!) touch the fence, they interrupt this circuit, and the power is all sent to the point of contact as a shock.

The type of soil and its moisture level are both important factors in achieving a good ground. Wet, heavy clay soils work best for good conduction. In excessively dry conditions or sandy soils, you may need to take additional measures such as burying the ground rods in a salt-and-clay mixture, and potentially adding more ground rods. If you aren't achieving a hot fence, check the system's grounding first. It's the most common problem for a poorly performing fence.

A final point is to never skimp on a cheap energizer. Always get one with a joule rating two or three times the recommended size, often expressed on the package as the number of miles of fence the energizer can handle. Keep in mind with netting that the length of the fence is actually the total length of the fence multiplied by the number of strands—this is effectively the length of material that needs electricity.

Moving Fence

In rotational grazing systems, moving fence can be a simple chore, or a daunting one. Know from the outset that, at first, there will be a steep learning curve, but over time the process gets smoother and easier. Fence moving is a skill, and skills take time to perfect. The reward for moving a fence is well worth it: healthier pasture and animals, more stocking per acre, and good soil-building practices.

Regardless of the type of fencing you use, get enough material to set up at least two to three

Figure 3.38. Once you get the hang of folding and rolling net fencing, you can do it relatively quickly. Photos by Jen Gabriel.

paddocks at a time, if not more. This gives you the freedom to move animals when it's needed, not when you have the time. A fully portable fence system has the distinct advantage of allowing you to mow the fence lines before setting them up, whereas permanent fence has to be mowed with the fence always in the way.

Strand, rope, and tape fences have the distinct advantage of being easy to roll up and move. It's worth investing in good reels and strong fiberglass poles to achieve this task. Net fencing can prove more challenging, at least until you learn the proper method of folding and rolling it for transport. Particularly in woody silvopastures, net fencing can prove to be a challenge, if not utterly frustrating. The material seems to find every branch and stump and get stuck on it as you move through the woods and brush. There are two approaches we have found useful in reducing this:

1. Make access points wider than you think. We aim for all our fence lines through hedgerows and through the woods to be 4 feet wide, so our small tractor and brush hog can fit though.

Figure 3.39. It's common to come across old fencing tacked to a tree, which eventually causes the tree to grow over it and decay early. A better design is to use a board with washers and nails, which can ease out from the tree as it grows and thus protect it. Photo by Boberger/Wikimedia.

TRADITIONAL FENCING FROM SWEDEN

In Sweden one traditional fencing technique is known as *gärdesgård*: Juniper trees are used as posts, and spruce as the rails. Interestingly, the best wood for theses fences is the worst wood for timber; slow-growing trees in the suppressed understory with tight growth rings are less susceptible to decay.

After you harvest pole-sized trees, lay long lengths of spruce branches in between the juniper posts and tie them on. You can heat the branches over a fire to soften them and then twist them to increase their strength and pliability.

These fences offer not only aesthetics but also a valuable use of the low-grade wood in the farmscape, and they bring woody debris from the forest into the pasture. Centuries of these fences, which decompose in 40 to 50 years and are then replaced, are often rebuilt on fertile mounds of the previous generation's wood. The system has a lot more appeal than barbed-wire and metal alternatives, which if left behind create a dangerous hazard. Visit www.Woodlanders .com to see a video about these fence systems.

Figure 3.40. These traditional Swedish fences are both practical and easy to construct, making use of low-grade trees from the forest.

2. Install temporary or permanent portions in difficult spots. Instead of struggling with a tricky place each time the animals rotate, just do it once. For instance, we use net fencing, but in many hedgerows we have more permanently installed eight strands of fence, which we then attach the netting to.

In many operations, some form of a hybrid among different fencing systems is what the farmers ultimately settle on. In silvopasture systems it's worth setting yourself up for success by thinking about approaches to fencing in hedgerows and woods. Our farm employs net fencing for both ducks and sheep, but we've started to install semipermanent fencing in the hedgerows. This gives us a fence to tie our netting into, without having to drag the net through the woods each time we move the animals in.

So Many Choices!

Beginners to grazing can feel overwhelmed by the vast array of choices at hand, from the animals to the water, fencing, and shelter infrastructure that supports them. So let's review the main points.

Choice of animal and breed. Make your decision based on the size of your land, the local markets, and the functions of the animal that best fit your landscape and management goals and style. Don't decide based on your emotions alone. Consider the smaller ecological footprint and lower cost of raising ruminants versus animals who need grain as a major feed source. Carefully consider the breed, and the breeder.

Shelter. All animals need access to shelter from wind, rain, snow, and excessive sunlight. The degree of protection depends on the species and breed of animal. Eventually, the trees in silvopasture can be an effective shelter, but you'll likely need some form of movable structure in the shorter term.

Water. There should be readily accessible water for animals that is as clean as possible. A portable option (tank on a trailer) is often worth the cost, to avoid the expense of buying pipe or having to move animals back to a central watering location. Winter demands good planning ahead of time.

Fencing. Choose a system that will keep your animals where you want them while also protecting them from outside threats. Weigh the benefits of permanent fencing (more money but less labor in the long run) versus portable options (lower costs, more labor, and greater flexibility).

Before diving into an animal system, visit as many farms as you can, and check out their choices. Ask a lot of questions and offer to help move some fencing. Most successful graziers rely on a community of peers as an ongoing resource to share ideas, commiserate, and trade gear and materials. Start building this network as early as possible, and before the animals get to the land, make sure you are all set up with infrastructure. Plan for a lot of extra time up front, but always seek ways to reduce the time you put into managing animals. There

is always another way to improve the system, if only you look at it with a critical and flexible eye.

The Final Piece: A Grazing Plan

At the center of all good grazing practices, regardless of the choices made in the previous section, is a plan. A good grazing plan consists of two parts: a paddock map and a grazing chart. The map acts as a reference to the physical spaces (paddocks) where you will move animals, and the chart keeps track of the dates you move animals, mow, and complete other activities in management. As with any data collection system on the farm, the information is priceless.

The remainder of this book actively engages you in the process of developing a grazing plan, starting with woodland and pasture assessment in chapters 4 and 5, then moving on to the development of maps and charts based on this knowledge in chapter 6. The process can be thought of as this: With animals and their necessary infrastructure in place, your role is to orchestrate their movement through a dynamic and complex landscape. Getting to know the habits and personalities of the animals is one part of the puzzle, with the minute details of the land and its vegetation equally important. To these aspects of silvopasture we now turn.

BUILDING A PIG SILVOPASTURE AT SPIRAL RIDGE IN TENNESSEE
By Cliff Davis

Southern Appalachia had a proud hog and tree crop culture for centuries. Entering the highland rim of middle Tennessee, you can sense this forest culture of yore. With a little investigation—say, chatting to the 70-year-old garbageman—you can get a glimpse into the past, where grand chestnuts and oaks filled the horizon and families raised pigs, reminiscent of the Euro-swine-tree-crop cultures mentioned in the influential book *Tree Crops: A Permanent Agriculture* by J. Russell Smith. Recent agricultural practices throughout the deciduous forests of the Southeast have left much of the highland countryside denuded of trees, only to create pasture for poorly managed cattle, fields of GMO corn or soybeans, and concrete-infested development.

Spiral Ridge sits within a majestic network of finger ridges, waterfalls, limestone outcroppings, pure springs, and intimate valleys, all clothed in a wardrobe of oak and hickory forests, accessorized with persimmon, black locust, mulberry, black gum, tulip poplar, plum, honey locust, and other perfect pig fodder.

Our pig adventures began a couple of years after we moved to our off-grid farm. A year prior to our move, the land had been clear-cut and the forested ecosystem destroyed. The temperate/subtropical nature of our ecoregion delivers 50-plus inches of precipitation a year, giving natural succession favorable conditions to grow with vigor and density. And it did!

Our end goal is to mimic habitat patterns of succession in the eastern deciduous forest, using the forest itself as our primary directive for a productive agroecosystem. Using the intelligence of succession and the high production of savanna ecosystems, we

Figure 3.41. The proper selection of genetics, a keen understanding of forest ecology, and an adaptive management style are the most important aspects of any farrow-to-finish outdoor pig silvopasture system. Photo by Cliff Davis.

set out to discover the past and future of farming: agroforestry. To do this, we employed animals to offset our manual labor and create a disturbance strategy that slowed down succession enough for us to plant fruit and nut trees with an understory of pasture. To re-create this highly productive savanna ecosystem, we put our faith and promise on the back of a herd of pigs.

The ecology of a pig has evolved to create a highly intelligent, self-reliant omnivore whose native habitat is woodlands. Pigs need simple shelter, shade, and a diverse diet. Add water, trees, and pasture and you have a resilient source of utilitarian delight for successful, small-scale, diversified farms.

Creating and managing a silvopasture system using pigs has been a lesson in animal behavior, forest ecology, holistic management, pasture establishment, strategic disturbance, and personal humility. This synthesis is mastered through a series of successes and failures, as an ecological farmer.

We have been researching and developing strategies and techniques particular to our holistic context, climate, management, and farm economy to create an adaptive agricultural system that baptizes the pig and tree crops as keystone species of a future renaissance.

Central to this work has been identifying compatible species of pig and tree crops, effective fencing options, strategic disturbance patterns, holistic management systems, regenerative pasture enhancement, and positive personal attitude/aptitude.

The selection of compatible species of pigs and tree crops has been time consuming and also costly. Tree crops were a little easier, as the native oak-hickory forest has a great selection of trees. Allowing the coppice regrowth of oaks, hickory, persimmon, and black locust, while planting species like apple, pear, chestnut, mulberry, pecan, hazelnut, honey locust, and pawpaw in the openings, helped initiate our silvopasture system. Fast-growing pioneer species, like tulip poplar and sumac, are being utilized as pig and goat fodder.

Figure 3.42. Acorn-finished craft pork. These craft pigs are being finished on chestnut oak (*Quercus montana*), pignut hickory (*Carya glabra*), and various omnivore delights. Photo by Cliff Davis.

We have been through several different breeds of pigs, from registered purebreds to hybrid crosses, from meat hogs to lard hogs. What we have realized is that the breed should be chosen for the ecology, the environment, the management, and utilization. All four of these criteria are very important when selecting a pig breed.

We have settled on creating our own hog for our own context. There are qualities and characteristics we favor and don't favor in each of the pigs we used. For instance, the heritage lard hogs are favored for their ease of management, quality of fat, and ability to graze pasture, but they took too long to grow, bore low numbers of offspring, and had too much fat—therefore they were hard to sell and make money from. The heritage meat hogs grew very fast and overly disturbed the forest/pasture ecology, but reached market weight in six to eight months. Creating a hog is no easy endeavor, but for the die-hard pig farmer, it is a viable option. Our High Forest Hog will meet all four criteria and give us a quality of pork reminiscent of the famed Iberico hogs of the Spanish dehesa ecosystem.

The most crucial infrastructure for pig farming is fencing. Our efforts to find simple, adaptive, low-cost fencing have allowed us to work with several fencing options. The most adaptable fencing we use is electrified poly net for pigs. This temporary system is fairly easy to set up, is low-cost, and can be used at different lengths and patterns. It also keeps all sizes of pigs from escaping, including the rowdy piglets. Another system of fencing we are working with is semipermanent 12.5-gauge aluminum wire. This is far cheaper than poly net, as you can use existing trees for posts and wire gates for easy rotation. Once you have moved pigs for several years using these simple fencing options, you can start to consider more permanent fencing options.

The dynamic nature of disturbance creates habitat niches for ecological adaptation to occur. When using pigs in any ecosystem, disturbance will follow, delegating the farmer to creating a holistic management

Figure 3.43. Two years of research using bamboo as a forage for pigs has proven to be a money saver and herd maintainer in the winter months. Pigs prune young fall branches and leaves from this *Phyllostachys nuda* timber bamboo. This bamboo will benefit significantly from the excess nutrients for spring forage production. After spring pole production, we maintain the groves using a cut-and-carry fodder bank system that we feed to the pigs throughout the summer. Poles get graded and selected for on-farm use. Photo by Cliff Davis.

system. This is a cautionary tale. With any slack in management of pigs, overdisturbance will occur. Over the years, we have seen soil ecology improved and destroyed, all caused by management decisions. We look at how much of the pasture has been trampled, eaten, and denuded. We like no more than 20 percent exposed, 50 percent trampled, and the rest eaten. Exposed areas are always covered with old hay, wood chips, or sawdust and reseeded with an appropriate seed mixture.

A mix of annuals and perennials makes the best pastures for pigs. We recommend a perennial clumping grass and a legume as a base. Seasonal annuals broadcasted over open areas and then mulched have worked very well as a feed supplement during the spring, summer, and fall. In early spring we broadcast oats, red clover, and field peas and undersow with brassicas, turnips, and other veggies. In the summertime we broadcast buckwheat, sorghum, Sudan grass, sunn

hemp, and cowpeas. In late summer we start to sow rye, wheat, vetch, peas, and crimson clover. During the winter, we move pigs less throughout the system because of the wet weather. We like to keep them in more mature woodlands in winter, eating acorns, grubs, and insects.

The key is to have a good attitude and know that you may mess up from time to time.

Management is no easy task. It takes years of study, observation, and adaptation to be a successful silvopasture pig farmer. As we move forward in our endeavors, we hope to learn from our mistakes, build more permanent fencing systems for ease of management, create ecosystem-fed craft pork that uses very little supplemental feed, and build awareness of the greatness of the pig and its native woodland habitat.

Learn more about this system and contact Cliff Davis through www.spiralridgepermaculture.com.

Figure 3.44. Research and development of select breeds for herd improvement, craft pork quality, management efficiency, and land disturbance is another important part of the work we are doing to develop bioregional regenerative agroecosystems integrating tree crops, earthworks, and pigs. This ancient Chinese breed shown here is called the Meishan. We are trialing this breed for litter size, mothering, and gourmet craft pork. Photo by Cliff Davis.

4 Managing the Woods for Grazing

While existing forestland in many ways offers a clear opportunity for silvopasture, not all forests should have animals in them. Farmers and landowners have a legacy of doing great damage to their woods, which is not a pattern that should be continued. Instead silvopasture practitioners should focus on forested lands that are marginally to severely degraded, in many cases. Healthy forests should be left alone, or perhaps limited to smaller animals such as poultry, or to only periodic (every 3 to 10 years) visits from animals to forage. Some disagree with this idea, which will be discussed later in the chapter. It's an open conversation.

In many cases land that is overgrown with undesirable (aka invasive) vegetation, hedgerows, and old plantations—as well as land where the forest overstory is already open enough to allow for grasses and other plants to grow—is the ideal, at least in the beginning. These areas can be cleared and renovated all on one dramatic pass, or you could work at them piecemeal, over time, even using the animals themselves to do a lot of the work.

Given that the trees in an existing woodlot are almost always the same age (see chapter 2), you can also arguably remove a great deal of the lower canopy trees, as well as make some strategic decisions in the canopy, to grow excellent timber species out of the remaining trees, all while capturing the shorter-term gains of grazing underneath.[1] This type of forestry must be done with careful planning and include a developed plan

Figure 4.1. These woods were high-graded for the best white pine about 20 years ago, leaving a collection of poor-quality trees and opening up the understory for opportunistic plants like multiflora rose, honeysuckle, and others. This type of woodland is abundant in the temperate climate and makes a great site for silvopasture.

for how forest regeneration (seeding trees for the next generation) will be introduced and protected.

It is likely that the vast majority of farms practicing silvopasture will engage in a diverse set of approaches, including converting some forested land to silvopasture, planting some open pasture with trees, and leaving some forest and some pasture as is. Ultimately, the more types of habitats you employ, the more you will be able to be flexible, adaptable, and prepared for whatever the weather, climate, and seasons offer. Over the next two chapters, we pick apart the nuances of these different approaches, and then in the design chapter (chapter 6) we will look at how all the parts all fit together.

Provided you've explored the basics of forest ecology (chapter 2), can competently read the forest, and/or are engaging the help of a forester, you can step into converting forestland to silvopasture. Always proceed with caution when cutting trees, of course, recognizing that decisions have consequences that can be both long-term and unknown.

Starting with this chapter, the information we present is going to be more applied. Follow along and, if you are actively planning a silvopasture system, complete the recommended activities. By the end of chapter 6, you will have a great start on a silvopasture design for your farm or homestead.

Climate Change, Woodland Conversion, and Managing Like Old Growth

The main disadvantage of silvopasture in the woods is that, while you're increasing usable productive land in the short term, you're also in most cases reducing the carbon-holding capacity of the forest. Clearing brush and trees means setting in motion the process of decomposition, and this initiates carbon release back into the air.

It's important to think about the life cycle, and about these impacts, when deciding on strategies. Consider the current carbon capacity of the woods you are managing, and what impact you will have in converting it to silvopasture. It all gets very context-specific. For instance, thinning a mature hardwood forest to 50 percent canopy cover would have a much greater impact on carbon than would thinning a woods or clearing brush from an area as depicted in figure 4.1. Further, both of these strategies are better than clearing the land completely, which still often happens for development or even (ironically) to make grazing land for animals.

While you might be looking at just your woods in this frame, it's important to note that the dynamics of carbon on a global scale are complex and interconnected. While cutting trees removes carbon from a woodlot, it all depends on how much carbon an individual tree is holding. For instance, cutting a very slow-growing, 90-year-old tree that is only a few inches in diameter would not have nearly the impact as cutting a 90-year-old tree that is several feet in diameter. Thus, thinning woods to support growth on the largest and most vigorously growing trees would have a lower impact.

In addition, if harvested wood can be put into use as a durable good (tool handles, buildings, furniture), this could even increase carbon storage, since those materials will likely last as long as or longer than a living tree, especially considering that wood has a much lower climate impact than many other building materials.

All in all, the multiple variables on many levels of scale should be considered, and landowners should seek to manage with win-win scenarios, including approaches that both retain as much carbon in the woods as possible (say, by keeping the largest trees), and also engage in productive use of the landscape.[2] And while the carbon dynamics of forest thinning is complicated, practitioners can rest assured that planting trees in pasture is always a win from the climate perspective (see chapter 5).

As we manage, we can also engage with the woods in a way that supports a more diverse forest structure and species composition. At the University of Vermont, a 15-year study of forest management techniques that mimic creating an old-growth forest has had promising results: Tree harvests can still occur while the woodlot maintains as much carbon as one that wasn't thinned at all.[3] The application of this concept to silvopasture is unknown, since the necessary harvest for this application is much higher, but it's some good food for thought.

What does managing like old growth mean? The technical term is *structural complexity enhancement*, and it refers to addressing the pattern common among current forests, which are largely single-age stands that feature a lack of diversity in the size and age of trees,

Figure 4.2. Cutting trees always has impact, but good understanding of the dynamics at play can mean a lesser impact. This tree is being cut to "release" surrounding trees that show greater growth characteristics, and the woods as a whole will retain the largest trees long into the future.

dead and dying trees in the canopy, canopy gaps, and various types of coarse woody debris decomposing on the forest floor.[4] This type of approach greatly enhances biodiversity, and even results in the residual trees putting on growth faster when compared with several more conventional forestry treatments.

The important takeaway is that converting a forest to silvopasture is not a benign activity, but neither is any form of land use. As land stewards, we should be aware that removing trees removes carbon, and that ideally those trees are put into a durable use. In our forestry practice, we should strive to leave the biggest trees in the woods, first thinning those that are growing at a slower rate. And where we have soils and conditions ideal for growing trees, we should pause before we decide to bring animals into the picture.

Different Types of Forests

For the purposes of this chapter, we will discuss four types of vegetation patterns that we want to assess for silvopasture conversion. Not every site will have these, nor will a specific stand of vegetation always fall neatly into these categories. But each requires a bit of a different approach to management, and often to assessment.

Figure 4.3. Plantations offer some of the best options for silvopasture, since these monocultures are low in productivity and habitat biodiversity. Trees can be strategically thinned, though may not need as much thinning as hardwoods since the high canopy can allow more light in. Photo by Jen Gabriel.

The categories we will consider include:

Forests are dominated by canopy trees (past the stand initiation phase described in chapter 2). They can range from very young to very old. Management varies widely based on the species composition. Some may be appropriate for silvopasture, while others should be left alone.

Plantations are often a monoculture of one or two species, most often intentionally planted in rows. These often were intended to be wholesale-harvested, but many such plans were abandoned as markets changed. These are often low-value forests that can offer great opportunity for silvopasture.

Hedgerows are strips or clusters of trees often left by previous farmers because they were either hard to harvest, hard to maintain with machinery, or desirable for their sheltering effects from wind and snow. Many hedgerows are also overgrown with thick brush, making them impenetrable. They offer some of the best places to get started with silvopasture, blending them into the edges of existing paddocks.

Abandoned ag lands / high grades / clear-cuts are all the result of large-impact management and, often, abandonment by previous landowners. They are among the most abused landscapes and often deemed as worthless. Yet they offer some of the greatest potential for silvopasture, which can support healthy forest regrowth and reclaim these lands for productive use.

Along this continuum, tree cover and canopy dominance of trees decrease as we move from top to bottom. Effect on carbon storage increases as you move in the same direction. And current value of the land (to wildlife, ecosystem services, and the economy) is often highest with forests and orchards, less so with the others. If we are to apply silvopasture most effectively, we should start with the most marginal lands, then apply what we learn to the more valuable places.

A LEGACY OF DISTURBANCE

Overall, it's important to note that, regardless of the forest type, we are working almost exclusively with lands that are disturbed in some form; and in almost all cases it's safe to assume the land is abandoned farmland that was once cleared for agriculture. The forest, then, is young, regrowing, and immature in its identity. In many cases the first approach is to recognize the tree species that are doing well on the site (if any) and act to remove any evidence of disease and defect, along with other species that are not doing well. And while it is tempting to remove dying or dead trees, in many cases those are some of the most important to leave, as they become critical habitat for wildlife.

Remember, from chapter 2, that almost all of these forests are even-aged stands, which means that it's pretty easy to determine which trees are faring better than others. If all trees in a stand are more or less equal in age, then those that are larger and taller are most likely going to be keepers, while others can be among the first to go. Think of your job as similar to that of a lawyer, where the forest and trees provide you with evidence for cutting: The answers are all there; they just need to be discovered.

Assessing a Woods

A good assessment process combines three things: measurements/data, observations, and clarifying the

Figure 4.4. Incorporating hedgerows into grazing paddocks offers immediate shade and shelter benefits to livestock, as well as better utilization of land. These areas are often overgrown and can be thinned over time, opening up the edges and bringing more forage into the understory.

management goals of the farmer or landowner. This is true whether you are examining biological systems such as forest or pasture (as we will discuss in chapter 5), or infrastructure elements such as a building or driveway site. In this section we will apply assessment to understanding the character of forests and woody areas, so we can determine their potential for a silvopasture system.

DEFINING A FOREST STAND

Within an entire forest, a *stand* is an area that features common patterns in vegetation, slope, hydrology, and microclimate. A stand is considered pure if at least 80 percent of trees in the main canopy are of a single species; a mixed stand is anything less than 80 percent.[5]

A stand map, then, is a map of the property of interest, with areas that are identified and marked as different from one another. Sometimes stand borders are obvious, such as alongside a gorge or creek, or they can be much more subtle. They can often be observed from aerial photography, but in most cases you'll walk the woods and mark stands on a map, or a forester may do this for you.

Think of this exercise as taking a piece of land and drawing a bunch of bubbles around it, dividing it up into chunks where the forest looks and acts the same.

Figure 4.5. A stand is delineated most often by the dominant trees in a given area. Stands can range from less than an acre to several dozen. Designating these areas helps prescribe management strategies that will differ from one stand to the next.

HELP WITH TREE IDENTIFICATION

One of the most basic aspects of forest management starts with being able to confidently identify tree species. This is not always an easy feat, as there are sometimes dozens of species in just a few acres of woods. Like learning a language, the process of getting to know trees takes time. The character of a given species is similar, but not always consistent, especially when looking at the bark of trees.

Consider going beyond simply "naming" a tree. Get to know it. Too often, we stop at learning the common and Latin names, and some properties, where trees offer so much more. Knowing when they leaf out in spring, when they flower, and when their leaves and seeds fall to the ground is especially useful when designing for silvopasture.

Start with just a few major species, and add others as you go. Some great resources to help your learning process include:

"Key"-based tree guides send you through a series of questions (the "key") and help you arrive at the identification of a tree while learning some of the ways they can be distinguished along the way.

- *Tree Finder* and *Winter Tree Finder* (Eastern US) by May Theilgaard Watts are pocket-sized books easy to bring along into the woods.
- *Bark: A Field Guide to Trees of the Northeast* by Michael Wojtech offers a great study of the different presentations of bark in trees, which is one of the more challenging variables in tree identification.

Some of the best desk reference books include:

- *Tree Identification Book: A New Method for the Practical Identification and Recognition of Trees* by George W. Symonds is an "oldie but goodie" book from the 1970s that has great keys and information on trees.
- *The Sibley Guide to Trees* by David Allen Sibley is a beautifully illustrated and informative text about families and species of trees.

The main factors that distinguish one stand from another include:

Dominant Vegetation

Forest stands are defined by the dominant tree species in the canopy. In some cases a single species can be almost entirely dominant; in others, several species might share the mix. What determines which species show up and persist are the environmental factors that each species prefers, along with the available seed.

It is not only a stand's composition that we should examine but also its age and the relative health of the dominant and co-dominant species. Some of these qualities can be observed or measured on-site, while others might require looking at historical imagery and/or consulting with a forester.

Soil Type

The factor that most limits trees' ability to grow large and tall is the soil type, which largely remains stable over long periods of time. The exception to this is the top 10 to 12 inches of topsoil, which in many forests has been radically altered through a history of tillage agriculture, disrupting the biological composition and structural makeup of the soil and depleting nutrients and organic matter in the process. In the end, the trees in a forest will only be as good as the soil, and there isn't a lot you can do in their lifetime to dramatically change this, unlike the scenario with grasses discussed in chapter 5. It's best to assess and understand the potential limits of a given soil, and design management practices accordingly. This is achieved through soil survey maps, as well as on-site testing, which is described on page 176.

Landform

The umbrella of landform includes elevation, slope, aspect, and position of a forest stand in the larger landscape, where:

Elevation is the number of feet above sea level. Certain species won't grow above or below specific elevations, due to the changes in growing conditions.

Slope is how steep or flat the stand is. This affects water runoff, soil erosion, and the angle at which sunlight hits the canopy.

Aspect refers to the direction the slope faces, which also dictates sun angle, soil moisture, and a number of other growing conditions.

Position is where the stand falls in the larger landscape. Generally, temperate landscapes are classified as uplands, upper slopes, lower slopes, or bottomlands. Each offers unique microclimates and sees different effects from water, wind, fire, and other forces of nature, including more subtle patterns of animal movement and tree seed dispersal.

Hydrology

The best way to understand the hydrology of a stand is to observe it firsthand and make a map. Of course, there are the obvious creeks, low spots, and the like, some of which can be discerned from a map. But even more important is to observe the dynamic set of changes that occur from one season to the next, and even from one year to the next. With increasing flooding and droughts occurring in much of the temperate United States because of climate change, and with rain events projected to increase in intensity,[6] paying attention to

Figure 4.6. A thorough assessment of our hydrology resulted in the deliberate channeling of water through our maple woods to address flooding issues. This created the added benefit of some periodic pooling of water that often coincides with our duck rotation.

water will become ever more critical to good farming practices, as we've discussed in chapter 1.

In addition to identifying the subtle flow of water in the forested landscape, identifying watersheds is often a good way to define forest stands. While people often think of watersheds in the largest sense, it's useful to look at the micro-watersheds of your land.

MAKING A STAND MAP AND CHART

The process of making an accurate stand map requires us to gather available data and on-site observations. Of course, the lines drawn on the map are not absolute, nor do they need to be absolutely perfect. Rather, in defining general patterns, we are then better able to define how we should manage one patch differently from another. Don't get hung up on the details.

To make a stand map, you will need to obtain an aerial map of the property in question and then either draw directly on it or use a computer program. Following these steps will prove worthwhile later on, as this map will become the foundation to which you'll add pasture stands and paddocks in chapter 5, and then turn into your overall grazing map and plan in chapter 6.

There are many online and computer tools that can help you develop a stand map, with some time spent out on the land to verify and fine-tune your work. Generally, there are two approaches for obtaining information throughout this assessment: One is online and with software widely available for consumer use, and the other is through your local soil and water district and cooperative extension offices, both of which are available in most counties in the United States.

As noted in the sidebar, this book encourages you to take advantage of Google Earth and online tools that are free and widely available, as well as to form relationships with actual humans in your local agricultural support offices. Both offer different advantages and disadvantages.

1. Obtain a recent aerial photo of the site.

You can get a good aerial photograph of your farm property online, on Google Earth Pro, or by visiting your local county soil and water office, which will print one off for you. The photographic record is updated at least every few years, so you should be able to get a photo taken in the past 12 to 24 months. Further, soil and water and/or your local library or historical society should have a collection of aerial photographs on file, as does Google Earth. These old images often provide valuable insight into the past land use and what effects it may have had.

With map in hand, determine how you will draw and take notes on it, so that you can use it over time. If you like computers, you can use any of a range of computer programs to draw on your aerial map (again, Google Earth proves among the easiest); or you can print out your map at home, or enlarge it at a local copy shop, and use tracing paper over the map to delineate stands.

2. Mark initial bubbles from previous knowledge/patterns obtained from the photo.

Before you go outside, there is a lot of information to collect and analyze, including your own previous observations, depending on how familiar you are with the site. The goal is to first determine the clearest boundaries that distinguish one stand from the next. Draw a line anywhere the species composition, slope, watersheds, waterways such as streams or creeks, and infrastructure, such as driveways and buildings, help provide clear boundaries (see figure 4.7).

3. Add soil and contour information.

Next, locate a copy of your soil survey and add this to the map. Your soil and water office can provide this, or you can use the national online explorer website (websoilsurvey.sc.egov.usda.gov), or import soil data into Google Earth (see the sidebar). The advantage of the soil and water explorer site is that you can get a lot of interpretations for a given soil type. The change from one soil type to the next may or may not align with given stands. Keep in mind that soil survey data are estimates, based on random sampling and some estimation. Some variability is normal.

After you've added soils to your map, the next step is to define any significant watersheds and note significant changes in slope, aspect, and position. To do this accurately, you need to obtain or construct a contour map of the property, which shows the land as it is in

FOR THE LOVE OF GOOGLE EARTH: A BRIEF TUTORIAL

Google Earth is a remarkable tool accessible to anyone with a computer and Internet connection (once it's downloaded, you can use it without the Internet) or even a smartphone or tablet. It provides landowners and farmers a powerful tool that is relatively easy to use.

There are many other ways to download an aerial map, including soil and water or extension office or other online mapping tools. There are also far more complicated GIS-based programs. We choose to provide some guidance specific to GE here in this book, since in our assessment it offers one of the most accessible ways to start working on landscape-level design.

Here are directions for use, step-by-step:

1. Download by visiting www.google.com/earth
2. Open and find your site by searching in the upper left corner with your address.
3. In a GIS (Geographical Information System) program, everything is based on the idea that we can mark a point in reference to its longitude and latitude. Thus, any "point" you make in Google Earth is actually linked to a point on the earth.
4. The basic functions you can use include marking a point, drawing a line (connecting two points), and making shapes (connecting multiple points).
5. When you click on one of these options on the top toolbar, a window opens that allows you to draw, label, and make notes about the point, line, or shape. Note that the new item also appears in a list to the left, which you can sort into a folder for easy reference in the future.

6. Be sure to save your work frequently by going to File ➔ Save ➔ Save My Places. It's also a good idea to click on Save Place As . . . under the same menu and back up your file in a folder on your computer.
7. Start by marking your property boundaries but using the shape feature to outline the borders. By default the shape is usually filled in, but you can change this under the "Info" window. This can always be accessed by right-clicking on the item you created and selecting "Get Info."
8. One of the big advantages of Google Earth is that once you define an area, stand, or paddock, you can measure the area and know its acreage. You can also easily measure distances. To do this, find the Measure icon at the top of the bar.
9. You can also use this to keep notes. Simply drop a pin where you want to make a note, and add any comments to the Description part of the info box.

Of interest to our discussion here on silvopasture, Google Earth is best used to delineate stands in the forest and pasture, determine paddock size and shape, and measure areas for fencing and water.

In chapters 5 and 6, we will outline the process of naming, labeling, and assessing stands. Later, in chapter 6, you'll use this map to make paddocks. You are encouraged to follow along and build your map as you read the text.

In addition to several sidebars in this book to help you get familiar with Google Earth, you can visit www.silvopasturebook.com and click on Resources to see tutorial videos teaching you the basic functions of Google Earth.

three dimensions. Reading a contour map takes a little practice, but it's a very important skill to develop. You need only a few pointers to get started:

1. Find the highest and lowest points on your map, indicated by elevation numbers. Now you know the general direction of the slope.

2. Lines closer together indicate steeper slopes, while those farther apart mean the slope is gentle.
3. Where a collection of lines "points" downhill, it indicates a rise in elevation (ridges), whereas pointing uphill means a decrease in elevation (valleys).
4. It's often good to define the micro-watersheds that exist on your property.

Figure 4.7. An example of an initial stand map for Wellspring Forest Farms' forested lands. In our case, the stands are relatively small.

FINDING SOIL DATA ONLINE

In addition to visiting your local county soil and water office, there are two main ways to access soil data online:

1. The USDA has a browser to view and assess your soil data:
www.gelib.com/soilweb.htm.
2. You can download and open soil map data in Google Earth through this link:
www.earthpoint.us/TopoMap.aspx.

With a contour map it's useful to get a copy you can draw on, then use a fat marker to define watersheds and draw arrows indicating the direction of flow. Like solving a puzzle, the first few pieces are trickiest, but it becomes easier as you add more information. Google Earth and other computer modeling software also offer the ability to pan around the map and view the landscape in 3-D, which coupled with a contour map really helps you learn the language. You can also easily calculate slope, and even display a side-view section of your land, known as an elevation profile.

4. Check your map against the land, and give each stand a letter.

Chances are you can create a pretty decent draft stand map with the preceding tools. Next, head outside to check your map against the land. Spend time especially exploring the edges between one stand and the next, and clarify any discrepancies. Then give each stand a letter or a number, useful for notes or record keeping later. Getting out on the land also allows you to take note of the more subtle features in the landscape that are impossible to discover from maps. It's especially worthwhile to note elements of water flow, the larger trees, and the timing of a stand's species or characteristics.

Creek   Vernal Wet   Wet Spots

Figure 4.8. One of the most useful maps to start with is a general site assessment of landform and hydrology that defines micro-watersheds, aspect, and landscape position. These are important factors in understanding the ways grazing paddocks and trees will differ around the site.

It's best to work on these observations over time, as it's pretty hard to get it all in one go.

5. Characterize each stand.

Back at the kitchen table, spend some time defining the character of your stand based on the criteria that set it apart from another.

For each stand, summarize the characteristics of:

1. Dominant species.
2. Structure, age, diversity.
3. Rare/interesting features and species.
4. Initial management goals.

An example for our farm is provided in table 4.1 As you can see, our farm contains little in the way of good forest; it's mostly abandoned shrubland, hedgerows,

and poor-quality woods. This means we are well suited for many silvopasture practices on much of the land.

You'll add the information on site index and basal area to the table once you get to the next stage, discussed on pages 129 and 133.

6. Articulate stand goals.

When assessment is done thoroughly, management solutions often arise naturally. Before taking action, it's important to reflect on what the stand needs to become a healthy ecology, and how this relates to any production goals. When in doubt, it's best to leave things be until a clear direction develops. But in many cases goals for the stand do become clear, such as:

- Thin to reduce the presence of disease and overcrowding in trees.

Table 4.1. Forest stand descriptions for Wellspring Forest Farm

Stand	Description	Dominant Species, Site Index	Age, Structure, Diversity	Rare/Interesting Features, Species	Basal Area	Initial Management Goals
A	Healthy, monocrop of sugar maple	Sugar Maple (100%), SI 70	~60 years old, mostly dominants and co-dominants	Oaks and hickories on edges, black cohosh patch	130, overstocked	Maintain as sugarbush, ducks only
B	North-sloped, steep, overgrown with some good mature trees	Mixed sugar maple, oak, hickory, and brush	~40 years, scattered structure	Butternuts, creekside (heavily eroded)	60, under stocked	Improve access, good silvopasture, creek restoration
C	Hedgerows, mixed species, abandoned	Hawthorn, ash, walnut, maple	~25–30 years, clogged with brush, some nice trees	N/A	N/A	Maintain as windbreak, feather into field
D	Riparian zone, creek with heavy erosion and deep gullies	Mostly overgrown undesired veg, some ash, willow, butternut, hickory, oak on drier edges	~25 years, poor structure and diversity, abandoned and neglected	Elderberry present	N/A	Restore with plantings and improve creekbed with settling pools, graze in dry times
E	Flat, floodplain-type zone with small outbuildings	Black-walnut-dominant, some red oak and ash	~50 years, good structure, low diversity (small patch)	Walnuts in very good shape, healthy	90, well stocked	Support healthy walnuts, maintain understory for silvopasture
F	Drier woods, pasture edge, mixed species with dense understory brush and low-growing trees, lots of water pooling in areas (drainage clogged)	Walnut, poplar, cottonwood, apple, white pine, mixed, brush	~40–50 years, decent structure	White birch	70, low end of stocking	Thin to release desired trees, open for silvopasture, address drainage issues
G	High-graded area with residual pines, choked with brush	White-pine-dominant, brush	~50 years, low structure and diversity	A portion of this blends into neighbor's wetland, lots of standing water	50, understocked	Address drainage flows, clear brush, plant new trees in clusters
H	Hedgerow, some hardwoods, mostly brush and low-growing trees	Hedgerow, brush, apples, hawthorn, buckthorn	~50 years, low structure, not a lot of desired diversity	Valuable windbreak for fields, excellent-tasting apples	N/A	Thin, establish forage
I	Mixed hardwoods, some areas mostly brush, pond	Hickory, basswood, poplar, willow, brush, hickory	~50 years, good structure and diversity	Pond, needs rehabilitation; some novel willow species, pignut hickory	60, understocked	Thin, establish forage

- Reduce thorny understory brush.
- Reroute water to reduce erosion after large rain events.
- Increase the diversity of species in the canopy.
- Thin canopy to support forage establishment.

Of course, a perfectly fine goal is to simply learn more and wait on taking action. And likely a portion of the woods, where hardwood forests are healthiest, should remain as is, at least until you have developed silvopasture in other areas.

At this point, use the stand map and table to jot down some basic observations and potential goals. This doesn't need to be complicated or take a lot of time. In essence, you are simply drawing lines to distinguish one space from another, and making some initial observations of the site. A qualified forester from your environmental protection agency or extension can help you get more detailed as you go, but be sure whoever is managing the silvopasture is involved in the process of mapping and understanding stands, as they will be the ones interacting with the site most frequently and making the decisions that matter. Even if the initial assessment offers more questions than answers, at least you know what to ask someone with more experience.

STAND-LEVEL MEASUREMENTS FOR FORESTS AND PLANTATIONS

After completing the overall stand assessment, the next step is to drill down into more detail, assessing each individual stand. For this, you may want to prioritize stands that you'd like to manage first, or those that would seem most clearly to benefit from silvopasture approaches. The goal of a stand assessment is to take some "eyeball" measurements and be confident in answering the following questions:

What is the tree density (aka stocking rate) of the stand? (Quantity)

A stand that is too dense means that trees are stressed to some degree, not growing to their full potential, and more vulnerable to pest and disease outbreaks. Ideally, forestry aims to thin stands to an appropriate density, concentrating the available sunlight to the best trees.

With silvopastures, the stand of trees must be thinned to a level that allows sunlight to reach the ground in order to grow productive forage and browse plants. This requires not only reducing the density of main canopy trees to an appropriate level, but also often removing trees and shrubs in the intermediate strata as well.

How are the trees structured in the stand?

Many woodlands offer homogeneous age structures, with a single layer of trees. As previously mentioned, forests need multiple layers, gaps, and a mosaic of structures in order to thrive.

How suitable is the stand for growing the dominant trees? (Quality)

The ability of a tree to grow well in a given stand is dependent on two things: how suitable the species is for the site conditions, and how fertile the soil is for growing good trees. If you have poor soil, you will never have amazing trees. It's good to learn this early on, before overcommitting time and resources to it. At the end of the day, tree growth is a combination of the site quality (known in forestry as site index) and competition. The site index determines height growth, and competition (partially) determines diameter growth.

Determine Stocking

Stocking is a term that describes the density of the trees in a stand, a value expressed as *basal area*. Technically, basal area looks at the total area in square feet of the tree stumps if we were to cut them down at the diameter at breast height, or DBH. The square feet of space those trees take up determines the density. As shown in figure 4.9, the larger the diameter of the trees, the fewer are needed to occupy the same density.

To calculate basal area, all you need is a penny, a string, and a notepad. The string should measure 25 inches long, and it's easiest to tape the string to the penny. You can also purchase a prism or basal gauge from a forestry supply company, which replaces the penny and string as your measurement tool.

Since it's unreasonable to assess each tree in a forest, foresters generally take a random sampling of 8 to

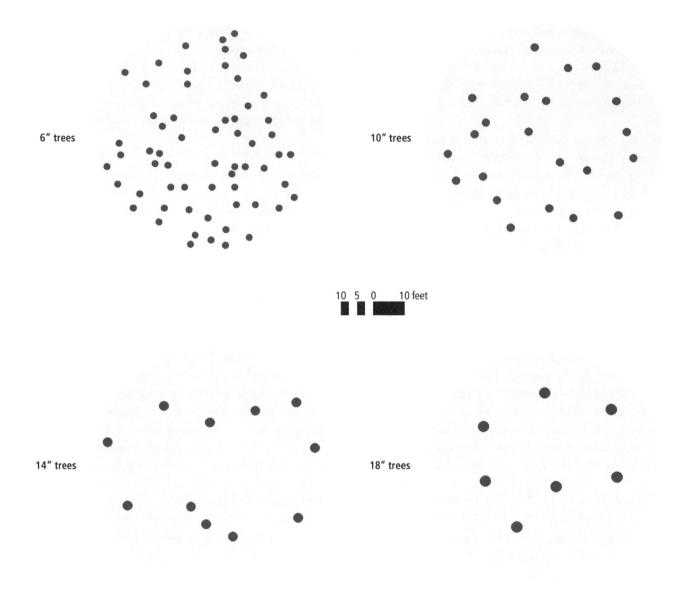

Figure 4.9. A representation of the basal area of 60 feet/acre and the number of trees at 6, 10, 14, and 18 inches in diameter. Image by John Gilbert, Auburn University.[7]

10 "plots" and project these findings on the stand as a whole. If your stand is over 20 acres, even more plot measurements may be needed to accurately reflect the whole.

There are various ways to establish a plot, but for this purpose we will use one of the easier methods, known as radial plot sampling. This involves randomly selecting locations to stand and then taking measurements from that reference point. It's useful to mark the point with a flag or stake to reference it again as you take measurements in the future.

Starting at your first of 8 to 10 plot samples, hold the penny 25 inches from your eye, using the string as the measuring length. Close the other eye and look forward. Count any trees that are the same diameter as or larger than the penny as "in" and ignore the others. Rotate 360 degrees and take a count of all the trees.

Figure 4.10. The cheapest tool for basal area calculation is constructed from a penny with a 25-inch piece of string taped to it. Use this to ensure the penny is the correct distance from your eye, as you count trees "in" that are wider than the penny.

Multiply by 10, and this is the basal area for the variable plot. Repeat this with each of the spots, and then average the total.

It's easiest to have someone else along to do this, as they can follow your eye and note the average diameter of the trees you count as "in." This can be achieved with the use of either a special diameter tape or a Biltmore stick (see the sidebar on page 134). This should be tallied in a notebook. If you are alone, mark the spot you are standing, and then after taking the count, flag the trees from memory; you can always check your work.

At the end of the exercise, you have two pieces of information. One is the average basal area factor of all the plots you took; the other will be the average stem diameter of the stand, determined by adding up all the diameters and dividing by the number of sample trees.

This sample method is appropriate in stands where the average tree diameter is between 3 and 15 inches as indicated on the stocking chart. If your average diameter is larger or smaller, a different multiplier is needed to get an accurate read.

UNDERSTANDING THE DATA: STOCKING CHARTS

A stocking chart is a useful tool to understand the dynamics of tree populations over time. This chart describes the 100,000 seeds to 100 trees phenomenon discussed in chapter 2, but here we will dive deeper into reading these charts to help plan management. Start by looking at the top of each curve, and note that the average tree diameter sets your first measurement. Start with that line, and then move down to where the basal area meets this, on the y-axis. This will give you a rating for the stocking of your woods. Note that the chart also offers a number of trees per acre, on the x-axis.

We can use this information and chart to know what to do next when it comes to thinning. Note that, when this chart refers to stocking, it is referring to the optimal density of trees to make most productive use of the woods, whether that is for timber harvest or for a healthy old woods. For silvopasture, the density will need to be much lower, in order to provide sufficient light to the understory so forage can be established. You may need the help of a forester or someone who has experience interpreting these charts to apply the information to your situation.

Looking at the chart, we might look at the A-line as the maximum stocking rate for timber, while for silvopasture, the B-line or perhaps a little above is going to be the upper threshold for density, as the recommended basal area for silvopasture is generally around 60 to 70. If you are very overstocked above (100 to 120), you may not be able to get to silvopasture-level density right away, as thinning too much, too fast can be detrimental to a woods. We will use this chart again later in this chapter to help prescribe thinning to the woods to this end. Keep in mind none of this is an exact science, but rather a series of guideposts.

WHAT IS THE STAND STRUCTURE?

As trees grow, and sort themselves out based on the site, growing conditions, and their genetics, multiple layers often emerge in the forest. It's useful to assess

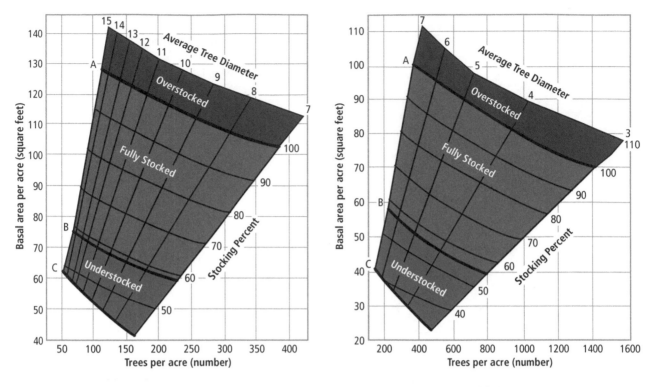

Figure 4.11. With your data on average tree diameter and basal area, you can assess where the stand falls on the chart above with regard to density. Illustration refined by Camilo Nascimento; data from USDA Forest Service.[8]

the character of a stand based on defining the different layers that are present. In some woods you will find all four, while others may only have one or two. The common ways that the layers are distinguished include the following characteristics:[9]

Dominant

- Crowns of trees extend above the general canopy layer.
- Crowns intercept sunlight across the top and along the sides of the upper branches.
- Tree diameters are among the largest in the stand.

Co-dominant

- Crown is within and helping to define the main canopy of the stand.
- Crowns intercept sunlight mainly from the top.
- Well-developed crowns but crowded at the sides.
- Tree diameters are among the upper range in the stand.

Intermediate

- Crowns might extend somewhere into the main canopy.
- Intercepts sunlight only at the top, but limited.
- Crowns are narrow and short.
- Diameters are within lowest range in the stand.

Suppressed/Overtopped

- Crowns entirely below the main canopy.
- No direct sunlight to the crowns.
- Small, sometimes lopsided, and sparse crowns.
- Diameters are among the smallest.

Some tree species fall into a non-dominant class because that is their evolved niche, so it's important to know the characteristics of each species you are working with. From a management lens, if the latter two categories contain trees that are adapted to be in the upper canopy, they simply won't respond much to thinning. In other words, they won't suddenly spring up and fill in

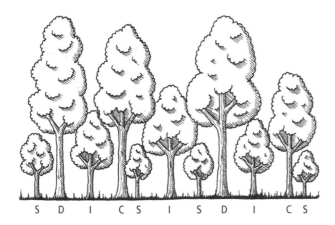

Figure 4.12. The position a tree occupies in the canopy is a reflection of genetics, site conditions, and sometimes plain old luck. Dominants and co-dominants are most likely to be the long-term canopy species, whereas intermediate and suppressed trees likely won't be able to catch up, even if the canopy is opened up significantly. D = dominant; C = co-dominant; I = intermediate; and S = suppressed. Illustration by Travis Forte.

the void. For conversion of woodlands to silvopasture, the reality is that many of the trees will be removed. In almost all cases the intermediates and overtopped trees will go, and then a relatively heavy selection of co-dominants and dominants may also occur.

An exception to this approach would be the retention of uncommon species in the lower layers for seed production to increase diversity, or the retention of lower tree species as coppice or pollard trees for animal fodder.

HOW WELL ARE THE TREES GROWING? (QUALITY)

While basal area is a very useful tool to sample the quantity of trees in a stand, you also want to look at how good the site is at growing trees of a specific species. Just because a species is there does not mean the circumstances for optimal growth exist.

Additionally, some sites are well suited for growing large trees, while others will never be great places to grow trees of any type. In this case, just because a site isn't ideal, it doesn't mean you should cut down all the trees. But the information you gather will help you make some long-term decisions about management.

This concept of site suitability for a specific tree species is known as *site index*, and charts exist for all major tree species. To use these charts, you need to know two pieces of information: the rough average age of the trees in a stand, and the height of the canopy. Many foresters don't bother going through the process of measuring age, relying on their experience with indicators such as tree height, relative vigor, species composition, indicator plants, and past history as ways to estimate the age of the stand. Historical imagery can even be utilized, as sometimes you can actually trace the timing of forest development using these resources. Regardless of the method, the goal is to answer the question, "How suitable is this site for growing X species?" so that you can determine how much management to put into the stand.

It's important to select good trees to use as samples for the site index, and for each collection of samples, only use trees of the same species. For instance, if the dominants in a woods are both sugar maple and white pine, then separate site index data should be taken for each species. The best trees to sample have good crown access to sunlight as dominant and co-dominant trees.

To determine the age, there are a few options. Remember that diameter does not equal age, which means you cannot assume age from the size of a tree. In most cases, since the stand is even-aged, most of the trees will be around the same age. So one strategy is to thin 5 to 10 trees in a stand that are chosen culls and cut them to determine the base age of the stand.

The other is to use a tool called an increment borer to take a core sample of a standing tree. A borer only extracts a small portion of the wood, so does minimal damage to the tree while getting a good sample if it's done properly. The tool can cost several hundred dollars, but often a local extension or private forester can provide one or take samples for you.

Determining tree height proves to be much easier than age. You can estimate the actual height of trees using a homemade cruiser stick, as shown in figure 4.14. This simpler approach only works on ground that is relatively flat or gently sloped. Sample five or more trees in a stand of the same species to confirm the average height. If your stand has more than one dominant

MAKE YOUR OWN "CRUISER" STICK TO MEASURE TREE DIAMETER AND HEIGHT

For measuring tree diameter, estimating height, and assessing the merchantable height (for timber), a tool known as a cruising, or Biltmore, stick can be very useful. While these can be purchased from forestry suppliers for around $60 or $70, you can make one yourself for next to nothing.

Get a thin piece of wood, about the width of a yardstick. Cut it to 36 inches.

Tree Diameter

On one side, mark measurements for tree diameter as shown in table 4.2.

To measure diameter, hold the stick with your arm straight out from your face and one eye closed. Line up the zero end with one side of the tree, and take the measurement from the other side, as shown in figure 4.13.

Height

Mark the other side of your stick with the measurements as shown in table 4.3. The labels indicate the height of the tree to the nearest 10 feet.

To use the stick to measure the height of the dominant trees, stand 100 feet from the tree and hold the stick vertically 25 inches from your eye, aligning the bottom of the stick with the bottom of the tree. Without moving your head, match the measurement with the top of the tree, and read the height to the nearest 10 feet.

The relationship is true only if the tree is significantly taller than you are, and the ground is relatively level. When you're working on more uneven ground, you may need to use a clinometer tool or smartphone app.

Figure 4.13. Using a Biltmore stick to measure tree diameter.

Figure 4.14. Using a Biltmore stick to measure approximate tree height. Photo by Jen Gabriel.

Merchantable Height

Traditionally, the Biltmore stick can also be used to measure what is called merchantable height, which is used to calculate timber values in trees by assessing the number of logs that can be obtained from a standing tree. This may be useful for some, but is a bit outside the scope of this book. These measurements can be marked instead of the height measurements. If you are interested in this application, visit the sources for more information.

Sources

"Building a Biltmore Stick" from the University of Purdue. www.agriculture.purdue.edu/fnr/stout woods/activities/Building%20a%20Biltmore%20 Stick.pdf.

"Simple Homemade Forestry Tools for Resource Inventories" from Washington State University. cru.cahe.wsu.edu/CEPublications/EM038E /EM038E.pdf.

Table 4.2. Markings for tree diameter

Mark lines at:	Label as:
5 7/16"	6"
7"	8"
8 7/16"	10"
9 7/16"	12"
11 3/16"	14"
12 7/16"	16"
13 11/16"	18"
14 7/8"	20"
16"	22"
17 1/16"	24"
18 1/8"	26"
19 1/4"	28"
20 3/16"	30"
21 1/8"	32"
22"	34"
23"	36"

Table 4.3. Markings for tree height

Mark lines at:	Label as:
10"	40
12.5"	50
15"	60
17.5"	70
20"	80
22.5"	90
25"	100
27.5"	110
30"	120
32.5"	130
35"	140

species, you will want to do this for each major species of interest.

For a more accurate measurement, the classic tool is a clinometer, which can be purchased from forestry suppliers. This tool makes measuring tree height simple. Most devices are calibrated to 66 feet, meaning that to take an accurate measurement you need to be standing that far from the tree. Check the user instructions, as some instruments are calibrated for different distances.

Once you've positioned yourself at the proper distance, simply look through the device, starting with level, where the two sides read zero. Then take two measurements, one tilting up toward the top of the tree, and the other tilting down to the base. Take measurements from the right side of the scale (the left side is for calculating slope) and add the two numbers together. This total is the height of the tree.

A third way to quickly get tree height is through a smartphone app. This uses GPS to determine your position and the phone as the angle gauge, completing the calculation for you. The accuracy of these apps depends on how good your reception is, of course.

With these data, you can use species-specific site index charts[10] (available for eastern US species at www. fs.fed.us/nrs/pubs/gtr/gtr_nc128.pdf) to determine how well your site is suited for a given species. You can find the point for a given stand by matching the average height of the sample trees with the average age. The site index number is expressed as a canopy height of the trees relative to the base age, which is often 50 years.

In other words, a site index of 70 on a 50-year base age means that the dominant trees in the stand will average 70 feet at 50 years. You can compare this with the other classes on the chart to see if the reading is good relative

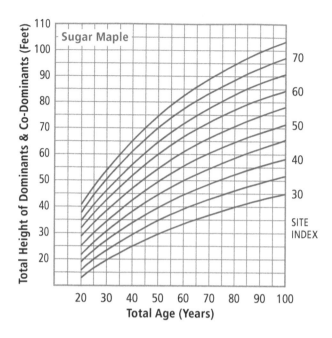

Figure 4.15. A site index chart for maple, which shows that, at a site index of 70, trees 50 years old are expected to be 50 feet tall. With your real-time data from site assessment, you can compare the growth of the trees you observe with the expected value. Image by Camilo Nascimento, data from US Forest Service.

to the potential of a given species. Obviously, landing on a higher curve in the chart means the site is better for growing the species at hand. This tool can be great to give you a sense of the growing potential of a site, or might be the determining factor.

PUTTING THE PIECES TOGETHER: BASAL AREA, CROWN CLASS, AND SITE INDEX

While collecting these three pieces of information may take some time, it's easy to see how useful it is to know them before you make decisions about what to leave and to cut. These three aspects of forest assessment help you determine three main things:

1. If the woods should be used for silvopasture at all.

Stands that offer good growing stock density, and are well suited for the tree species of interest, are best left to do what they do best: grow trees. So stands with a high basal area or ideal stocking rate, along with a

good site index, should make you consider not using them for grazing integration. Low-stocked woods, and/or woods with a poor site index for species are the better places to implement silvopasture. It should be noted that not everyone involved with silvopasture agrees with this, believing that the "best" sites being responsibly managed for timber gain from the income, which helps offset the cost of establishing a silvopasture in the first place.

2. The nature and type of thinning.

Overstocked stands, regardless of end use, should be thinned to concentrate sunlight on the best trees. The crown classes help us determine the species composition and potential for supporting a healthy woods, while the site index often confirms the observations of forest structure while defining the actual potential for a given species on a given site. Stands with a low average diameter might be too young for you to determine their future potential, while those with a higher average diameter are often more clear.

3. The best species/communities to favor on a site.

Observations of forest structure, along with the site index, paint a clear picture of the best species and groups of species to favor when thinning. If the site is poor for all the species present, then your focus can be more on supporting species diversity, giving more consideration to using the trees in relationship to grazing animals.

These assessments should help give you patterns to apply to the woods, and build your confidence in making the case for your approach to thinning, to which we now turn.

Thinning the Woods

Thinning is the process of opening up more light in the forest. In more conventional forestry the goal is often to improve stand density and concentrate the growth on the best trees. The main goal in silvopasture is to choose which trees are cut, based on opening up more light for forest, while also leaving the best possible specimens for future tree regeneration and/or harvest. During this

Figure 4.16. Many people think of silvopasture like a plantation, with evenly spaced trees and a "clean" forest floor. In actuality, the best stands are a mosaic of open areas, dense clusters of trees, and everything in between.

type of manipulation for silvopasture, several dynamics are affected, including[11]:

- The shade available for livestock.
- The way the thinning affects a future harvest.
- The impacts of thinning on future regeneration.
- Creating advantages for some species over others, among both trees and forages.

In the process of thinning for silvopasture, we seek to allocate sunlight in a balanced way among trees, fodder trees/shrubs, and ground forages. Along the way, we also want to make each stand unique in its composition and avoid uniformity over the range of the stand. Many people picture a forest of equally spaced trees with a clean ground and grasses growing. In actuality, a mosaic of spaced trees, clusters, and a ground mixed with both

open areas for grasses and woody debris is the ideal picture (see figure 4.16). There should be standing dead trees (snags) and downed trees, which serve to support soil microbiology, and provide a food source or cover for wildlife.

In general, space between trees is important so that good light can get to the ground. We just don't want complete uniformity. It's also useful to consider the edges of paddocks; in these spaces we might thin fewer of the small-diameter trees, choosing to leave them if it facilitates fencing systems or provides wind protection. In very dense forests it's best to thin over time, not removing more than one-third of the trees at a given time. Opening up too much light too quickly can lead to issues such as sunscald with shade-loving trees, or epicormic branching. Too few trees on a windy site can also lead to premature windthrow of trees, so it's best to leave a few extra standing as an insurance policy.

The process of thinning is when decisions are made over which trees stay and which go. This requires the decision maker to zoom in and look at individual trees, as well as zooming out to view the larger impact of those deliberations on the stand, and even the forest as a whole. Making wise choices is something that takes time and practice. It's highly recommended that you join a local woodland owners club, take a cooperative extension training if available, and/or work with more experienced foresters in choosing trees. Go slow, and take risks in the least vulnerable parts of the woods.

SOME PATTERNS TO CONSIDER

There is no specific recipe or prescription that applies to the task of marking trees, as each place in the woods is a unique variable plot. For silvopasture practitioners, we can offer a few forestry strategies to consider, each of which gives a pattern for us to approach the challenge of stand thinning from a slightly different perspective.

Basal Area

With regular forest management, the goal of a thinning will often be to take a stand from full stocking down to a density that optimizes the growth of residual trees; allow the woods to regrow to a high density (toward A); and then cut again, repeating this pattern. Over time the tree diameter should increase, along with the basal area, as illustrated in the previous figure 4.11 on page 132.

With silvopasture, the goal at the end of our work in most cases is to reduce the basal area of a stand down to between 60 and 70 square feet. It's important to remember that stocking is a tool for even-aged stands only, and that it focuses on the mature canopy timber trees alone.

The two main differences between a more traditional thinning and one for silvopasture are:

Thinning occurs from the main canopy level all the way to the ground. When managing for timber alone, often the trees growing underneath the main canopy are less of a concern. In some cases it's even beneficial to maintain a lower canopy to shade out undesirable plants. With silvopasture much of the understory will be cleared by default, to enable enough light to establish forage on the ground.

Lighter and more frequent thinnings are needed to keep stocking lower. While more traditional forestry might seek to maximize regrowth up toward full stocking so as to maximize returns on timber, with silvopasture the focus is going to have to be different. The "B-line" serves as a rough guideline for a stocking that helps establish shade-tolerant cool-season forages, and thinning keeps stocking in the lower portion of the green zone. This is a guideline at best, because other factors like species composition, stand structure and age, and the density of tree crowns also have a great effect on the outcome.

It can be helpful to keep this pattern in mind as you think through thinning the woods. But as we will see, this is much more an armchair analysis than other approaches we explore in the following sections.

Diameter Times Spacing Guide

Another pattern to consider is the relative spacing that might be desirable for a given target basal area, known as diameter times spacing.[12] This formula expresses ideal spacing of trees in a given stand based on the tree diameter as well as the desired basal area. For any given basal area, the ratio between tree diameter in inches and spacing in feet is a constant, giving us a formula of:

$$\frac{\text{spacing}}{\text{constants}} = \frac{15.4}{\text{square root of desired basal area}}$$

For a desired BA of 60, for instance, the spacing constant would be:

$$C = \frac{15.4}{7.74} = 1.98$$

Then we can calculate the spacing between trees relative to their diameter as:

spacing (in feet) = DBH × C

If the crop tree diameter averages 12 inches, the calculation would read:

12″ × 1.98 = 23.76 feet between trees

This calculation is useful to get a general sense of the spacing we might want to aim for, in order to have a relatively even spread of trees. This information could be used in a few ways. When identifying keeper trees (or crop trees; see below), we can consider this by taking a tape measure from the trunk and seeing what trees would be cut in order to maintain such a spacing. You could also use this number to check your work after marking trees, to see if your decisions are consistent with an overall target.

What is limited in the use of stocking charts and calculations is that the forest is not in fact consistent or standard, but highly variable and ever-changing. This is where, while these two factors help frame our thinning practice, we ultimately need to rely on an approach that can be exercised on the ground on a tree-by-tree basis.

Crop Tree Management (Leaving the Best)

As figure 4.17 depicts, a stand is a mix of trees that are in various states of health, growth, aging, and decline. The trees in any given plot offer the story of stand development, where outcomes are determined by tree genetics, environmental factors, response to disturbance, and just plain luck.

For most of us working on the ground, the easiest approach to good forestry is known as crop tree management, where the best of the trees are identified, with management focused on supporting their growth. This approach results in a healthy forest, and it keeps our task simple: Identify the best trees and give them room to grow. Another way to think of this is: "Thin as Mother Nature would thin." The forest is, in fact, thinning itself with or without us intervening. Good forestry, then, can be looking for the evidence of trees that have either reached their maximum potential or are on their way out.

Figure 4.17. No stand is a perfect set of trees. Some of the choices here are obvious candidates for thinning (*darker shading*), while some are harder choices (*lighter shading*) to support the residual stand (*no shading*). Illustration by Travis Forte.

Of all the forestry approaches out there, it's arguably most useful for beginners to learn about crop tree management, at least as a starting point. See the sidebar on page 140 for references for some excellent (and free) resources to consult and learn more from. In addition, this is a topic best learned from working with experienced people on the ground. Seek them out in your local community.

Before selecting trees, it's important to know what you are aiming for. Are crop trees those that are most beneficial to wildlife? Or best for timber markets? Sometimes a crop tree is simply one you find aesthetically pleasing, or one that is providing an ecosystem service (holding the edge of a streambank, for example). In most cases a mix of objectives will likely drive the decision-making process.

Once goals are in place, the objective is to mark the best trees, then perform a crown release to give those trees more room to grow. Research has found that this release affects tree development in four growth characteristics: height, diameter, crown width, and the length of the "clear" stem (timber-quality wood free of lower branches). Trees that are touching the crowns of crop trees are especially important to fell or girdle (see pages 142 to 144), and removing trees on the south (sun) side of the tree has a greater impact than removing those on the north side.

The crop tree management approach has the distinct advantage of being useful for both even- and uneven-aged stands, as well as for many stages of forest

RESOURCES FOR CROP TREE MANAGEMENT

Gary W. Miller, Jeffrey W. Stringer, and David C. Mercker. USDA Forest Service. "Technical Guide to Crop Tree Release in Hardwood Forests" (2007).

Arlyn W. Perkey and Brenda L. Wilkins. USDA Forest Service. "Crop Tree Management in Eastern Hardwoods" (1993).

Dave Swaciak and Peter Smallidge. Cornell Cooperative Extension. "Crop Tree Management: Getting the Most from Your Forested Land" (2003).

You can download these at www.silvopasture book.com and find links to more from their authors and agencies.

development, from sapling to polewood to maturing tree stands. It has emerged as one of the more common recommended strategies, partially because the concept is simple to understand. The nuance comes from actually standing in the woods and deciding which trees to mark.

MARKING TREES

It's important to balance the overall patterns discussed previously with a determined focus on looking at the individual trees in a stand and assessing their health and potential. This is a constant back-and-forth assessment between the stand as a whole and the individuals within. In approaching the marking of trees, it's best to deliberate. The worst possible time to decide which trees to cut is when you're standing in the woods, chain saw idling, staring up at the canopy. It's important to be of a good mind and not feel distracted when marking trees. Separate the functions of marking and the actual work of cutting.

I personally like to visit a stand that I intend to thin three times.

The First Visit

The first visit, I observe trees and mark both crop trees and those I think I want to cut with different-colored flagging tape. For a silvopasture treatment, I might also bring paint, since many of the intermittent and suppressed trees will be thinned regardless. The flagging is useful for those trees in the co-dominant and dominant strata.

One good place to start is to mark the crop trees or "hub trees" in the woods. Once these trees are identified and flagged, you can look to release them from any crowding on any side. This starts a process of elimination, favoring the most exceptional trees in the woodlot.

When looking at an individual tree, ask the following questions:

Crown health. Is the crown (top of the tree) full and healthy? Stressed or limited by other trees?

Structure, branches, and stems. Does the tree have a split crotch? Is there excessive lean? Any signs of premature decay, including cracks in the bark, dead limbs, or insect or woodpecker holes?

Vigor. What is the diameter for the height? Is the tree straight? What is the height to the first branching? How tight is the bark?

Evidence of disease. Are there fungal cankers, staining, or signs of heart rot?

Effects on neighbors. Would thinning this tree help other, more desirable trees?

Best use. Is this tree's best use as a live tree for a future harvest, or seed? Would it make a good snag or nurse tree on the forest floor for wildlife? Is there a good market or productive use for this tree?

Strive to justify marking a tree with at least two or three good reasons. A good question to ask is: Over the next 10 years, is this tree likely to increase in growth, stay the same, or decline? The goal is to first try to remove or allocate declining trees to harvest or for wildlife snags or debris on the forest floor. Keep the best, and remove those Mother Nature is already thinning. Once a stand has been marked, allow yourself to take a break and then come back for round two.

Figure 4.18. Forester Mike DeMunn points out some of the characteristics that help him decide to cut a tree or not. Reading the forest and making good decisions is a skill learned over many years, and often it's best to have a good forester help in the process.

The Second Visit

The second visit, the goal is to be critical of those first choices, and check your work. It's also time to revisit marked spots, recalculate basal area, and visualize what the forest might look like after the work is complete. When you're satisfied with the choice of a given tree, you can remove the tape and mark the tree with paint: a slash on both sides for firewood, a dot for a wood product (such as mushroom log), and a ring for timber.

The Third Visit

In the third visit, I come with felling gear, ready to focus solely on safe and efficient cutting and moving of trees. The practice of tree selection and cutting should never be a mixed activity.

FELLING AND MOVING TREES

Once the decisions have been made on tree selection, it's time to shift to the activity of safely cutting and removing trees from the woods. This section provides some strategies for those who are doing this work on a smaller scale, on their own time. At larger scales, it's often worth contracting out this work to a professional outfit. Sometimes funding is even available to help. For more discussion of these possible opportunities, turn to the end of chapter 6.

The work of removing trees needs to be done with three goals in mind: safety, efficiency, and minimal damage to the trees and forest left behind after the work is finished. Most of this work is going to be done with a chain saw, and then some sort of vehicle to pull out any stems that have a next use. In particular with silvopasture, there is also the matter of dealing with the likely considerable amount of brush and debris from tops and wood.

Equipment

Running a chain saw is a dangerous task. It should not be done without both the proper safety gear and training. When working with the tool, a good head space and awareness are also critical, as one mistake can be a painful lesson, or worse. The same goes for equipment for moving logs: Winches, chains, and tractors are all helpful, but mistakes are costly and can damage equipment. Trees are heavy and hard to move, and a lot of time can be wasted if you don't have the proper equipment and experience on the jobsite.

For chain saws, it's critical to maintain a clean engine and always cut with a sharp chain; if you have to strain and push the saw through the wood, it isn't sharp enough. The saw should be a newer model and

have a working chain brake, anti-vibration technology, and a catch to protect hands in the instance of a broken chain. Anyone running a chain saw should be wearing protective chaps and a logging helmet, which combines head, ear, and face protection in one unit.

The simplest way to move a log is to simply hook it to a tractor, ATV, or other vehicle and drag it out of the woods. Winches allow for a much greater ability to pull logs from a distance without having to get a vehicle close, though this is not as critical in silvopasture thinning, since the low density should allow for easy navigation through the woodlot. A logging arch can be used to minimize the damage when moving logs, though some scarifying of the soil from dragging logs can actually be beneficial to prepare the soil for seeding forage.

METHODS OF CUTTING

The purpose of cutting trees is ultimately to remove them from the canopy of the forest, which creates more space for the residual trees, supports tree regeneration, and offers the harvest of material for use. This action can be done quickly, by cutting and removing the whole tree (felling), or much more slowly with girdling or coppice/pollard systems.

Felling

This refers to cutting a tree until it falls over and can be moved. The goal is to remove most (about 90 percent) of the wood, all while controlling how the tree falls. This is achieved by cutting a wedge and leaving a small strap of wood connected, referred to as the hinge. With proper momentum the tree falls, the hinge controls the fall, and the wedge provides the space for the tree to close on itself.

Most people learn the basic felling technique of cutting a face wedge in the side of the tree that points toward the desired direction of the fall, opening up a space, and removing around 50 percent of the wood. A back cut then finishes the job and (hopefully) the tree falls the right way. This chase-felling is dangerous and risky, leaving too much up to chance. You're running the saw at full speed right up until the moment the tree falls, putting yourself in danger.

Figure 4.19. While it takes some practice, directional felling is by far the safer way to fell a tree. It involves a proper face cut as well as accuracy when boring through the tree. It leaves the tree connected in two places, and thus stable until you're ready for the final cut, allowing it to fall in the desired direction.

A better method for working trees is called *directional felling* or *bore cutting*. This method allows you to remove most of the wood while maintaining more control over when and how the tree falls. The basic method is to make a face cut, as in the previous method, except at a much steeper angle. Then place the saw about 1 inch behind the face cut and bore directly through the tree level with the bottom of the wedge. You'll remove the majority of the wood, leaving only a small portion connected at the back. Depending on the lean of the tree, you can insert wedges; when you're ready, you cut the backstrap and better control the release of the tree. This method is covered in more detail in *Farming the Woods*.

It's important to get proper practice and learn felling techniques directly from experienced chain saw users.

There are numerous chain saw safety courses, and one that teaches directional felling is an international outfit called the Game of Logging—not to imply that logging is a game, but the way the course is run is that different aspects of the practice are scored and taught in a fun way. Regardless, lack of experience running the saw and cutting the trees means that much more time can be eaten up in the process, especially if and when a tree becomes hung up on another tree.

Girdling

Girdling is a useful way to thin trees without having to deal with felling. It's of particular benefit in creating snags, or standing deadwood in the forest, which are a boon for wildlife. The best snags are mostly of larger-diameter trees that have wood on the softer end of the spectrum, so that woodpeckers and other animals can bore into them and make use of the wood. All trees make decent snags, but pine, poplar, basswood, red maple, tulip poplar, and cottonwood are among some of the best in the temperate woodland. The softer the wood, and the bigger the tree, the better.

Girdling severs the cambium layer in the wood of the tree, effectively killing it. It's always surprising how long a tree can remain standing after it's died; some snags will persist for several decades before falling over. The best practice is to sever the tree all the way around in one ring, then repeat 2 to 4 inches above or below and chip away the outer bark in between (see figure 4.20). This will ensure that the tree won't repair a single cut and continue to grow.

Girdling effectively thins the woods, and can achieve the same objectives as felling: The tops and upper

branches are usually the first to go, opening up light in the canopy, while the tree remains a viable home for a diversity of wild creatures. Ideally, it's best to aim for 5 to 10 snags (standing dead trees) per acre, whether they occur naturally or are stimulated by girdling.

Coppicing and Pollarding

A third option for woodlot conversion for the purpose of silvopasture is the timeless practice of coppicing and pollarding. This essentially works off the concept that all deciduous leafy trees will resprout following cutting, provided they are given access to enough light. This is potentially, then, a very good practice to partner with thinning for silvopasture, if you'd like to use the cut trees as fodder for animals to consume.

Coppicing is the practice of cutting down to the bottom of the stump, while *pollarding* cuts above browse height. Depending on the vigor of the species, both of these methods can be employed, and in the context of silvopasture can be excellent strategies for providing animal fodder. The key to coppicing or pollarding is opening up enough light so that stump sprouting can occur, and then protecting the initial growth from browsing animals.

We will leave further discussion of how coppicing can integrate into silvopasture for the next chapter, where several scenarios and ideal tree species are discussed.

Abandoned Pasture / Cropland and Hedgerows

Shifting gears from established forests, lands where thick, impenetrable brush persists are the forgotten spaces in the farm landscape. In the cultural pursuit (at least in North America) of colonizing the once vast forested land, a stark division was drawn between forest and field. As farmers walked off the land, the least valuable places were abandoned, and a mix of trees, so-called invasive plants, and thick vegetation often covered the land. These are often viewed by landowners as a problem, a thorn (pun intended) in their side. How will this ever change? The good news is that silvopasture provides both the tools and the incentive to reclaim and shift the ecological succession of these lands.

Figure 4.20. The proper way to girdle a tree is with two cuts about 2 to 4 inches apart, and a full removal of the outer bark in between the cuts, to ensure the tree doesn't grow back together. Photo by Wikimedia/Lamiot.

Figure 4.21. This hedgerow was poorly valued by the previous farmer, used as a dumping ground for trash and soil. After decades of only being mowed along the edge with machinery, the border was thick with thorns and impenetrable. Fortunately, through a combination of grazing with sheep and hand and machine pruning, it is coming back nicely, serving as a critical windbreak for our house and many acres of land, as well as providing shade and shelter for the animals. Photo by Jen Gabriel.

Of all the places to start silvopasture, these spaces offer one of the best opportunities; they lack the level of risk involved in thinning a woods, and they don't require the patience of waiting for trees to grow in the pasture. They can be a great place to experiment and reap the satisfaction of reclaiming land for productive use. And they can become some of the first spaces to use as a learning laboratory, observing animal behavior and interactions with vegetation, long before translating this information to forests and field. Silvopasture can be a powerful incentive to reclaim these lands and more immediately reap one of the basic benefits it offers to livestock: shade and shelter.

Another invitation of these marginal spaces is the opportunity to interact with a woody landscape in a more intimate way, meaning not just with chain saws and large vehicles. This harks back to an older way of working the woods, as for too long farming and land use have been dictated by the machines we've become reliant upon to do the work. Working these spaces, then, provides the opportunity to rediscover tools like the billhook, machete, and scythe, and to spend time pruning and shaping spaces, all while improving habitat for our animals, and wildlife too.

In the progression of forest succession, we can generally categorize these spaces as falling somewhere in the middle

THE DUST BOWL AND A FORGOTTEN AMERICAN LEGACY

When speaking of hedgerows and shelterbelts as a part of US agriculture, we cannot fail to remember the largest tree planting endeavor in the nation's history: the so-called Great Plains Shelterbelt of the 1930s.

Reeling from the impacts of the Dust Bowl, President Franklin D. Roosevelt sought to alleviate the problem, and in the process supported one of the largest environmental efforts ever put forth by the United States. He initiated the project soon after taking office in March 1933.

The plan called for the planting of millions of trees in an area 100 miles wide, stretching from the Dakotas to Texas, including a wide range of climate conditions. It was initially incredibly ambitious, as Roosevelt signed an executive order devoting $15 million to the project, of which only $1 million was actually authorized to be spent. Congress rejected several proposals, and threatened to kill the project in 1936, but it was saved by an allocation from the Works Progress Administration (WPA), which changed the name to the Prairie States Forestry Project. It remained active until 1942, when the demands of World War II quelled its efforts.

The public gave the plan mixed reviews, though many eventually came to rally around the project.

The first tree was planted in Oklahoma in 1935, part of a planting that still exists today.[13] Trees were usually planted in long strips at 1-mile intervals within the 100-mile-wide belt, with WPA and the Civilian Conservation Corps doing much of the planting. It's estimated that the project successfully established over 200 million trees and shrubs planted on 30,000 farms over a span of 18,600 miles.[14] A survey of the Prairie States Forestry Project in 1954 determined that about three-quarters of the original shelterbelts were in fair or better condition than when they were established, and only 8 percent had been destroyed.

Unfortunately, for a range of reasons, many of the shelterbelts have been removed in the last several decades. Many of these plantings fell victim to economics; farmers saw greater value in clearing them for planting crops than in the potential savings from preventing soil erosion.[15] How soon we forget the lessons of the past.

While the practice was not related to silvopasture, we can only wonder: If livestock had been integrated into these shelterbelt systems, would more value have been seen, and more of these remain today?

---

of a pioneer or climax forest. They often offer the widest range of species diversity, as well as the most textured habitat. In a landscape where we've crafted only forest and field, these middle-ground spaces offer critical habitat supporting the widest range of bird and animal species.

Research has more than established the multitude of benefits that hedgerows and scrublands provide, including:

- Depending on the size (height/width/volume) and species composition, many beneficial habitats for birds[16] are found in these mid-succession environments.
- Cropping systems adjacent to these types of vegetation see an increase in crop production because soil

erosion is reduced, creating variations in soil moisture and moderating microclimate.[17]
- They can serve as critical wildlife habitat and affect many important dynamics at the landscape scale.[18]

When we encounter these spaces, it's important to discern their history. Generally they have been cleared and farmed, then abandoned; or the trees were highgraded to the point that light was open, and then the land was abandoned. These patterns might tempt us today, but we must resist. The old mind-set says that these parts of the landscape are wastelands and should be cleared as soon as possible. Hedgerows are still commonly ripped out by farmers, despite clear evidence of their many benefits.

However, hidden inside any scrubby mess of brush is value; we just need to see and discover it. The process should not be seen as an overnight activity, but something that happens in pulses, over time. We can approach renovation of hedgerows and scrublands as a task in ecological transition, where pulses of activity and disturbance happen over a period of several years. Some of these pulses may be more dramatic, as we will discuss, such as shredding vegetation with machinery, while others are more fine-tuned, such as pruning residual trees for access.

All told, these spaces are ripe for silvopasture. They offer a great starting point to implement the practice and reap a clear and immediate reward: transitioning space into something both more functional and more beautiful.

Understanding Hedgerow Design

The words *hedgerow*, *windbreak*, and *shelterbelt* are often used somewhat interchangeably, though technically a hedgerow is vegetation that defines an edge or boundary, while windbreaks and shelterbelts indicate a more intentional planting to protect sites from the effects of wind and blowing snow. Regardless, the effects of these plantings on the immediate environment and the larger ecosystem depend on the size, shape, composition, and density of the species. In most cases we might expect to encounter unintentional hedgerows and abandoned fields, which might exhibit similar patterns to hedgerows except for being much wider and larger in size.

For silvopasture specifically we might be interested most in the shade and shelter effects this type of composition can offer our animals, though there can be many secondary benefits. For instance, existing hedgerows can help buffer the effects of weather that can stress young trees we are seeking to establish in an adjacent field for future silvopasture. We can use our base map to assess current and future possible plantings with some of the basic concepts of design in mind[19]:

- Shelter is most effective when plantings are oriented perpendicular to the prevailing and seasonal wind.

Table 4.4. Wind-speed reductions at various distances on leeward side of shelterbelt

	5H	10H	15H	20H	25H	30H
Single Row Deciduous	50	65	80	85	95	100
Single Row Conifer	30	50	60	75	85	95
Multi-Row Conifer	25	35	65	85	90	95
Solid Wall	25	70	90	95	100	100

Note: Expressed as percent of open wind speed (where H is the height of the canopy) (100%).
Source: Brandle, James R., Laurie Hodges, and Xinhua H. Zhou. "Windbreaks in North American Agricultural Systems." In *New Vistas in Agroforestry.* Dordrecht: Springer Netherlands, 2004, pp. 65–78.

- Protection of animals and buildings is most effective with multiple rows and densities of trees.
- The line of trees needs to be extended far beyond the area to be protected, to avoid infiltration of side winds.
- The effect of windbreaks on both the windward and leeward sides of the hedge or shelterbelt is a combined function of the composition, depth, and height of the vegetation (see table 4.4).

We can use the above concepts and figures to think about our silvopastures as multifunctional, by ensuring we maintain those that already provide valuable functions as a shelterbelt, and perhaps expand or enhance them further by widening, elongating, or reshaping them as parts of the overall farm landscape.

SHELTER AND SHADE VALUE ASSESSMENT

Before heading out with our tools to begin the process of eradicating plants, then, what we want to do is map and recognize the benefits of existing windbreaks and hedgerows on a landscape (if applicable), as well as identify opportune places to plant in the future. We can call this a "shelter and shade value assessment." The steps to this process are:

1. Map seasonal and prevailing winds on-site.
This can be done with a combination of research and on-site observation. In the United States there are two websites that might prove useful, one from the NRCS/

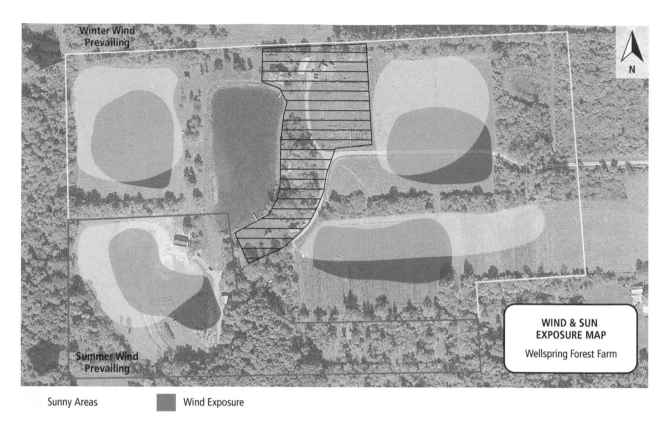

Figure 4.22. Mapping existing and potential areas of shade and shelter on a landscape.

USDA that offers "wind rose" maps from historical data (www.wcc.nrcs.usda.gov/climate/windrose.html), though only for major cities, and the incredibly handy windhistory.com, which provides data mostly from local airports. While these are great tools to explore, note that the conditions at your specific location are likely to vary somewhat from these data.

2. Depict protected zones on the map.

Based on the seasonal and prevailing winds, and considering the path of the sun during the hottest months, begin by identifying sheltered areas on your property by labeling them and calculating their value, at least in rough terms. For example, a hedgerow can be assessed for its general height, using the methods mentioned on page 134 with the Biltmore stick, then some projections can be made as to the protected areas that extend the benefits.

Windbreak calculations can be made based on table 4.4, in relation to the prevailing and seasonal winds. Since wind is always changing, these assumptions will be rough, but you can at least start to distinguish areas that are most protected from those that are less so.

As you move farther from the equator, the dynamics of sun and shade become more and more complicated. Depending on the time of year, and the time of day, the sun is in a different place both on the horizontal landscape and vertically in the sky. So there is no easy way to map the entirety of these patterns, but we can at least get a sense of it, by focusing on the hottest times of the year.

Shade can be calculated for a given time of year using the handy calculator at planetcalc.com/1875; those who are more ambitious can map and play with shadow

Figure 4.23. Once seen as a wasteland, this hedgerow was a process of discovery as it was grazed and opened up to support more forage on the ground. Included in the hedgerow are white ash, hawthorn, sugar maple, and serviceberry. Photo by Jen Gabriel.

dynamics at www.findmyshadow.com. If you are curious to see how the sun arcs across your landscape, a good online tool is available at suncalc.net.

For instance, by using the PlanetCalc tool, it can be determined that for a 30-foot hedgerow in Central New York, the shade cast at noon in the hottest summer months ranges from 10 to 21 feet, whereas in September through November (when the sun's angle is lower in the sky) the range is about 30 to 45 feet.

With a smartphone you can also literally stand at a spot and use a number of apps to display on your phone the path of the sun, effectively showing you areas that will be shaded or sunny. Two good apps are Sun Surveyor and Sunseeker.

With all these free tools at hand, it's useful to make some notations and sketches on your site map to get a sense of the value of existing hedges and plantings, as well as where more could be added to improve the overall environment. It's also important to consider this when thinning a woods or clearing a brushy patch of land, as it could be a major error to apply a uniform treatment to the entire stand rather than leave a buffer of thicker and denser vegetation on edges more exposed to wind and/or sun. An example of our sketch map for this can be seen in figure 4.22.

VEGETATION ASSESSMENT

Unlike the forest measurements and thinning strategies discussed for forests and plantations, there are no well-developed metrics for assessing abandoned fields, cropland, and hedgerows, in part because each composition is unique, and also because, as mentioned, there is little value seen in these environments. Often these woody-dominated landscapes offer opportunity for some clearing, some thinning, and some tree planting, too. So assessment will be a bit more of a case-by-case scenario.

The approach, then, is to assess species composition and then determine what stays and what goes, much as we discussed in the previous section. This is going to depend largely on what is there, and on your goals as a farmer. Resist the temptation to label plants "good" and "bad" based merely on their species type. We've found that some of the plants most hated by people are favorites of animals. In addition to a general assessment of vegetation, note whether any species appear to be thriving, particularly valuable hardwood and softwood trees. These give clues to future potential for the site.

For example, we discovered some hidden birch thriving in a brushy (and somewhat wet) area, which inspired us to plant more in the future. Some of the other species we have found include hawthorns (great for medicinal use), basswood, maples, hop hornbeam (some became the railings in our new home), hickory, and several delicious old wild apple trees.

All in all the assessment of scrubby lands can be an enjoyable process of discovery, alongside the work to reclaim and put acres back into good production. In some cases the assessment and clearing may happen together, if the site is such that the beginning state doesn't allow any access at all. But as with marking trees, discussed on page 140, it's important to separate the assessment and clearing functions. It's easy to strap on the brush cutter or get behind the wheel of a skid steer and quickly do more clearing than you like.

One useful strategy is to flag vegetation you want to keep so it is obvious, keeping in mind the methods you use to clear—a person standing on the ground versus an operator in a machine will have a different vantage point. With machinery it's not a bad idea to leave some "buffer" vegetation that can be thinned later by hand—precision is harder with large equipment.

It's also not necessarily beneficial to thin wholesale, as research has shown that a more complex edge ecology is beneficial. The practice of "edge feathering" is often done for birds, insects, and other critters, but could be a useful pattern to consider for silvopasture, too.[20] This approach, whether by thinning trees to leave scattered ones, or planting in a similar pattern, offers a better edge than just a stark line. All this should be done in consideration for how you will potentially plant trees, move equipment, and retain access to make your farm work reasonable.

Methods of Clearing

In most cases practitioners will find that transitioning brush to silvopasture will necessitate a few different approaches. We can divide these into mechanical, animal, and chemical, and all carry with them the possibility of doing both good and harm. In addition, it's best to take a multistage approach to the work. This allows you to patiently explore, make discoveries, and shape the space into its best use. A helpful approach we have discovered is:

1. CREATE ACCESS FOR FENCING, YOU, AND THE ANIMALS.

The first step is putting enough effort into clearing the space to offer access for fencing and for the animals to get to most parts of the paddock. Prioritize removing plants that are, literally, a pain, most notably those with thorns and those that are taking up lots of space without producing a ton of edible forage. The clearing work is often easiest when leaves have dropped (and before/after substantial snow)—except that it's harder to identify plants at this time. Get to know the space and what is there.

The animals, given the opportunity, will assist in this endeavor, making trails as they seek out foods. Depending on the species and their ability to access the existing vegetation, they may be able to assist in browsing, stripping, trampling, or otherwise impacting an area.

It's nice in the first few visits to bring along a pair of loppers and a folding saw, and use the excuse of feeding the animals to help open up pathways and more access points. Once you've determined where fencing will go, make it really easy on yourself by cutting larger pathways along the fence line.

Figure 4.24. We found access points in our hedgerows and wove in the fences, quickly realizing that in the long term we'd be much less frustrated with wider passages. It's much easier to assess the situation and do some additional clearing once the animals have been through.

2. IDENTIFY AND MARK THE MOST VALUABLE SPECIES.

As you explore, carry a roll of neon flagging tape and be sure to mark any trees that are rare, unique, or desired. You will be amazed at what you find, and at the incredible persistence and ability of trees and shrubs to survive and thrive in such challenging circumstances. It's important to clearly mark these trees, and even keep them

buffered with some undesired vegetation if equipment is coming in to clear.

3. IDENTIFY SPECIES TO REMOVE; CLEAR IN A MOSAIC.

After the most important specimens have been marked, focus on clearing other vegetation, whether with tools, machinery, or, if absolutely necessary, chemicals. (More on these options later.) It's great if you can time thinning work so that it follows the residency of the animals in the space, which often opens up the vegetation, allowing for an easier job. Some find it useful to also mark vegetation they want to remove, in case they don't get to it until later.

Try to do clearings in a mix of patterning. Clear some gaps in places, while perhaps just removing a plant or two in others. Cut some species to the ground or pull them out by the roots, while also leaving some that you know the animals like cut higher. Leave good shade and shelter trees alone at first, even if the species isn't your favorite. You can always take more, but it's harder to replace vegetation once it's gone.

4. PULSE THE WORK.

Depending on how thick and overgrown things are, for the first year or two the paddocks will mainly serve to feed the animals. It isn't worth trying to establish forage until the space is more or less how you want it, from a density perspective. In the end animals should have access to the space, and mostly what is left provides either food, shade, or shelter. It's wise to spend time in the summer grazing and thinning some, marking the brush to keep and remove, and then returning in the winter when there is often more time to get the bulk of the work done.

All along, think of ways to incentivize your work. Feeding animals as you prune, freeing up an old apple tree that will bear fruit, and the satisfaction of making

Figure 4.25. We've been working on this hedgerow for over a year. In summer the sheep grazed back the shrub vegetation, stressing the plants. We then fenced them in for part of the winter, finding that our sheep spent time stripping the bark from each shrub, effectively killing it back to the roots. We've been able to literally drive the tractor over these shrubs, or step on the vegetation to remove it, and the sheep then graze the young shoots.

LAYING A HEDGE

Hedge laying is an old practice commonplace in England and other parts of the British Isles, traditionally used to create livestock-proof barriers while maintaining aging hedgerows by encouraging them to put on new growth. The hedges are also remarkably beautiful; laying one is a true skill to master.

It's easiest to lay a hedge when trees are all in a row. Hawthorn is one of the more common trees, and was often planted for this very purpose. Trees are first pruned to feature a single leader stem, which is then cut on the backside using a tool called a billhook, or a chain saw for larger stems.

Rather than cutting all the way through, leave the cambium and outer bark on the side where the tree will be laid intact, effectively forming a hinge with which you can bend the tree over to the side. This also keeps the tree alive and encourages it to throw up new shoots of growth.

The trees are laid down and left at a height determined by the function of the fence. Then, depending on the style, the pleachers, as they are called, are woven through vertical stakes and finished off with a binding along the top to hold it all together. Generally the process is repeated every 8 to 15 years.

I mention this process as food for thought while you examine your hedgerows and consider the possibilities. Of course, you could also plant trees with this intention, to eventually lay them and form a robust and time-tested living fence for both function and beauty.

Figure 4.26. A laid hedge of hawthorn and buckthorn by Karl Liebscher of www.shropshirehedgelaying.co.uk.

beautiful spaces in once unapproachable parts of the land all help.

MECHANICAL METHODS

The use of tools—from pruners, saws, and loppers to brush cutters and any number of larger machines—is the most common approach to achieving a more desirable level of vegetation. The quietest, simplest option would be using pruners, saws, and loppers to clear, which is most often slow going, and better reserved for clearing really errant branches and to make way for brush cutting, or to clean up in the wake of larger machinery. These tools are also wonderfully appropriate for harvesting brush as feed for animals.

Of course, there are also a number of more traditional tools that should find their way into the cache of a silvopasture farmer. Machetes, billhooks, brush hooks, and scythes are all wonderful tools that have a learning curve but ultimately offer a high degree of satisfaction in their use. There is something special about using only the body, and not only fossil fuels, to get the job done. For some, there might be a time and a place for this; whereas an initial clearing might use machines and gasoline, maintenance can then be more easily carried out with hand tools once the vegetation is under management.

The next step up would be a handheld or walk-behind brush cutter. One of the more common models is basically a high-powered weed whacker that is worn with a harness to limit stress to the body (see figure 4.28). The unit is fixed with a circular or triangular blade, and models with dual handles and a

Figure 4.27. It's highly recommended to combine functions of pruning as a harvest event for your grazing animals. It's not only entertaining but also more rewarding than just clearing brush for hours on end. This can be done slowly, a bit more with each time the animals rotate through.

harness are much easier on the body than the more typical one-handed weed whacker. It also often has a more powerful motor, necessary for thick vegetation. Rather than "saw" vegetation, whack it with a back-and-forth swinging motion. It's important to keep the blades sharp, as they quickly become dull. Plan on sharpening each time before use. Protecting your head, eyes, and body is also a critical aspect of using these tools. Generally, handheld models are easier to use than walk-behind, though there are some really beefy examples of the latter, which can do quite a good job.

Chain saws are not recommended for clearing brush, as they are very dangerous, and the work is rough on the chain. Swinging a saw in brush is very unsafe: It's easy to hit rocks and dull your chain in the process. The

Figure 4.28. A brush cutter is basically a high-powered weed whacker with a saw blade on the end. It can be a very effective tool, but it's tiring to use! Photo by Jen Gabriel.

exception would be when you prune a thick shrub back to a stump or cut a multistemmed trunk that can't be efficiently dealt with another way. It's useful to have the saw on hand for these moments.

From these handheld tools we move to machines. Many landowners and farmers already have tractors and a brush hog or mower, which can be used to some effect. You can use a bucket to push and compact brush before backing over it with the mower.

A number of other PTO and hydraulic tractor attachments also cut and shred brush, as well as chains and stump grabbers that latch on and can be used to pull roots. While effective, these tools take a lot of time to hook up and drive back and forth—and a tractor can't always offer enough traction to get the job done. A rear-mounted backhoe is nice if it's already purchased, but for less money you can rent a mini excavator, which can do a more thorough job.

Many local companies will rent and drop off mini excavators, which are great tools for clearing, though inexperienced operators will find there is a decent learning curve to get the hang of one. The advantage of an excavator is that it can be easier to dig out a plant and remove most or all of its root system, reducing the chance of it returning. Rentals are also sometimes available for skid steers, both those with a driver cab and walk-behind models. Bulldozers can often be rented and are very effective, though they may be harder to use if your goal is to remove only some of the vegetation. The cost of renting these machines can arguably be worth the time it saves versus more manual methods of clearing.

The third level of approaching the work with machinery is to hire it out. Contractors who have both the equipment and time can be pricey, but effective. They are often used to clear acreage entirely, so you will need to shop around to find someone who will do careful work not to damage what you want to keep. Depending on the tools they have, they can do minimal or extensive damage to the land. Ask a lot of questions, and try to visit a previous jobsite if you can. Hiring out the work means you will likely do a much more dramatic clearing of your site. The other drawback is it removes the opportunity for you to explore

Figure 4.29. A Fecon brush cutter opens up a landscape, leaving good ground for seeding. The third image shows how it looks after two seasons. For the right site, this, or similar front-mounted hydraulic tools, can create the conditions for an excellent silvopasture. Photos by Brett Chedzoy.

and discover the hidden gems in overgrown areas. Still, sometimes it's necessary.

ANIMALS

We talked extensively about animal behavior in the previous chapter, mostly as we considered how to provide them a diverse diet. Another angle to come at this is to also employ the animals to do the work of creating beautiful silvopastures from hedgerows and abandoned farmland. From our perspective, the interplay of working with animals on landscape rehabilitation is a great experience. It is not, however, necessarily a turnkey proposition, and your goals are invariably different from those of the animals. Ruminants are going to be the most amenable, with goats and many sheep breeds ready and willing to engage in the task, though some breeds of cows aren't so hard to convince, either.

As previously mentioned, the selection of animals and breed, along with their previous experience browsing, will have a big impact on their willingness to eat what you want help clearing. Some training is going to be necessary. It's important to know what plants you have in the area and ensure that none are dangerous. Start with the most vocal and curious animals in the flock or herd, as their behavior will be easier to transfer to others. And for particularly stubborn animals, set up a routine in which you make a sound and use a specific bucket to feed them something novel each day, starting with the most palatable and adding a different food each day, until you work through problem plants. Many animal caretakers note that building a trusting relationship with the group is also key to getting this to work. Our sheep follow us around willingly, always curious and eager for the next thing we prune for them.

A well-known innovator in this area of animal management is Kathy Voth, who offers a plethora of information on her website, livestockforlandscapes.com, which is useful for identifying those plants that the animals *won't* touch. While there are many details, the basic steps to training animals to eat weeds, according to Kathy, are:

1. Know your plant.

Begin by finding out about the nutritional value and the toxins in your target plant. Many weeds are very nutritious, but like all plants they contain toxins. Prevent illness by knowing your toxins.

2. Choose the right animals to train.

Young animals are more likely to try new things; females stay in the herd longer and teach their offspring. Train only as many as you can handle. They will teach everyone else for you.

3. Reduce the fear of new foods.

Setting up a daily routine of feeding animals something nutritious but unfamiliar gives them positive experiences with new foods and makes them comfortable trying other novel foods. Feed them something new twice a day for four days. When you introduce your target weed on the fifth day, they'll eat it because it's just one more new thing in their routine of new things.

4. Practice in pasture.

Each new plant requires that your animals learn a new grazing technique. Give them a day or two to practice in small "classroom"-sized pastures. Then when you send them out in the world, they'll have the skills they need.

In addition to getting the right animals in place and perhaps engaging in some training, pay attention to the browse height of the brush, and make it easy on the animals. I like to employ two techniques when helping animals do the work of clearing brush. One is to not place clippings on the ground or in a pile, but rather to use a shrub as a "basket" where I can spread out the food right at their comfortable eating height.

Figure 4.30. No animal will completely clear vegetation, but they can help along the process, and if you're raising them anyway, they are the most cost-effective tool. You may need to leave them longer in a paddock to get them to eat less desirable plants, but eventually at least goats and many breeds of sheep will do it. Photo by Scott Bauer, USDA.

Another strategy that works well is to snap or saw branches almost all the way through, but not quite, so they bend but do not break, and the remaining connection keeps them off the ground. In just a few minutes, I can walk through a brush area and harvest pounds and pounds of good forage for the animals, with little effort. This is a much more satisfying task than just clearing and moving brush without the animals enjoying it.

It's important to follow after animals move through an area, doing a bit of weed whacking or clipping behind them. This allows you to choose what will be knocked back, as well as what you might leave to resprout and provide food for the next time. A huge revelation for us was shifting from the idea that we would remove 100 percent of the brush, and instead leaving as much as 40 to 50 percent to regrow. If the food wants to grow, why not let it? We favor plants that are favorite foods based

Figure 4.31. This large honeysuckle bush has been pruned, grazed in summer, and then stripped in winter by the sheep. The resprout from the stump will continue to serve as fodder for several years, until better forage is established.

on what we observe, and target those that we know are prone to spreading or causing problems (thorns and burrs) for the sheep or us humans.

CHEMICAL METHODS

The use of chemicals in woodland management is less common than in farming; they are more effective in a limited and targeted application than in being sprayed extensively over an entire swath of land. Since much is unknown about the persistence and problematic effects of herbicide use, it's not a recommended approach in silvopasture, especially since much of targeted spraying of plants could end up in the mouths and stomachs of animals.

One of the more common go-to herbicides is, of course, glyphosate, or RoundUp, which has been the subject of debate for decades. Claims that it's biodegradable are currently under dispute, and California recently listed it as a known cause of cancer.[21] With the aforementioned options, and the plan to graze the land, it seems like chemical treatments aren't often necessary, and probably more useful when landowners or farmers aren't looking to keep livestock, which is a tool that can do the job just as well.

If you do consider chemicals, it's important to get proper training and read all the labels to ensure you are using only the required dose. Too often folks just head out to the fields and overapply the material, which is what in many cases can cause issues for people and the environment. Treat chemicals as a poison, and take the necessary precautions.

What to Do with All That Brush?

It's almost a guarantee that your work will quickly generate brush and branches, which can really pile up. Those who have approached this problem often start by sending material through a wood chipper, only to discover how much time and energy this takes, for little yield of a usable product. If you do decide it's worth it, be sure to make it easier on yourself by laying material in a single direction; otherwise the brush gets all intertwined and is a big mess to untangle.

It's perfectly fine to leave the brush—or, more specifically, to pile it. One strategy useful in high-deer-pressure areas is to use it to build brush walls to discourage the deer, especially where you plan to plant future trees. Try to avoid cutting more brush than you are able to move and pile (this is another good reason to pulse the work as discussed previously). Brush can also be piled in one place as a habitat pile for wildlife.

Another option if you're using machinery is to bury or shred the brush. With an excavator, it's feasible in an open area to dig a hole, place the brush in it, and cover it with soil. Some call these mounds *Hugelkultur* and tout their benefits for planting. The science on this is still unclear. Brush can also be shredded, which is in many cases the best solution, as it immediately becomes

Figure 4.32. Cornell's Arnot Forest is experimenting with "brush walls" as a use for brush that will ideally keep out deer and support forest regeneration. Photo by Brett Chedzoy.

a mulch layer and can quickly integrate back into the soil. The very action of some clearing machines (Fecon and brush eaters) will create this product, while other people pile material and then use a brush hog mower on a tractor to shred it in place.

Brush can also, of course, be burned. This can release a lot of volatiles into the air, and definitely is a quick release of carbon, though on the global scale it is not a significant climate change contributor. If you can find a purpose for it, though, it's better to leave the biomass in the landscape to build soil. When you burn

it, you lose it. But sometimes it's necessary. Making biochar from the brush is another option, and much more benign on the environment. See *Farming the Woods* for more info.

Whatever the decision, don't get ahead of yourself. It's easy to get excited and tear out a whole bunch of brush, tossing it behind you, only to realize that you've now created another task with an equal amount of work. Better to clear less and be done with it than to go overboard. Have a plan in place before you start removing material.

CLIMATE CHANGE, DROUGHT, AND HEDGEROW GRAZING
AT WELLSPRING FOREST FARM

The drought of 2016 was tough on farmers, notably in the central part of New York State, where we almost felt cursed as storm after storm passed just to the north or south of our little bubble. Literally—we were in the red portion of the map in figure 4.33, which shows the conditions in the state as of September 2016.

This extreme drought arguably began the previous fall, when rainfall began to drop noticeably, followed by an abnormally dry and low snowfall during winter, and then a devastating gap in rainfall coupled with extremely hot temperatures during the summer. All this led New York to declare portions of the state a natural disaster area, opening up loans and funding for farmers struggling with the conditions.

It was a stark difference, after record rainfall the previous summer and record winters the two years before that. We were seeing the effects of a changing climate unfold before us, and as farmers we knew we'd better get in line. Agriculture has always been at the whim of the weather, any farmer's favorite discussion topic and the thing they most enjoy complaining

about. Rarely are conditions "perfect" in their timing, duration, and intensity.

Yet the northeast United States has enjoyed an amazingly consistent pattern of precipitation for hundreds of years. We usually get just over 3 inches of rain, on average, each month of the year. Or perhaps we should instead say we *used* to get this, since we may soon find this pattern to be a thing of the past.

Climate change is here, and it's something we've thought a lot about in our farm plan, though mostly we have considered the impacts of *too much*, rather than too little, rain. We are actively planting trees along contour, on swales that catch and store water from heavy rain events. These swales slow down and allow the maximum amount of water to infiltrate into the soil, with overflow filling ponds or slowly trickling out of our streams. Indeed, we saw the fruits of this effort during the 2015 growing season, where our water systems dampened any negative effects of the excessive rain events that occurred.

But 2016 was different. We realized we were not as prepared as we had hoped. The most noticeable aspect of this came into play with our sheep grazing. We started out all right in the spring, rotating the sheep through about 35 paddocks we'd carefully designed, moving the flock every two or three days. We made our first trip through these paddocks starting in mid-April and coming back around to the first paddock in late May. No grass. We couldn't believe it, but the paddocks, which usually were begging for grazing with tall, bushy grasses abundantly packed in, were full of short, dry, and browning stubble.

The heat, which seemed to always be over 90 degrees F (32 degrees C), coupled with the realization that we were out of grass, quickly took its toll on our nerves. I literally had dreams of rain, which didn't come *at all, in any amount*, for over a month. Our concern weighed heavily on our emotional well-being. We had no backup plan for this kind of event.

While we started searching for hay (also in short supply because of the rain) and priced out purchasing grain (our feed cost went from $0 per day to $30),

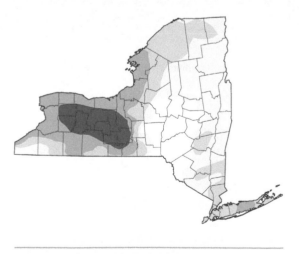

Figure 4.33. Our farm is in the bottom right corner of the red bubble, indicating extreme drought conditions as of September 6, 2016. Sourced from the US Drought Monitor.

Figure 4.34. Perhaps because they were hot and hungry, the sheep took right to the hedges, stripping every last available forage. We replaced time spent moving fence with time pruning and giving them fresh limbs every day.

we fell back on our permaculture thinking, which encourages us to consider how a challenge can become an opportunity. Enter the hedgerows, and the concept of hedgerow grazing.

Our farm, like so many, has three main land types; open field, forest, and scrubland. This third type is a kind word for areas simply abandoned, neglected, and unvalued by the previous farmers, left to grow in a tangled mess of shrubs, thorny plants, and trees suppressed with vines. A quick calculation on Google Earth revealed that grazing these areas was equivalent to adding almost 4 acres of pasture. But would the sheep agree?

The answer: yes. We used a brush cutter to cut paths through the vegetation, which in some places

was a full 6-foot wall of thorns—and set up our fence. The sheep went right in and set to work. We quickly realized that doing this wouldn't actually make more work; the level remained about the same. Instead of moving them every few days from one half-acre paddock to the next, we instead fenced in larger portions of the hedges and left them there for a week or more. The time we normally spent moving fence and their shelter was eliminated, replaced by time pruning the vegetation each morning to give them access to food above their browse height.

The sheep readily stripped honeysuckle, buckthorn, multiflora rose, and privet, along with the native trees we pruned and fed them. This pruning became enjoyable; we weren't just feeding the sheep but also engaging in the hedgerow clearing that we'd had on our to-do list for a long time, since most areas were so overgrown we couldn't even get in there to see what was happening. This new system shifted our perspective, as we moved from seeing a painful chore of clearing these plants change to the pleasant task of harvesting food for the sheep.

The hedgerows not only provided us food for the sheep for almost 40 days (how biblical), but also shade and shelter for them from the brutal heat of 90-plus-degree days. Remarkably, the sheep continued to maintain and even gain some weight. And the hedgerows began to open up. And so we joyfully found ourselves accomplishing multiple objectives, expanding grazing land and providing a more comfortable climate for our animals.

We will continue to work on these hedgerows in the coming seasons. It's not that the sheep were 100 percent effective, since in some areas the vegetation was so thick neither sheep nor person could get through. But they've made inroads, and we can build off that. With winter coming, we will be able to get in there and improve access for next year. Over the next few seasons, the vegetation will change, improve, and become a valuable asset to our grazing system.

A final paradigm that shifted for us was that we now see food value in the species that before we saw as noxious. Rather than fully eradicate honeysuckle and buckthorn, we have decided to shift toward managing them. The fact that they grow so vigorously is annoying if we don't want them, but a good thing if we think of ourselves as grass and fodder farmers.

Our plan moving forward is to bring these hedgerows fully into the rotation. This means doing some clearing of brush, mainly for access, and some thinning of trees, all to allow more light into the hedge while still maintaining shade. This will be a bit of a balancing act, but ultimately will allow us to not only manage the shrubs and trees as sheep food, but to establish grasses, too.

Because these hedgerows edge many of our grazing fields, incorporating them into the grazing system means that suddenly over 60 percent of our paddocks have tree and shrub fodder, and natural shade. For many farms, hedgerow grazing may prove to be the easiest entry point into silvopasture. It can offer four benefits to the farmer:

1. An increase in available food for the animals.
2. Diversifying the sources of food for the animals.
3. The addition of natural shade and shelter to paddocks.
4. Increased utilization of land.

It's certainly offered all these benefits to us.

Establishing Forage After Thinning or Clearing

We've spent a lot of time in this chapter talking about the assessments and choices involved in thinning woods and clearing brushy land in preparation for silvopasture. This is no mistake: Proper analysis prior to action, along with deliberate activities that are at once safe, effective, and efficient, makes up the majority of the work converting spaces to silvopasture.

Nonetheless, after opening up land in this range of ways, you must also quickly establish forages in order to produce more food for your animals in a timely way.

Many landowners and farmers thus take the wait-and-see approach after thinning and clearing, deciding to save their dollars and learn what Mother Nature has to offer in terms of forages. This is perfectly fine, and in many cases it might work, as there is likely a decent mix in the seed bank of the soil. But remember, if the ground has been disturbed, a niche has been opened, and introducing seed at that moment means likely increasing pasture diversity. In reality, on acres and acres, very few have the time and resources to do a complete reseeding.

Ideally, good pasture establishment (or renovation) is best planned well in advance, though we are well aware that the ideal doesn't always happen. We like to have on hand a mixture of red and white clover, triticale, annual ryegrass, perennial rye, and timothy in our cool-season mix. At the right time of year, we might also add field peas and tillage radish (spring and early summer) and buckwheat to the mix (anytime there is no danger of frost). This way, we can always be ready to throw down seed when the opportunity arises.

In a perfect world, the following steps ensure good pasture forage establishment:

1. TEST THE SOIL AND AMEND TO BALANCE IT.

Carefully follow the soil sampling instructions in chapter 5 (page 176) to determine if your pH is balanced and/or if there are any major nutrient deficiencies. The pH for grasses is best between 5.8 and 6.2; for legumes, 6.0 and 6.8. Old farmland often needs a lime application to balance the pH, which should be done 6 to 12 months before forage establishment so that the lime can integrate into the soil. Severe acidity can take years to remedy.

It's important also to have good phosphorus levels at seeding, which encourages root development. High potassium levels are important for legumes, and for all plants when under heat and water stress. Nitrogen should be given only in small doses so as not to overdo it.

All this is to say, if the soil test shows relatively good levels of nutrients, and a pH in the 6-ish range, you should be good to go.

2. PREP THE SOIL FOR GOOD SEED CONTACT.

Some of the methods of clearing previously discussed, such as the Fecon, provide nice seeding conditions on the soil surface. In other cases, it's important that the soil is loose and soft to promote good seed-to-soil contact for germination. This can be achieved by hand, or with light tillage, though you must use extreme caution in woodland settings not to affect the shallow feeder roots of trees. Strategically placing chickens, turkeys, or pigs for a brief time would also accomplish the task.

3. SELECT AN APPROPRIATE MIX.

It's important to choose species that will do well in your climate, provide a balanced diet for your animals, and offer good volume as the seasonal temperatures fluctuate. As previously discussed, different geographic locations are dominated by cool- versus warm-season grasses. Legumes tend to persist throughout, though variations within a species are important to pay attention to as well. A mix commonly recommended for cooler temperate pastures, for example, is all legumes and cool-season grasses: orchardgrass (6 pounds/acre), birdsfoot trefoil (6 pounds/acre), ladino or red clover (1 pound/acre), and Kentucky bluegrass (4 pounds/acre).[22] Limiting the mix to cool-season grasses is especially warranted in silvopasture settings in cooler climates. Consult with your local NRCS or extension office to get the most accurate species recommendations for your area and soil type.

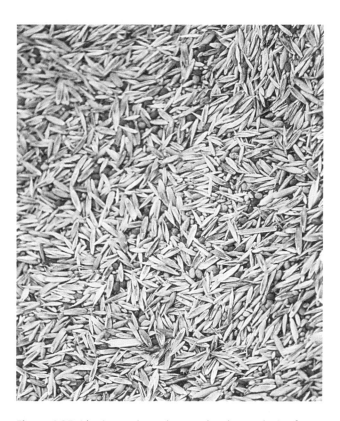

Figure 4.35. It's wise to always have on hand a good mix of annual and perennial grass seeds, along with legume seeds, on the farm. Photo by Jen Gabriel.

4. SEED AT THE PROPER RATE AND TIME.
Seeds can be drilled, cultipacked, or broadcast by
machine or hand. Drilling uses an implement that cuts
a thin line in the soil, deposits seed, then rolls over
it. Cultipacking involves seed being dropped from a
hopper, then pressed into the ground with a toothed
roller. Broadcasting is simply spreading the seed over
the surface, then packing or raking it in.

5. ALLOW TIME FOR
ESTABLISHMENT BEFORE GRAZING.

While it's tempting to let animals on that new pasture,
it's best to wait. Let the grasses grow at least 8 to 12
inches tall, and consider mowing the first time, or
doing only a quick graze at most. It's important not to
overstress plants as they are getting established. Some
choose to wait a whole season before letting animals at
the new forages.

These steps can, of course, also be used to renovate or
establish forages in open pasture—the topic of the
next chapter. A good rule of thumb is to assess pasture
first, and only renovate areas that have less than 40 percent of your desired species. Anything above that can
best be improved through good grazing management,
over time.

Regeneration:
An Event in the Life of a Forest

No matter what type of land you're starting with, the
goal with a silvopasture system is to maintain a forested
environment long into the future. As mentioned in
chapter 2, this means that at some point there will need
to be a period of tree regeneration, which is the establishment of the next generation of trees in the forest system.

People first learning about silvopasture often com-
ment that regeneration seems impossible, because the
grazing animals will prevent it from happening. This
is true, so long as we keep grazing them in the same
way, and on the same spaces, year after year. But for
silvopasture to work long-term, we have to have a plan
for regeneration. The hard part is that for many silvopastures, the timing of regeneration may not coincide

with the lifetime of those of us who are managing these
systems. Talk of regeneration quickly leads to questions
of landownership, tenure, and transfer, which are a
whole other set of challenges facing farms today.

If we reach back to the discussion of the stages of
forest development from chapter 2, we recall that most
of the woods involved in silvopasture are either in the
stand initiation phase (1) or stem exclusion phase (2).
Regeneration tends to be a characteristic of phases 3
and 4. In many cases, we won't reach this edge for at
least several decades. Further, this question of how to
regenerate stands successfully is one that many forest
managers are already asking, in a time where high deer
population density is having a significant effect on the
ways forests are regenerating naturally.[23]

So we have time to figure this one out—but not
forever. It's important to plant the seed (pun intended)
at the outset of our exploration of silvopasture, because
there are still things we can do to support regeneration
now, for the future:

1. **Natural regeneration** is the natural seeding from
   parent trees in a given area, or the emergence of seed
   from the ground under the right conditions. Birds
   and animals contribute a significant amount of seed
   to a forest ecosystem, with a big impact on what's
   available for the next round of trees. This is in part
   why it's so important to maintain habitat, including
   snags, brush piles, and large logs on the floor of our
   silvopasture woodlands.

   We can also identify hub trees and create conditions for them to regenerate, based on their specific
   needs as a species. For instance, we might open gaps
   in the woods and promote squirrel, crow, and jay
   habitat around oak trees we wish to see regenerate.
   These animals are the acorn planters.

2. **Artificial regeneration** is the introduction of seed
   from desired tree species, along with the conditions
   for it to germinate and thrive. This strategy is particularly important where species diversity is low.

3. **Planting** can take the place of regeneration through
   seed stocks, and has a particular advantage of getting
   a jump start by using stock several years old, which
   increases the likelihood of survival.

Figure 4.36. Natural regeneration of eastern white pine in a forest in New York State. Photo by Jen Gabriel.

In areas where deer pressure is high (much of the climate zone this book focuses on), it will be essential to exclude deer to a partial or great extent in order to support regeneration. Indeed, many woods, silvopasture or not, suffer from deer overbrowse, which greatly impedes natural regeneration.[24] The most effective measure to address this challenge? Add exclosures that keep the deer out, often in several smaller patches spread throughout a forest. It's much more economical to fence an area than individual trees, and it's more effective, too.[25]

Summary: Converting Woods to Silvopasture

In this chapter we've spent a lot of time diving into the details of converting existing forests, hedgerows, and overgrown land into silvopasture. Many areas provide ample opportunity to do this, but we should consider starting with those that are deemed most marginal. Healthy and mature forests are a precious resource, and because we don't yet have clarity around the effects of silvopasture long-term, many of these spaces might be best left as is.

Thinning trees and brush, and establishing pasture, should occur only after a good and thorough assessment has been done. The tools in this chapter will help you get started, and it's also a good idea to seek help from local services and professionals (see chapter 6). The concept of bringing animals into the woods is just one direction silvopasture can head, so let's contrast this with the approach of bringing trees into pasture, the subject of the next chapter.

SILVOPASTURE DEVELOPMENT AT WILHELM FARM
by Ann Wilhelm

At Wilhelm Farm in North Granby, Connecticut, we are in the early stages of developing a silvopasture unit. We have seen silvopasture practiced in other parts of the world, but it wasn't until 2013 that we began to seriously plan for a silvopasture on our farm. At that time we were concerned that a stand of mature pine trees in our forest was at risk of windthrow. The trees had been planted 70 years earlier by my grandfather, father, and uncle.

The site is favorable for white pine, and under the nurturing care and good management practices of my father the stand contained mature trees of exceptional size and quality. We feared that strong winds of a hurricane or nor'easter could topple the tall trees, putting years of investment in jeopardy.

We contacted a Connecticut licensed forester to conduct a forest inventory, update our forest stewardship plan, and mark and tally timber for harvest. The forester segmented our 35 acres of forest into seven management zones, which included a strip of land just over 2 acres in size. The forester was unable to inventory this zone, "due to the density of invasive shrubs." This area had once been an upland pasture, sloped and spotted with large rocks, but over time had become an impenetrable jungle of multiflora rose, Japanese barberry, grapevines, and other opportunistic, invasive plants. We envisioned converting this wasteland into a silvopasture area to serve as a transition zone between existing hay and pastureland and the managed forest.

We discussed this management goal with the forester. In his final report he recommended the removal of invasive plants and prescribed the NRCS practice code for brush management for this zone.

Figure 4.37. Dense growth of native grapevines and exotics made 2 acres of a former upland pasture virtually impenetrable. Photo by Ann Wilhelm.

Based on this recommendation we applied for and received an NRCS EQIP grant, which helped fund the initial steps of creating a silvopasture. In fall 2014 a Tigercat with a powerful, mounted mower cut through the tangle of invasive brambles and vines. Later that winter the loggers removed trees that were of low value or that posed a hazard to livestock, like wild black cherry.

In the spring and fall of 2015, we sprayed the area with chemical herbicides using backpack sprayers. Through 2016 we continued to combat the invasives with spot-spraying of herbicides and the use of a walk-behind mower. White pine seedlings were planted in spring 2015 on the more steeply sloped areas, and naturally regenerated oak seedlings have been protected in tree tubes.

We have made good progress in reclaiming this wasteland, but challenges remain. Japanese barberry is remarkably persistent, and preventing the wild rose and other brambles from recolonizing feels like fighting an incoming tide. However, the area is now cleared to an extent that we can get equipment and vehicles to the area, and use of portable fencing is feasible.

We purchased two goats in early summer 2017, starting small to gain experience in managing goats on browse. Pasture grasses were successfully established on several small, cleared areas. We will add more goats next season to continue clearing the resprouting brush and will seed more pasture as the land is prepared. One-year-old oak seedlings, grown last season in PVC-pipe containers, will be planted into the silvopasture area in spring of 2018.

We see silvopasture as a beneficial land management tool, not only for our farm, but as part of a regional vision for food production in New England. By bringing land that has been abandoned back into a productive state, we hope to improve the economic prospects of our farm. We know there are others in the region who face similar land-use challenges, and we plan to have a model silvopasture unit established

Figure 4.38. The first pass of the Tigercat mower cut through a tangle of vegetation and began to open the silvopasture zone. Photo by Ann Wilhelm.

Figure 4.39. Two years after the initial clearing by the Tigercat mower, sprouting brush is cut in early spring with a walk-behind DR Field and Brush Mower. Naturally regenerated oak seedlings are protected in tree tubes. Photo by Ann Wilhelm.

Figure 4.40. The Tigercat completes a second pass in the silvopasture unit, clearing along the edge of the pine forest. Photo by Ann Wilhelm.

by the summer of 2019 that can be used for demonstration purposes.

We also believe it's important to play a role in improving food security. A number of factors put the regional food system at risk, including climate change and an overwhelming dependence on food that is produced outside of the region. *A New England Food Vision*[26] proposes a number of changes to the region's food system, with the intent that by 2060 the region produces half of its own food. The report envisions "a future in which food nourishes a social, economic and environmental landscape that supports a high quality of life for everyone, including generations to come."

It is the fervent desire of Wilhelm Farm to be actively involved in making this vision become reality.

Connect with Wilhelm Farm at www.facebook.com/ WilhelmFarm.

Figure 4.41. The silvopasture zone in early spring after mowing brush sprouts. The next steps are to introduce goats to continue to combat regrowth of brush, establish forage grasses, and plant oak seedlings growing in our home nursery. Photo by Ann Wilhelm.

5

Bringing Trees into Pasture

While the conversion of appropriate lands already covered in trees and brush to silvopasture often both reclaims abandoned land and improves productivity, the prospect of bringing trees into pasture is a longer-term task on the part of the farmer or landowner. On the other hand, the climate benefits of planting trees into pasture make this a clear win-win situation, unlike the more complicated questions with clearing woods, as previously discussed. The main challenge is logistical, as many farmers are resistant to planting trees in the pasture and, in fact, often have spent considerable time and energy *removing* trees and hedgerows. Why would we want to put them back?

If you have read the earlier chapters of the book, you know that there are a multitude of reasons. Once we get beyond the challenge of establishing trees, we

Figure 5.1. Sheep grazing among four-year-old black locust trees at Wellspring Forest Farm in New York. With fast-growing trees, silvopasture can take shape quicker than you might think.

will be farming systems that are better for the climate, better for the animals, more resilient to weather and extreme events, and more productive on the same acreage. Adding trees to pasture also brings diversity to the farm ecosystem; there will be more species of birds, more diverse soil biology, and more options for fodder for livestock. And it's important to note that the landscape will be more aesthetically beautiful, interesting, and enjoyable for animals and humans alike.

Yet in order to reap these rewards, farmers must be willing to dive into the complex variables that silvopasture presents. Certainly one of the advantages of open pasture is a more uniform and simple ecosystem, one that is easier to manage. Straight lines and open fields have a purpose, no doubt. Silvopasture challenges and invites practitioners to think outside the box, and with time it does become easier, and the benefits justify the increased time and costs up front.

Still, many pitfalls can come along the way, including planting the wrong species for the microclimate of the site, experiencing high tree mortality, planting trees in patterns that make it harder for equipment, fencing, and animals to navigate the pasture, and potentially drastically increasing the amount of maintenance (and stress) of trying to keep trees alive, healthy, and resilient. This is where planning and thoughtfulness enter the picture. With a little understanding and careful species selection, a fully functional and beneficial system can be achieved.

In this chapter we will start with an assessment process similar to that discussed in the previous chapter for wooded lands. Then we examine what is known about the dynamics of trees and pasture, as the foundation of a good plan. Finally, we look at strategies and techniques to improve the pasture, along with establishing trees and selecting species that are good for silvopasture systems.

Assessing Pasture

Before digging holes, planting trees, and setting up paddocks, it's important to assess the current state of your pasture. We can think of open pasture much like a forest, where the grazing plants form a canopy, though instead of towering over our heads this canopy lives beneath our feet. As with the forest, what is growing is a product of the conditions of the given place, the availability of seed, and previous management decisions.

We will begin by differentiating among pasture patches, and then walking through an assessment process similar to that used for a forest—albeit using different tools and measurements, given the significant differences in the way we manage pasture versus forest. This exercise gives you the opportunity to build upon the map you created in the previous chapter, adding descriptions of your pasture patches alongside those of your forest, hedgerow, and abandoned cropland.

Assessment can seem overly tedious, and many skip the process, opting for the "meat" of what they want (pun intended). Instead, we challenge you to work through the process of assessing your current pasture before jumping to the often-made error of deciding what is, rather than discovering it. In other words, the process is most successful if you take the time to look at your pasture and define what is unique about each part, which gives direction to the choice of species and strategy for planting.

DEFINING PASTURE PATCHES (AKA STANDS)

As with the forest stands discussed in chapter 4, we will define *pasture patches* with the same four criteria: vegetation, soil, hydrology, landform. The word *patches* is used to differentiate this process from forest assessment, and because often, in pasture, we might look at even smaller areas where vegetation dynamics change. If you haven't read the descriptions of these characteristics in chapter 4 it's worth taking the time to review them, as a way to compare how they are similar and different depending on the context.

At this point we also want to differentiate between a patch and a paddock. A patch is an area of your pasture that contains similar environment and vegetation. As with forest stands, areas with different characteristics are divided up and assessed as separate patches. In this case patches can be rather large or small, whereas paddocks might match up with or might encompass many smaller patches. As with a

Figure 5.2. This paddock is sized in relation to the number of grazing animals, seasonal conditions, and desired time frame before the animals move on. A paddock could contain several patches, could match up with a patch, or might encompass only part of one patch.

stand, the purpose of identifying a patch is to prescribe a certain management approach. The purpose of a paddock is to create the right space for grazing animals to graze an area in a given time frame. As with forest stands, it's up to you to define patches based on what designations are useful for management. Paddock design will be discussed later, in chapter 6, after assessment is completed. In summary, paddocks are for managing animals, and patches help us manage the pasture.

Dominant Vegetation

While many forest stands can be delineated based largely on the dominant tree species, with pasture the nuances are likely more subtle. Since many forages are annuals and short-lived perennials, succession works differently. Often the entire field may have at some point been cleared, tilled, and planted to a specific species, or a mix.

As with forest stand assessment, species ID is an important component, but it's arguably even more useful to look at the vegetation from the "pasture scoring" framework, which evaluates a number of variables in the pasture. The USDA and other extension agencies offer a range of worksheets and approaches to monitoring, many of which are so detailed that they likely discourage farmers from actually completing the assessment. Instead, it's better to keep things

Figure 5.3. The vegetation of a given patch can say a lot about the site conditions. For instance, the high population of plantain and wild carrot here, along with clover, suggests a more disturbed soil that is recovering. Over time the goal for this patch will be to increase the presence of grass species. Photo by Jen Gabriel.

simple and achievable, by focusing on the step-by-step instructions below.

Plan on (ideally) walking your pasture at least three times a season (spring, summer, fall) for the first few years, and then on an annual basis thereafter. Make this activity fun, taking a morning or evening stroll, along with a clipboard and a beverage of choice. This allows you to get to know your resource and target areas for renovation or improvements over time. While it might feel like a challenge, as we all have limited time, the consistent interaction with your land will prove very beneficial over the long term. We can integrate this type of pasture assessment with our grazing chart, as discussed in chapter 6.

Soil Type and Health

As noted in the forest assessment section, a particular tree species' ability to reach its maximum potential is directly related to the soil it is growing on. And while there's little you can do to manipulate tree growth in

the big picture, you can do a lot to improve a pasture's ability to grow high-quality forage.

Using soil survey data is adequate for assessing forests (as discussed in chapter 4), but for pasture plants it's wise to do more thorough sampling and get specific. See page 175 for the process, and more resources.

Following good sampling practices from the start saves time, money, and frustration, because balancing pH and learning of any deficiencies early on will save you money in the applications of seed, fertilizer, or other inputs, as well as avoiding tree establishment problems related to soil deficiencies.

Landform and Hydrology

As mentioned in the previous chapter, landform includes the elevation, slope, aspect, and position of a forest stand in the larger landscape. These factors, along with the way water moves and flows, will greatly affect the microclimate of a given pasture stand, the dynamics of seasonal growth, and the appropriate tree species. For instance, higher points on the landscape with southern exposure and even a moderate slope will offer warmer, drier conditions, whereas lowland areas with slopes facing north or east will be considerably cooler. One of these isn't better than the other—just different.

Paying attention to these differences, and their effects on vegetation growth and post-grazing recovery, is important. It may mean the difference in days of available forage; as we will discuss later in this chapter, overgrazing is the fastest way to ruin good pasture.

Making a Pasture Patch Map

Rather than a separate map, this activity takes the map you started in the previous chapter, building upon the assessment framework to fill in the pasture portions of the site you are designing. As with forest stands, the way you divide up space will be partially determined by the vegetation patterns you observe, and partially directed by the plans you have to manage the different areas.

1. GET OUT THAT AERIAL PHOTO AGAIN.

Using the aerial photo you acquired for your forest and woody stand assessment (see page 124), you'll now add a pasture stand layer. This should fill in the gaps from the forest and woody maps you previously completed, leaving you only with ponds, buildings, infrastructure, and other un-grazable areas on the farm unmapped.

2. Mark initial bubbles from previous knowledge and from patterns you glean from the photo.

Each bubble should contain similar vegetation patterns, soil type, topography, and microclimate. Don't let the patterns of fields and hedges the previous farmer left sway you in where you draw the boundaries. A few drafts with pencil may be necessary to get it right.

While a single open pasture that has a similar slope and aspect may seem uniform, keep in mind the effect of edges and hedgerows. For instance, research indicates that a well-managed windbreak can offer protection for

a distance of 20 times its height on the leeward side and 5 to 10 times the height of the tallest tree on the windward side.[1] The sheltering effect is, of course, related to the distance from the trees, with the more significant benefits occurring the closer you are. So a 25-foot windbreak offers benefits up to 450 feet, though the most significant ones are in the first 150 to 250 feet.[2]

There is no "right" or "wrong" delineation; use your best judgment, and have a farmer friend check your work and offer feedback. The process of taking time to notice and document difference is a big step forward. You won't get it all the first time; think of your map as a living document, something you can refine over time.

3. Ground truth with pasture scoring and soil sampling.

Once your pasture patches have been defined, it's time to understand what's there and how healthy the pasture

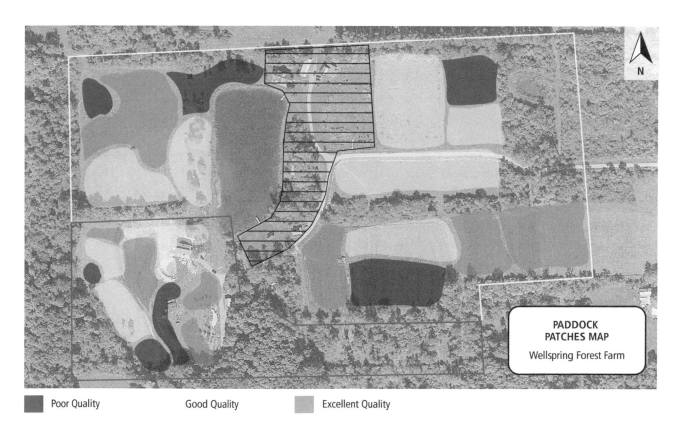

| ■ Poor Quality | Good Quality | ■ Excellent Quality |

PADDOCK PATCHES MAP
Wellspring Forest Farm

Figure 5.4. While pasture scoring will help further articulate the conditions of a patch, you can start by simply identifying the patches and rating them as poor, good, or excellent. This will help you determine which to focus your assessment and management on.

Figure 5.5. Scoring a patch requires getting down and dirty, noting the types of vegetation, density, presence or absence of bare soil, and diversity of pasture species. By comparing patches, practitioners can better learn and prioritize the parts of their pasture that need the most attention.

SOIL TESTING RESOURCES

Jerry Lindquist. "Proper Soil Sampling Depth Is Critical for Pasture Soil Testing." Michigan State University, 2015. Available at msue .anr.msu.edu/news/proper_soil_sampling _depth_is_critical_for_pasture_soil_testing.
John Lory and Steve Cromley. "G9215: Soil Sampling Pastures." University of Missouri Extension, 2005. Available at extension .missouri.edu/p/g9215.
Fred Magdoff and Harold Van Es. *Building Soils for Better Crops.* No. 631.584/M188b. Beltsville: Sustainable Agriculture Network, 2000.
Kristen Stockin, Jerry Cherney, and Quirine Ketterings. "Nutrient Management for Pastures (Factsheet 17)." Cornell University, 2005. Available at nmsp.cals.cornell.edu /publications/factsheets/factsheet17.pdf.

is. As with forestry, it's ideal to start from the pattern level with observation, and then fill in the details with identifying specific species composition.

What are the important factors for assessing pasture? There are three essential things to observe:

1. Does the pasture stand have a good mix of plants?
2. Is the stand growing as much food as possible?
3. How does the stand recover and regrow after grazing?

As with forest assessment, the revelations often come from comparing one pasture patch with the next over several visits. A recommended process for each visit is to take a notebook and a brightly colored ball (like a tennis ball) out to each patch. Toss the ball 5 to 10 times within one identified patch, and visit the spot where it lands. Take your assessment, using table 5.1 and marking a 1 (excellent), 2 (good), or 3 (poor) for each criteria. Add up your total to get an overall score for the patch. This is the method we used to generate our map (figure 5.4), after several draft versions.

In addition to a vegetation analysis, a soil sample will prove useful, especially in the initial years. Pasture soils are tested for the presence of any major deficiencies in pH, phosphorus, potassium, organic matter, or other nutrients or minerals, which can limit the productivity of the forage and/or limit the species present. A proper sample is critical to getting accurate results. Proper soil sampling includes:

1. Defining different sample areas of interest within fields or patches. These may align with your patches, or might encompass a larger area if you think the results might be similar. For instance, fields with a relatively similar slope that are in the same soil type are likely to be similar in their composition. Noticeable or dramatic changes in slope, soil moisture, vegetative cover, or past use are all-important considerations when deciding how to draw the line. So is cost. Taking 100 samples might give very precise results, but at $10 a sample that can be costly, and management for pasture isn't likely to be that complex. We usually send in about 8 to 10 samples for 30 acres of pasture.

Table 5.1. Simple rating system for scoring pasture patches

Characteristic	Excellent (1)	Good (2)	Poor (3)
Overall diversity of species			
Mix of grasses (70%), legumes (20%), and other forbs (10%)			
Density of forages (are there bare spots?)			
Palatability for livestock			
Regrowth rate of forages after grazing			
How well was forage utilized during last grazing event?			
How decomposed is manure from last grazing event?			

Note: Rate each patch with the above chart, then total up your score. A total of 7–12 is excellent, 13–18 is good, and 18+ is poor.

2. Visit each sample site, bringing with you a 5-gallon bucket and a hand trowel (or a bulb planter, which works great). At least 5 and as many as 10 random spots within the patch should be sampled to a depth of one trowel blade, and mixed in the bucket. This becomes a composite sample to be sent to the lab. It's important to sample at roughly the same time each year so you can compare results. Soil temperature and seasonality can alter the results. Avoid sampling if there has been excessive rain and the soil is oversaturated.

3. At each site, cut a sample 4 inches deep, and no deeper. Sampling too deep can lead to inaccurate readings that might mean a misapplication of minerals or fertilizer. Shake the soil from the core and add to the bucket. Repeat and mix as you go, so the final sample is a cross-reference of the soil. The more samples you take, the more your results will account for the variability that can occur in pastures, especially where animals feed or loaf, or where human activity may be more pronounced.

4. Send the sample to a lab, following its specifications for volume, moisture, and so forth. Your local soil and water or extension office can help you interpret the results and consider the benefits of adding amendments against the cost of doing so.

4. IDENTIFY SPECIES.

Many farmers (myself included) have a low level of confidence when it comes to identifying pasture vegetation. Partly this is true because pasture plants are hard to identify, and partly it's due to being busy farmers

RESOURCES FOR PASTURE ID

These sites offer a variety of free downloadable resources for help in identifying pasture plants:

- *On Pasture.* "Identify Your Pasture Grasses." onpasture.com/2016/05/02/identify-your -pasture-grasses.
- *100 Native Forage Grasses in 11 Southern States* (USDA). plants.usda.gov/100_native_grasses.pdf.
- www.midwestforage.org/pdf/326.pdf.pdf.

who don't make the time to learn the wide range of species that potentially inhabit a pasture.

Obviously, different stands will have repeat vegetation, as well as some novel species. Aim to dig up samples and bring home three to five species you can't name each time you visit the pasture for assessment. Take pictures and use a guidebook (see the sidebar)—or ask another grazier, an online grazing forum, or your local extension for help. Over time you will learn a few species here and there, and suddenly know more about your pasture than you ever thought possible.

5. IDENTIFY KEY ACCESS AND CIRCULATION ROUTES.

At this point in the assessment process, it's important to label critical routes for people, equipment, animals, and materials to access and circulate around your pastures.

For instance, if you have a movable shelter, you'll need the space to move it from paddock to paddock, turn it around, and so on. It's important, too, to leave space for future needs, including those of a farmer who might buy or lease the land in the future.

Knowing these needs will help you set gates and fence lines, if you are going to install permanent fencing. For the assessment phase, simply mark access and circulation routes on your map as another overlay of information. While the focus so far has been on defining different spaces, it's important to note these access and circulation points so you don't block them with trees.

6. SET SOME PASTURE PATCH GOALS.

Set one or two goals for patches, starting with those you rated "poor." These are likely areas that need problems addressed, such as soil pH or fertility, water drainage, or combating a species that is dominating. You might be able to address some areas through grazing practices, while others may need a complete overhaul. The ultimate goal is to maximize the production of quality pasture across the board, a process that takes several seasons. It's also good to identify patches that might be best for tree planting, which often can be coupled with disturbances that will improve the overall pasture quality.

Once you've completed this assessment, your map will serve you well as we further explore the process of deciding what trees to put where.

Designing for Trees: When to Add Trees to Pasture, and When Maybe Not

Ultimately the decision to add trees is of course up to the farmer. The right species, spacing, and patterns emerge from the process of assessing the pasture, formulating

Figure 5.6. Placing trees in pasture is a balancing act. Getting the right species in the right place, at the proper spacing, is something that is part design and part learned from trial and error, accepting that some trees will die and a design won't work perfectly the first time out. These black locust trees were planted close together with the intention of pollarding some and leaving others in the overstory to mature. Photo by Jen Gabriel.

production goals, and applying the resources and energy to plant trees successfully. This activity will certainly benefit the system long-term, but you can see positive gains in just a few years, especially if you're working with faster-growing trees.

Over the years experienced graziers learn the little differences among their paddocks and can exploit these to their advantage. For instance, grasses and forage in upland areas might come on quicker than those in lowlands during the spring thaw, whereas bottomland paddocks are often excessively wet in the muddy spring season. Some paddocks likely already offer unique benefits at different times of the year, by protecting livestock from wind and intense storms. The flexibility to utilize different pastures at different times of the year and in different seasonal conditions is ideal for managed grazing systems for many reasons.

Adding trees to pasture offers variability and the benefits of diversity to the farm. Likely, many farms will end up with some paddocks in silvopasture, and some as open pasture. There are, of course, advantages to open pasture, depending on your location and goals. And since silvopasture is a steep learning curve, it's better to start slow and grow, perhaps planting at most one-quarter to one-third of your open pasture with trees to start, then expanding from there.

Let's look at the three major design concepts we should consider when adding trees to pasture:

1. The first is to **ensure that productivity is maintained**; that is, that the quantity and quality of forages are maintained or enhanced, and thus overall animal performance (pounds gained, milk produced, health) is equal to or better than with open pasture.
2. The second is to **design and plant in patterns** that both enhance the natural character of the land and ensure the efficient movement of animals, equipment, fencing, and water.
3. The third important aspect is **tree species selection**, along with developing a sound process to establish trees efficiently and increase their survivability.

In the coming sections we will explore each of these topics.

Pasture Productivity

The story many of us are taught from a young age in biology class is that plants compete for sunlight, water, and nutrients, so planted trees will inevitably shade out forages and lower their productivity. Unfortunately, this greatly oversimplifies the dynamic. The reality is that *some* trees will inhibit the capacity of *some* forages. Careful species selection, timing, and management decisions all have an important role to play.

Inherent in the above assumption is also the idea that pasture management is merely a volume game: that the amount of edible biomass produced per acre is the most important factor. In fact, good pasture management balances quantity, quality, and animal performance. Let's pick apart each of these aspects in order to better understand the whole.

Quantity

Ongoing research has established that silvopasture at 50 percent canopy is around 20 percent less productive than open pasture in terms of the total forage biomass produced. Overall forage production continues to decline as trees grow and the canopy closes in.[3]

Let's zoom in to understand the effects of shade on the growth of specific forage species. A number of studies have compared forages under various light conditions, most often potted samples under synthetic shade cloth. Some of the more recent trials compared grasses at full sun, at 50 percent shade, and at 80 percent shade over two seasons, finding that, unsurprisingly, plant response varied given the level of shade, species, and growing season.[4] In other words, "It depends."

More to the point, in summer and fall months, research trials found that six cool-season grasses (Kentucky bluegrass, 'Benchmark' and 'Justus' orchardgrass, 'KY31' and 'Martin' tall fescue, and timothy) and two legumes (Cody alfalfa and white clover) did not show significant reductions in 50 percent shade. All of the warm-season grasses declined under 50 percent shade, and all plants were noticeably affected at 80 percent shade. Many of the grasses and legumes performed worse during the spring and summer months. This, along with other studies,[5] suggests that species selection and a canopy cover of less than 50 percent are critical aspects of a good silvopasture.

Important to note in these studies is the questionable use of shade cloth to mimic forest shade. Cloth casts a uniform amount of shade—unlike silvopasture, which offers patches of sun and shade at any given time, depending on tree spacing and arrangement. Another study found that a structure with wooden slats was a much better representation of the dynamics at play,[6] though nothing beats the real thing, of course. Ultimately, more research needs to be done, but what we have offers some key learning points.

Planting pattern, orientation, and species choice also all greatly affect the amount of sunlight that actually hits the forages. Research suggests that tree impact on forage growth can be improved with careful selection of tree and forage species and the spacing and orientation of trees.[7]

QUALITY

Another major effect of trees in pasture is the way they change forage quality. There is plenty of research

Figure 5.7. Shade can maintain or even enhance the quality of pasture forages, especially in the hottest months and during extended droughts. This is dependent, of course, on the species and conditions in the overstory and on the ground.

showing that many grasses grown under more shade exhibit higher crude protein concentrations—as much as 20 to 30 percent more under shade.[8] A high amount of crude protein content means that animals get more protein while consuming less grass.

Among other indicators of digestibility, including acid detergent fiber (ADF, a measure of the least digestible components) and neutral detergent fiber (NDF, which indicates how woody a plant is), the change was small, from 1 to 4 percentage units.[9] Many of the specific grasses mentioned on page 179 can maintain quality under shade, while other research on specific systems found that pastures with a dominance of annual ryegrass offered better quality than those of open pasture in a pine-walnut silvopasture.[10]

Also observed in many studies is the effect of trees on stabilizing forage quality, especially during later parts of the summer, when ambient temperatures tend to be above the optimum for cool-season grasses. Well-designed silvopastures can also moderate early frosts on forage, thereby extending the grazing season in colder climates.[11]

ANIMAL PERFORMANCE

The concept of "performance" generally refers to how well dairy animals produce milk, or how efficiently meat animals maintain or gain weight. A third parameter is sometimes offspring mortality or birthing problems. These measurements fail to recognize the value of a happy animal, a factor that is difficult for scientists to measure objectively but readily observed by any astute farmers building relationships with their animals (see chapter 3).

From a forage perspective, livestock production from silvopasture is unsurprisingly equal to that for open pastures during the first several years of tree growth,[12] though that can decline as trees mature. For instance, a trial in New Zealand found performance (in this case, weight gain) of sheep reduced by 50 percent in a maturing 15-year-old pine plantation with 200 stems/hectare.

This makes sense: As the canopy closes in, pasture productivity can decline. Thus it's important to balance trees against light penetration, as discussed in chapter 4.

The above results stem from looking purely at the decline of forage as shade increases, but we cannot

Figure 5.8. Trees offer dynamic patches of shade throughout the day and support animal performance, especially in hotter climates. Photo by Gabriel Pent.

forget the moderating effects of silvopasture, which is documented to reduce heat stress in animals. In a review of heat stress and shade research, Blackshaw and Blackshaw (1994) noted that an animal's ability to maintain a comfortable temperature depends on a combination of radiation, air temperature, air movement, and humidity. The effect of radiation is notably further impacted by the orientation of the animal to the sun and the reflective characteristics of its coat (color). When the air temperature exceeds the animal's body temperature, then the animal must either move to shade or sweat to cool itself down.

Sweating equals an increase in the demand for energy and a loss of water (respiration), which accumulate to about 15 percent of heat loss, while the majority of loss occurs through conduction, convection, and evaporation from the body.[13] Heat stress also leads to a reduction in feed and water intake, and the rate at which this occurs has been shown to correlate strongly with the breed and coat color of the animal.

Many research papers have established that ruminant heat stress lowers milk production and decreases growth rate, while shade increases average milk production consistently.[14] For meat breeds, it has long been recognized that animals are adversely affected by high temperatures. One four-year study found that shade increased the summer gain of yearling Hereford steers by 8.6 kilograms per steer, thereby increasing profits for the farmer.[15]

Since many of these trials are considering the effects of extreme heat, you could argue that, under normal conditions, shade isn't a critical factor in cool temperate grazing systems. While historically this has been true, one of the most likely effects of climate change on temperate ecosystems that have been traditionally more moderate is the increase in extreme heat. The availability of shade, along with increasing forage diversity to include both cool- and warm-season grasses, will likely prove to be an essential practice for cooler-climate farmers in the coming decades.

SILVOPASTURE RESEARCH AT VIRGINIA TECH
FINDS BENEFITS OF TREES IN PASTURE
By Gabriel Pent

Growing up, summertime consisted of me rising early every morning to help my dad milk our small goat herd before spending the day mowing the grass under landscape trees on our tree farm. It never occurred to me that anyone would find it desirable to combine the two operations into a single system until I heard about a silvopasture study taking place at Virginia Tech. I joined the study as a graduate student, interested in finding out if the shade of black walnut (*Juglans nigra*) and honey locust (*Gleditsia triacanthos*) trees had any positive impact on the lambs grazing beneath them.

We used time-lapse cameras and audio recorders to document lamb grazing behavior throughout the summer months in Blacksburg, Virginia. Some researchers have worried that animals provided with shade could get lazy and spend less time grazing, yet the lambs under the well-distributed shade of these silvopastures grazed as much as or more than lambs in the open pasture. Most important, the lambs in the silvopastures spent more time lying down throughout the day. The lambs in the open pastures spent more than two hours longer each day standing up than those in the silvopastures, indicating how much more comfortable

Figure 5.9. Lambs grazing and resting under the shade of a black walnut tree during a hot summer day. Photo by Gabriel Pent.

the lambs were in the silvopastures. In fact, with small vaginal temperature loggers, we found that lambs in the black walnut silvopasture were 0.7 degree F (0.4 degree C) cooler than lambs without shade during the hottest hours of the day.

Animal well-being and animal productivity are inextricably connected, and we have found that lambs in silvopastures gained weight as fast as or faster than lambs in open pastures. While black walnut silvopastures produced 30 percent less forage than open pastures, improved individual animal gains in these systems compensated for the lower stocking rate. Thus we found that our silvopastures produced the same animal output as our open pastures.

One of the primary problems facing sheep or goat farmers is parasite control. While conditions under silvopastures may seem like an ideal habitat for parasites (moister and cooler than open pastures), we found lower fecal egg counts in manure samples collected from lambs in the silvopastures than from samples collected from lambs in open pastures during one year of our study. One potential reason for this could be the regrowth of residual stumps in the silvopastures. This highly preferred forage keeps lamb intake further from the worms at ground level.

This browse also contains condensed tannins, which can inhibit the growth of internal parasites. Woody shrubs may prove an invaluable feed and anthelmintic resource in an integrated parasite management plan.

One of the trees established in these silvopastures is a grafted honey locust cultivar (*G. t.* cv. Millwood). The large, sugar-filled pods produced by these trees are a perfect complement to the cool-season grass species in our pastures, which can contain high levels of protein during the winter months. During one year, these trees produced a particularly large number of pods (about 4,300 pounds per acre from under 27 trees per acre).

Unfortunately, while other animals, including cattle, sheep, goats, and deer, have been observed to relish the pods, we had to train the young naive lambs in our trial to consume them. Once the lambs finally began to eat the pods, lamb gains in the honey locust silvopastures were greater than those of lambs in open pastures for a short time, indicating the potential value of this fodder source if animals are willing to use it.

While some work at other sites has indicated that forages may be more nutritious in silvopasture systems, we have found that any small improvements in forage characteristics under silvopastures are likely tied to species differences in the sward. In particular, lowered levels of neutral detergent fiber (NDF) seem to follow increased proportions of clover (*Trifolium* spp.) in honey locust silvopasture swards. In black walnut silvopastures, warm-season species such as nimblewill (*Muhlenbergia schreberi*) and foxtail (*Setaria* spp.) are associated with greater NDF levels. However, higher total nitrogen in silvopasture forages seems to be independent of species present.

Tree species selection can have a large impact on forage species composition over time. Although weeds were more of a problem in the thinner swards of the black walnut silvopastures, there was no horsenettle (*Solanum carolinense*) in these systems, likely due to juglone—an allelochemical secreted by black walnut trees.

This particular site was developed to support grazing lambs as a model for cattle.

However, sheep may be more heat-tolerant than cattle. In addition, this mountainous region rarely experiences a temperature humidity index greater than 72. Joining our network of experimental silvopasture sites across Virginia is a new silvopasture research and demonstration site in Blackstone. This 40-acre site will enable us to study the impact of silvopasture on cattle under conditions more similar to those found throughout the southeastern United States. Preliminary data indicate that cattle gains may be 50 percent better under woodlots that have been thinned for silvopasture establishment than in open pastures.

Even with lower forage productivity in some silvopastures, research such as ours has consistently shown that silvopastures can produce at least the same animal output as open pastures during the summer months, in addition to the products provided by the trees. Improved animal well-being in silvopastures likely compensates for any reduction in forage productivity.

Gabriel Pent is the ruminant systems specialist at Virginia Tech's Southern Piedmont Agricultural Research and Extension Center in Blackstone, Virginia.

ECOSYSTEM BENEFITS

Finally, it's important to zoom out from these specific components, which attempt to describe the complexity of a set of dynamics and reduce variables in order to craft data. To the credit of almost all agroforestry researchers, there is plenty of acknowledgment that the data must be taken in context, as seasonal variables, timing, and weather all play a hand in results.

When comparing silvopasture with open pasture, some of the most important factors are that wind speed is reduced, solar radiation modified, and temperature moderated.[16] In addition, humidity tends to be higher, and there are lower rates of evapotranspiration and higher soil moisture levels.[17]

All these factors combine to potentially support forage growth, maintain quality, and protect animals from stressors in the environment. As we consider the breadth of scenarios that a changing climate will continue to bring (seasonal drought, excessive rainfall, increasing heat waves, magnified intensity of storms) to our pastures, it's clear that trees will create a more stable, moderated ecosystem for our animals to graze. Of particular concern are periods of drought and water stress, coupled with intense heat, as many areas of the northeastern United States experienced in summer 2016. This combination can be deadly for pasture production.

Ultimately, it is a combination of factors that will stress our livestock-based systems, and so silvopasture appropriately offers a combination of solutions to mitigate them. Anyone farming knows that the only guarantee is change, and as good farmers the best we can do is hedge our bets and set ourselves up to be as flexible as possible with our options. Rather than applying the same ecosystem type and pattern across all acreage, it would be wise to have a range of habitat types, as this offers the most options for management.

PRODUCTIVITY: SUMMARY

To summarize the entire picture of productivity in pasture-to-silvopasture conversion, we can say that, generally speaking, while the quantity of forage for all warm-season grasses and some cool-season grasses is definitely lower in silvopasture, with many cool-season grasses the loss is insignificant, and the quality of those forages can be equal or greater. Animal performance has been found to be generally equal or greater in silvopasture, especially given the increase in heat stress that is likely as the climate changes.

Threaded though all these factors is the need to design carefully with respect to species selection, tree spacing, and orientation or patterning of plantings. It's important to recognize that each paddock is unique in these considerations. For example, a paddock on the north side of a field with tree rows orienting north-south will have a very different microclimate than the same configuration of trees in a paddock on the south side of the same field. When assessing pasture, one of the main challenges is to recognize these and other subtle differences that aren't always apparent at first glance.

Tree Spacing and Patterning

Before getting into the nitty-gritty of arranging trees, let's pause for some ecology. How do young trees come into a forest system? For many species, a large amount of seed is broadcast throughout the woods by wind, gravity, birds, and animals. This seed hopefully encounters a soft layer of soil and leaf litter to help nurture it as it gets started. And trees emerge in thick, dense groupings, sometimes of one species, and often in mixed-species clusters. There are not individual trees, but a whole community at work.

All too often we don't follow this pattern, instead taking trees out into the field as isolated individuals that we plant alone, then watch as they suffer. We should strive to plant not trees, but communities, in the pasture. It's important to begin a discussion of tree planting in this mind-set: We are orchestrating not a collection of individuals, but a community of them. Some will live a short time, while others may outlive us. A few will become favorite perch spots or nesting sites for birds, while others may succumb to heavy wind or snow. The goal is not to have a uniform pasture with each tree looking perfect in its form. Instead we are seeking to create a mosaic, a patchwork. In this case perfection is certainly the enemy of the good.

While you might be eager to establish trees in any open space, it's important to strategize and prioritize the spaces to start in. Consider the pasture areas that would most benefit from trees, and start there. These may include:

- Stands with large southern exposure that don't have shade for animals.
- Areas that don't offer good protection from seasonal winds.
- Steeper slopes that would benefit from the soil-holding capacity of trees.
- Seasonally wet areas and riparian areas.
- Areas identified for pasture renovation anyway.
- Pasture stands abundant in cool-season grasses and shade-tolerant species.

Additionally, one of the fears farmers express in planting trees in pasture is the complications that arise from getting equipment, animals, and materials into, out of, and around the pasture. Ensuring good access up front

means less labor installing trees that have to be removed later. It is always dangerous to assume that just because you don't access an area today, you won't need to drive to it in the future. Leave some extra room to be sure.

Most grazing operations require mowing at some point, though good management can reduce this need greatly. Plan to allow access not just for your current vehicle, but also for potential future ones. For instance, while we currently do all our field work with a small tractor, I still plan my access lanes 10 to 12 feet wide to accommodate my truck (especially for when the tractor breaks down) and a possible larger tractor someday. Turning radius is also important to consider. Think beyond your lifetime as well; just because you don't use a particular size of equipment doesn't mean the next farmer won't.

PATTERNS OF PLANTING

When considering how to arrange trees spatially in a given area, it's best to design from bigger patterns to

Figure 5.10. Many typical silvopasture images depict trees in straight rows with wide spacing, which is just one of the possible arrangements for planting, best suited for flatter, more homogeneous landscapes. Depending on the site and goals, trees can be in rows, on contour plantings, in clusters, or planted more randomly. Each configuration offers different benefits and outcomes. Photo by Wikimedia/panoramio.

smaller details. Decisions around the patterns, densities, and canopy sizes mature trees will reach all lead in a thousand directions, resulting in different canopy densities, trees per acre, and site qualities. The following patterns, diagrams, and tables will hopefully provide you some ideas and help you figure out the numbers as part of your tree planning process.

Rows, clusters, and scattered trees all have advantages and disadvantages, and likely the overall pasture will benefit most from a mixture of patterns. There are literally thousands of variables and combinations, but the following diagrams and table should help get you started in planning.

Single Rows

Single rows are ideal for steep lands and areas, along contour, or in rows in flatter spaces. We usually think in terms of the spacing of trees within the row, and the spacing between rows. A common spacing might be 6 or 8 feet between trees, and 40, 60, or 80 feet between rows. For many species, we like to plant even closer, at 3 to 4 feet, which encourages straight growth. This does mean you'll need to more aggressively coppice and prune trees earlier on, but this just results in fodder for the animals.

Of course, you could plant double rows, too. Beyond a double row, growth may be affected for middle trees if the density is too close and timber is the main goal.[18] Figure 5.10 shows examples of three row spacing scenarios for rows, to give a sense of the finished canopy cover and number of trees per acre.

If you're planting in rows, one question that arises is: What orientation? The general guideline for best light to the pasture floor is to plant rows north-south at mid- to higher latitudes, and east-west at low latitudes. This of course also depends on considerations for hillsides and sloped pasture, which would benefit from trees being planted along contour. A third consideration is whether trees would be helpful as a windbreak, meaning their orientation is roughly perpendicular to the prevailing winds. In some cases, one, two, or all three of these can align together. For instance, much of the contour of our farm's aspect is south to southwest, which allows for trees to be both planted on contour and remain roughly north-south in orientation. These blend well into an east-west hedgerow that is a critical windbreak from the north-prevailing winter winds. This hedgerow is on a flatter portion of land than the other rows. In this sense, there likely isn't a single strategy to seek; rather, look for a combination that integrates existing and new vegetation patterns on the farm landscape.

Offset Rows

Offset rows are plantings where the trees have relatively even spacing all around. This promotes more even shading and open growth for the trees. Depending on the spacing, trees could be fenced in rows for protection during establishment, or individual trees may need their own protection. Figure 5.12 shows a few examples assuming tree canopies at 25 feet. For more detailed numbers, table 5.2 offers the calculation for various canopy sizes and tree spacings and was designed to help users visualize what different canopy percentages mean and look like with different-sized trees. This is helpful because a 35 percent canopy cover is harder for many people than visualizing fifty 20-foot-diameter trees per acre. The table can help you decide how many trees to plant based on your desired light conditions and mature tree size. It can also help you determine when to coppice or thin plantings to ensure that light conditions don't cause too much decline in forage production.

Scattered or Random

Scattered or random trees mimic a savanna-type ecosystem. Relatively even spacing between trees will result in a uniform 50 percent canopy. This pattern offers even light distribution throughout the paddock while giving trees ample room to grow. It's ideal spacing in many ways, but problematic from an establishment perspective, as each individual tree will need robust protection from curious animals.

Clusters

Like oases in a desert, tree clusters can be valuable "living barns" for winter shelter and a cool space for loafing in the summer. They can be densely planted and more easily protected from browse, though you'll need to

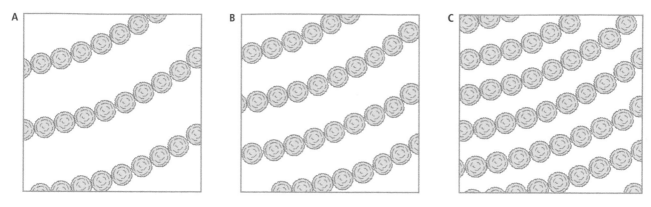

Figure 5.11. Curved rows (assuming some contour) at 80-, 60-, and 40-foot spacing, resulting in different densities and trees/acre. Each square in figures 5.11 through 5.13 represents 1 acre of planting space. *A* has 80′ row spacing, 18–22 trees/ac, and 20–25% canopy cover. *B* has 60′ row spacing, 28–32 trees/ac, and 30–35% canopy cover. *C* has 40′ row spacing, 40–45 trees/ac, and 45–52% canopy cover. Illustration created with data from Connor Stedman.

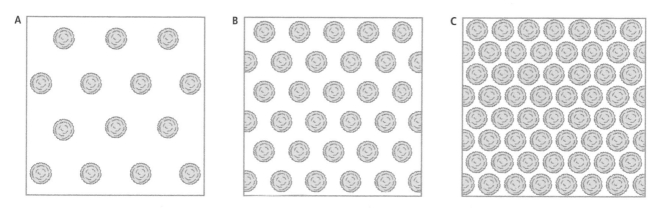

Figure 5.12. Offset rows at 60-, 40-, and 30-foot spacing, which provide different canopy cover densities with trees that have 25-foot canopies. *A* has 60′ row spacing, 13–15 trees/ac, and 15–18% canopy cover. *B* has 40′ row spacing, 30–34 trees/ac, and 32–38% canopy cover. *C* has 30′ spacing, 48–54 trees/ac, and 54–62% canopy cover. Consult table 5.2 on page 188 to help make further decisions about the impacts of spacing at various canopy sizes. Illustration created with data from Connor Stedman.

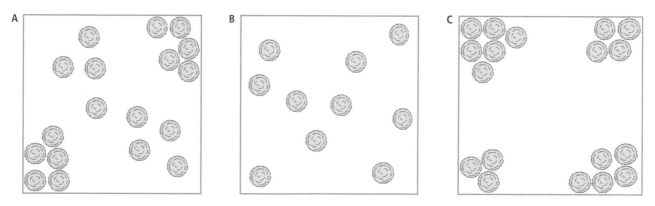

Figure 5.13. Various scatter and cluster patterns can be planted in a silvopasture. *A* has 16–20 trees/ac, 45–55% canopy cover in clusters, and 10–12% canopy cover in open pasture. *B* has 10–12 trees/ac and 10–14% canopy cover. *C* has 17–20 trees/ac, 45–55% canopy cover in clusters, and 0% canopy cover in open pasture. Illustration created with data from Connor Stedman.

Table 5.2. Silvopasture spacing chart: woodland/savanna spacing patterns

Spacing (Feet)	3	5	8	10	12	15	20	25	30	35	40	45	50	55	60	65	70	75	80	85	90
Trees/Acre	4,840	1,742	681	436	303	194	109	70	48	36	27	22	17	14	12	10	9	8	7	6	5
3	79%	28%	11%																		
5		79%	31%	20%	14%																
8			79%	50%	35%	22%	13%														
10				79%	55%	35%	20%	13%													
12					79%	50%	28%	18%													
15						79%	44%	28%	19%	15%	11%										
20							79%	50%	35%	26%	19%	16%	12%	10%							
25								79%	54%	41%	30%	25%	19%	16%	14%	11%	10%	9%	8%	7%	6%
30									78%	58%	44%	36%	28%	23%	19%	16%	15%	13%	11%	10%	8%
35										80%	60%	49%	38%	31%	27%	22%	20%	18%	15%	13%	11%
40											78%	63%	49%	40%	35%	29%	26%	23%	20%	17%	14%
45												80%	62%	51%	44%	37%	33%	29%	26%	22%	18%
50													77%	63%	54%	45%	41%	36%	32%	27%	23%
55														76%	65%	55%	49%	44%	38%	33%	27%
60															78%	65%	58%	52%	45%	39%	32%

Canopy Diameter (Feet)

% Shade Calculator — Woodland / Savanna

Note: Thanks to Connor Stedman for developing and sharing this table.

exclude them from grazing until an acceptable number of trees reaches sufficient size and girth to be grazed under. This means taking some land from productive pasture for three to five years. Existing hedgerows can be good areas to expand into clusters.

Regardless of the patterns and density you determine, it's important to accept tree mortality as normal. While you might have a desired final density, plan on a 25 to 50 percent loss in the first three to five years. Thus it's best to overplant, spacing trees closer together, or planting two or three trees where you might eventually hope to have one. Since most plantings will be from seedlings, choosing from a dense planting means a better likelihood that a healthier, desirable tree will exist in a given spot long-term. This is only affordable if you're working with unselected seedling trees (see more on this on page 189). If you're working with more costly cultivars or selected stock, it's best to interplant them with less expensive seedling trees.

PLANTING IN SUCCESSION

Tree patterning and spacing also need to factor in the restricted paddock size that will be inevitable during tree establishment. In other words, an open paddock of 1 acre may be reduced to strips 40 to 80 feet wide, with fencing keeping the trees out until they can be integrated with grazing animals (three to eight years). It's important to consider the grazing habits of animals in this regard. For example, while cows are more adept at grazing individually, sheep like to graze in a flock, often moving together. On our farm we have observed that row spacing closer than 40 feet creates uncomfortable grazing patterns for the sheep.

This presents a challenge, because in the long term it might be desirable to have tree/row spacing closer together. One strategy to overcome this is to plant successionally, with a wider spacing of trees for the first 5 to 10 years fenced out from grazing. Once it's safe to integrate those trees into pasture, you can move in fencing between established tree rows and plant new

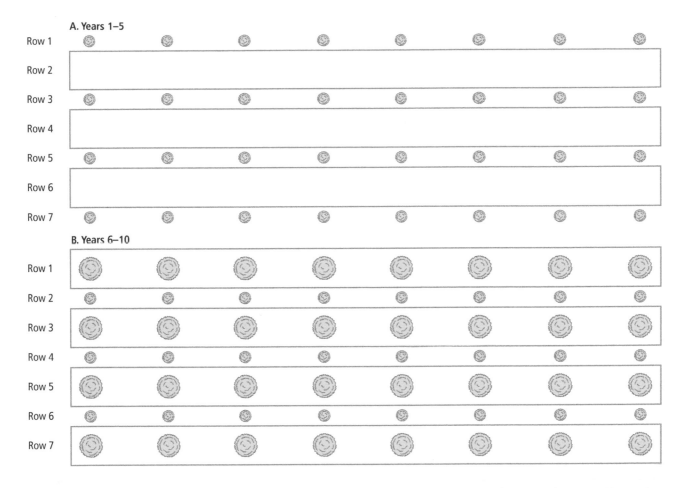

Figure 5.14. This concept for succession planting involves two phases. For Phase 1 *(A)*, in years 1 to 5, plant rows of trees at wider spacing (80 feet) than you intend for final rows. Portable fences are arranged to exclude the newly planted trees. Phase 2 *(B)* kicks in when the first succession of trees is large enough to be included in grazing paddocks, where paddocks are shifted to exclude a new row planting. This approach allows for denser final spacing, without creating bottlenecks for grazing animals with multiple thin rows to graze between.

seedlings. This pattern in effect maintains a good-sized paddock while eventually filling in the void and achieving a silvopasture canopy of around 50 percent tree cover. This concept is potentially important for many sites, where a planting plan can be designated as the final spacing but created in stages given that fencing, tree protection materials, and the cost of planting can all be more economical when implemented over time.

Selecting Trees for Silvopasture

As with many decisions on the farm, selection of tree stock is a balancing act among goals, time, and money.

We encourage you to shop around and discuss the philosophy of a given nursery or provider, or seek out your own materials for propagation.

The main tree options on the market are:

Seedlings. These are trees started from collected seed. Their genetic variability will be high, and characteristics may not closely match those of the parent. Seedling stock is usually either container- or field-grown, and then dug during dormancy and sent to customers as a bare-root seedling. Seedlings usually cost between $0.50 and $1.50 per tree and often range from 6 inches to several feet in height.

"Selected" seedlings means some amount of attention was paid to identifying desirable characteristics. Depending on the species, this may or may not mean that those qualities are transferred from the parent to the offspring. Seedlings cost between $2 and $4 apiece, or as high as $10 or $20, depending on the species and value of the seed stock. These come anywhere from 4 inches to several feet tall, with the price proportionate to the size and girth of the seedling.

Cultivars. Depending on the species, cultivars can be named selected seed (see above), or cloned material, which takes the form of grafted rootstock, cuttings, and/or tissue culture. These latter approaches offer an advantage in that the material is the same genetic makeup as the parent, so traits are almost guaranteed to express themselves. This is a good approach for trees where yields of fruit, nuts, or pods are desired, or where the goal is straight wood or high-sugar sap, for instance. Cultivars can be expensive, often $10 to $50 per tree.

It naturally follows that, given the difference in investment for the options above, you'll want to baby cultivars and will worry less about planted seedlings. Cultivars would be planted at their mature spacing, with the assumption they will survive, while seedlings are often planted very close together, to mimic the natural seeding of the forest, along with the assumption that many will die or be selected out over time.

For silvopasture, the likely approach is going to be to plant many trees out over many acres. This leads to a default preference toward the seedling option, with perhaps some cultivars thrown in for good measure if affordable. A hybrid approach appropriate for some species is to plant cultivars in the ground and then later "top-work" those that survive with selected scionwood from cultivar trees of the same or a compatible species. Generally, bare-root stock should be ordered and planned for either an early-spring or late-summer/early-fall planting.

All these points of information are species-dependent. It's critical to get to know the nuances of the specific trees you want to plant. Too often, farmers

Figure 5.15. For most broadacre plantings, bare-root tree seedlings are the likely choice, mostly because of their much lower cost and ease of shipping. These trees are either started from variable or selected seed, or grafted onto rootstock, then sent without soil, usually in spring or fall at the end/beginning of the dormant season. Photo by Jen Gabriel.

get excited about a particular species and don't pay attention to its unique needs, or seek to determine if it will do well in their soils and climate conditions. The more you plan ahead of time, the better your long-term outcome will be.

The selection of a particular tree species is a balance point between the site conditions and the goals for productivity. Rather than seeing the limits a site might present as constrictive, it's helpful to see the benefit of having some of the decisions already made for you. In the context of silvopasture, we can generally group the desired functions of trees into the following categories.

SHADE AND SHELTER

As has been said throughout this book, a primary reason for placing trees in a silvopasture is often shade and shelter. This goal, along with fodder, arguably offers the clearest return on investment, a wide range of species to choose from, and the ability to focus on fast-growing trees that are adaptable to site conditions. Any tree offers shade and shelter, but some do it better than others, due to their structure, leaf mass, and adaptability to high winds.

While all trees cast shade, consider when the trees leaf out in the spring, when they drop leaves in the fall, and the density of shade they cast. For instance, maples are early to leaf out, whereas black locust and black walnut leaf out very late. Additionally, maples cast a denser shade, whereas the other two species mentioned let more light through.

Shelter is sometimes an opposing feature, since the trees that offer the most protection also block the most light. The best shelter trees are conifers, including pines, spruce, and firs, which offer year-round protection and are sometimes referred to as living barns (see page 103) when planted in dense clusters. An old plantation or planted windbreak the previous farmer left behind can often be renovated for this purpose. We discuss the roles of conifers in silvopasture in more detail starting on page 206.

BROWSE (FODDER)

Overlapping in many cases with good trees for shade and shelter, many species can be utilized for feeds that diversify the forage diet of grazing animals. Once trees are established, the animals can self-harvest the material, or you can make the material available for them through pruning, coppicing, or pollarding.

As discussed back in chapter 3, animal nutrition is achieved through a diverse diet of forages with differing amounts of proteins, fibers, and nutrients. The percent composition of forages changes with the growth stage of the plant. Woody plants and trees tend to hold value as food for a longer portion of the season. Fodder is also high in secondary compounds that offer nutritive and even medicinal benefits to animals, most notably ruminants. Planting trees densely as we've been suggesting means that, after the establishment phase, you can leave some trees to grow above browse height while managing others as animal feedstock.

If we believe in the natural wisdom of the animal body, we need only look around to see the value of fodder. When we let animals into a pasture with available tree forages, they immediately head to them, mixing their consumption with that of grasses and forbs. What is most exciting is that, once they're old enough, trees can be managed and maintained by the animals while they harvest food for themselves. The ultimate satisfaction is seeing animals voraciously consume woody brush, while cleaning up weeds underfoot and fertilizing as they explore.

As trees mature further, you can leave lower branches to be stripped and enjoyed by the animals each time they visit the pasture. While it may be tempting to prune these after the animals leave to improve the aesthetics, leaving them will likely regrow another round of fodder for the next visit. In addition, a visit to the paddock can be an excuse to do some tree pruning, letting the branches fall or, better yet, piling them up for the animals to consume.

A portion of the trees in a row or cluster can be coppiced (cut to ground level) or pollarded (cut above browse height). Doing this in the dormant season encourages an explosion of new shoots the following growing season. With coppice, the animals will browse the whole plant when rotated into the paddock, whereas you'll have to harvest and drop pollarded material for their use. You can cut or break branches almost all the way through, and then bend them downward to give animals access.

Tree fodder can also be harvested and stored for later use, sometimes referred to as tree hay.[19] This differs from dormant coppice or pollard as tree limbs are harvested during the growing season anytime after they are fully leafed out. The tree is generally cut just above browse height (a pollard); if the tree is young, it will begin the process of balling, where a hard fist of calloused wood forms over time, providing a stable place for new growth that is less prone to breakage in the wind. Late June and into the middle of July appears to be the best time to cut, so that nutrients are retained in the material.

As with regular grass hay, tree hay needs to be bundled tightly or else it will succumb to fungus and mold contamination. Traditionally in Europe, these bundles were sometimes known as faggots, and they were hung or stacked in shelters to dry out and await their feeding to animals during the cold winter months.

Some have also experimented with fermenting trees as silage (willow, most notably), by wrapping harvested fodder in plastic and burying it in the

Figure 5.16. Trees can offer nutritious feed for animals while also providing shade and shelter. Black locust is highly desirable—essentially the nutritional equivalent of alfalfa.

THE UNDERUTILIZED UNDERSTORY: PERENNIAL HERBS
By Jonathan Bates

When the misty vapors start to freeze and the sheep and goats do sneeze and wheeze, Then, quick! With herbs ward off disease: Hollyhock root & raspberry leaves, Peppermint and chamomile if you please.

So you've got your animals grazing the grass under and between the trees. You might not think about it much, but there's room for an herbaceous layer, too, and every silvopasture animal could be taking advantage of these plants.

Why is this important? We want to think of all the grass, forbs, shrubs, and trees holistically; each has multiple functions in the system. We can also frame, in terms of general animal care and welfare, that the overstory trees provide shelter, grasses are the core food source, and herbs are a valuable medicinal component. Herbs are chock-full of vitamins, minerals, and other important phytochemicals.[20] Just as an herbalist might give a prescription of herbs for your well-being, pasture-raised animals learn (by themselves, and from other animals) how to cure and prevent ailments by way of their mouth.

Many of our pastures include wild and naturalized plants not considered trees, shrubs, or grasses. Botanists put these "non-grasses" in the category of forbs. The Old World shepherds called fields that included these plants herbal leys. Common herbs that might already be in your field include clover, dandelion, Queen Anne's lace, field mustard, wild onion, and peppermint. It's important to have each group of herbs—legumes, asters, umbels, mustards, onions, and mints—for a healthy, diverse silvopasture.

Not only do these diverse plant families provide a more complete buffet for your animals, but your overall living system benefits as well. They provide habitat to countless insects, birds, and small animals, as well as healthier soil biogeochemistry, water conservation, carbon sequestration, and pest control.

Some forage plants are more palatable to or preferred by the livestock.[21] Likely tender woody plants, flowers, and non-aromatics are their most exciting foods. They might choose specific herbals for specific purposes. Some mint family plants, for example, can either increase or decrease milk production.

There is history between herbs and pasturing animals. Our pasture forebears were the pioneers with their herbal leys,[22] and we can glean from their knowledge and experience. What we still don't know, however, includes which herbs are of benefit and will succeed when we put trees into the silvopasture. Factors like shade, variable grazing periods, competition, soil diversity (fungal versus bacterial), relationships among system elements, and vegetation maintenance need to be better understood.

This sidebar is here to inspire all of us to pay attention to herbs.[23] Observe them in your system, encouraging them where you can. What are the animals eating, and why? Which herbs are already among your trees or might be missing? Are you struggling with animal health problems? If so, are there herbs near your trees that could help? Wormwood, or sagebrush artemisia, can be offered as a powerful dewormer, for instance, as can other plants like birdsfoot trefoil and sericea lespedeza.[24]

Herbs can be best planted in communities, known as polycultures. A couple of sample polycultures anyone could adapt to their system are:

More heavily/routinely grazed after three years of protection, able to be more heavily grazed—say, under black locust: Purple prairie clover; wood aster; mountain mint; yarrow; Turkish rocket; perennial leeks.

More lightly grazed, as ground cover under tree crops such as under elderberry: Illinois bundleflower; chicory; anise hyssop; sweet cicely; perennial sylvetta; bulb garlic.

Also included are a few other underutilized/unknown herbaceous plants for inspiration, organized as "Best Bets," tiers 1 through 3. Most were chosen for their multiple functions, which could include nutrition, dense clumping habits, high regeneration rates, hardiness, indigenousness, human uses, or beneficial insect qualities. (When creating your herb design, consider plants that are useful and probably in the pasture already, such as red clover, dandelion, plantain, yarrow, and nettle.)

Best Bets

Tier 1: Purple prairie clover; Illinois bundleflower; lespedeza; giant Korean celery.
Tier 2: Lupine; blue wild indigo; comfrey; garlic chives.
Tier 3: Wild senna; bee balm; lemon balm; yarrow.

These "best bets" are general recommendations. Many factors determine how animals use these plants, as well as how the surrounding environment and management style affect them: moisture, soil health, grazing period, parasite pressure, plant growth stage, shade, and so on.

Jonathan Bates is a designer and teacher who owns an edible plant nursery, Food Forest Farm, and developed the famous "Paradise Lot" gardens in Holyoke, Massachusetts, with Eric Toensmeier. Learn more and get in touch with him at www.foodforestfarm.com.

ground for one month or more to ferment slightly. In Bhutan this practice was recommended in late October/early November, with the trees being packed several feet deep in plastic and covered with heavy rocks to create compression.[25] Some limited work in the UK verified the potential of willow silage as a decent-quality feed, but noted that a lot more research should be done.[26]

All this is to say, the potential of fodder is great, but we know less than we ought to. We know the food value of many trees, and so should give our animals access to them. The potential to harvest and store fodder as hay or silage is also intriguing, but several unknowns and logistical hurdles currently exist. Silvopasture practitioners are needed to experiment and share their experiences. Universities and other research institutions would do well engaging in more research to develop more robust species lists and nutritional information for a wide range of potential tree species.

TIMBER, POLEWOOD, FIREWOOD

The overwhelming focus of academic studies on silvopasture economics is the value of trees as a timber crop. While this yield can certainly be viable, timber markets are ever-changing, and a valuable tree today will likely be different in the future. Timber can be thought of, then, like an investment account, focusing on species

that are likely to hold their value and not succumb to emerging pest and disease challenges. (Black walnut is a good example.) Timber is distinct from other harvestable woods: The goal of growing the wood is to eventually mill it into dimensional lumber—the most common use of forest resources globally in the modern era. Trees at the time of harvest are often at least 40 or 50 years old. In this case, one might want to prune younger trees, which will result in less knots and deformities and can drastically increase the value of the timber down the road.

Polewood is a term for timber that is intended for harvesting left in the round, which might serve as a fence post, hop pole, or outdoor structure, or in roundwood construction projects. Often the wood is harvested on a shorter rotation, maybe 20 or 30 years. Certainly you can select from a dense planting over time to favor the best trees for this purpose (or even for future timber). The advantage from an ecological perspective of growing timber or polewood in silvopasture is that harvested wood is likely to be put into durable building projects, which is essentially a form of long-term carbon sequestration.

The economics of firewood are highly dependent on your locale and the demand. The advantage firewood offers is that, unlike other wood products, the quality of the material (so long as it isn't rotten) isn't as important as

Figure 5.17. Harvest of black locust posts from a silvopasture, which is likely the fastest-growing and most profitable tree in the Northeast, in high demand for fence posts and hop poles. Prices range from $1 to $2 per linear foot. Photo by Sean Dembrosky.

its heat value. Many of the species discussed later can be harvested for good-quality firewood, and this type of harvest can also be a byproduct of good forest management.

The challenge with each of these yields, which differs from the others, is that we are often talking about harvesting whole trees. Depending on the planting, this might mean losing a tree's shade, shelter, and fodder benefits. If you're interested in these products, a good approach is to plant densely, recognizing that only a portion of the trees will be harvested eventually for firewood, polewood, or timber. A dense cluster or multi-tree row will allow for this type of management and won't risk losing the benefits of standing trees, whereas harvesting trees from a single row might cause undesired effects on the pasture and animals.

For an extensive discussion of the nuances and uses of these wood products, readers are encouraged to consult our previous work, *Farming the Woods*.

MULTIPLE YIELDS

Ultimately, we can manage silvopasture plantings with more than one of these yields in mind, giving us both flexibility and staggering yields in the short and longer term. A nice reality from an ecological perspective is that trees that wouldn't last to become timber or polewood can be managed for animal fodder, and the multilayered nature of short bushy trees interplanted with taller trees contributes to a more robust form of shade and shelter. At this point you want to prioritize which of the preceding outputs are the most important to you, as this will help you zero in on specific species and on the amount of time, energy, and skill you must put in to get your desired results.

Many species overlap and fit in more than one category, and those that offer more than one potential yield can be arguably the most beneficial, since their utility is more diverse. You may have your heart set on a given species or a given goal. Each type of tree and yield comes with its own set of management needs, which generally increase in both the time and the skills they require as you move from the top to the bottom of the following list.

A wonderful progression of silvopasture, no matter the yield focus, is the transition in labor that occurs over the course of the first 10 years. Initially much of the work focuses on soil preparation, tree planting, and protection from various threats to survival. Over time this shifts to pruning, thinning, and maintaining. Eventually you'll need to remove wood in order to keep the system light levels optimal for understory forages. This succession of tasks is at the heart of a long-term perennial system. Perhaps someday our grandchildren won't need to plant many trees, if we leave them lands abundant with forests to steward.

Best Bets for Silvopasture Trees

While literally any tree could be chosen for a silvopasture, we can certainly make an argument for a list of characteristics that make a tree ideal for this practice:

- Fast-growing to get above browse pressure.
- Adaptable to many different ecotypes and microclimates.
- Low pest/disease pressure.
- Good for fodder, at least as an option for use.
- Multifunctional yields for future harvests.

To this end, instead of naming every possible tree we could plant, this section will profile some "best bet" species for silvopasture. We will start with deciduous trees, then work through conifers, and finish with more specific situations where livestock could be integrated into orchards and Christmas tree operations.

It's easy to get overly complex with species selection, choosing many different trees and having one of each in a given paddock. For management reasons, however, it's often wiser to choose one to three species in a row or cluster, seeking diversity on the site as a whole.

THE TOP FOUR TEMPERATE DECIDUOUS SILVOPASTURE TREES

We are in urgent times as members of the global community, where ecological destruction, social inequality, and climate change are every day amplified in many places around the world. Because the choices a generation makes in its lifetime have great effect on future generations, we must put our attention to the tasks of increasing the productive capacity of forested lands and providing for the needs of people and other living things. These "top" trees are presented in the context in which we are currently operating:

The climate is rapidly changing. We need to enact solutions to both mitigate the effects of current changes, as well as pull carbon from the air and store it back in trees, roots, and soil. These trees have properties that can help us get there, and fast.

The five-year silvopasture. Focusing on the top four species presented here offers landowners the opportunity to get silvopasture up and running in well under a decade. These species can be established quickly, thereby reducing costs in time, management, and lost grazing lands.

A gentle learning curve. These trees are among the most forgiving in terms of successful cultivation. They require fewer inputs, can handle damage from browse and recover, and offer several flexible options for usable yields.

This list assumes interest primarily in achieving the benefits of silvopasture for the animals, and getting trees growing quickly on pasture, therefore improving the planet's carbon sequestration capacity. While this may not be your personal goal, these trees are still worth considering at some level in your silvopasture system. They are great entry points.

Note that most of these trees have an extensive range, many into areas beyond just the cool temperate climate. Many are potential players on the global stage.

Willow: High Biomass and Anti-Parasite Medicine

If all trees were as easy to propagate as the willow, agroforestry would seem much less daunting. The amazing quality of willow is its ability to be live-staked—wherein a section of branch from any portion of the tree that is ½ inch to 1½ inches in diameter is cut and driven into new ground where a new tree is desired. This works almost any time of year, but is best done if the staking material is harvested prior to spring leaf-out. Material can be left soaking for months until planting. All this translates to one of the cheaper ways to get trees into silvopasture systems. It is also a tree that rapidly offers food *and* medicine to grazing animals.

This genus (*Salix*) includes some 400 species of trees and shrubs that exist primarily in the moist soils of the temperate regions north of the equator. Once established they support a large network of strong, tenacious roots that are extremely tough and fibrous. They also provide an important food source for early-spring pollinators, as mentioned in the sidebar back in chapter 1.

While it is common knowledge that fast-growing trees with dense root systems such as willow are good potential storage sites for carbon, this isn't an overnight phenomenon. Carbon can, of course, be stored in both the aboveground and belowground portions of the tree, with the most stable forms in the roots and soil. Research has found that willow begins to be a significant carbon sink five years from planting,[27] and that at least 20 percent of carbon remains in the underground portion of the plant even when the entire willow tree is cut down.

While willow is highly adaptable, keep in mind that these trees grow quickest through loose soils. Willow loves loamy floodplain soils and wet feet, and is easily established under these conditions. In old fields and compacted soils, the ground should be loosened prior to planting through mounding or tillage. Another strategy is to grow trees in pots before planting out, to develop a more thorough root system. Once trees are established, growth is quite rapid, and they recover very well from animal browse.

Figure 5.18. Willow is easy to start from 12- to 18-inch cuttings taken in the late winter before bud break. The tree produces abundant food, and its high tannins can be thought of as a digestive aid for grazing ruminants.

Stakes should be cut to 12 to 18 inches long and can be pointed at the end with a hatchet to help facilitate getting them into the soil. Ideally, when pounded, more than 50 percent of the material should get into the ground, though this isn't always an easy prospect depending on the soil type and previous compaction.

There are hundreds of different cultivars, and all will have value as fodder. Selecting the right cultivar depends on your site conditions, desired form, and available plant stock.[28] Willows can be divided into three types:

1. **Tree willows** form the tallest single-stem types.
2. **Osier (basket) willows** form a medium-high shrub.
3. **Sallow (shrub) willows** are low growing with multiple stems.

A few species recommended specifically for fodder include:

- *Salix purpurea* and other osier willow—are productive and highly adaptable.
- *S. matsudana* × *alba* (aka hybrid willow) is a tree willow that will produce the most biomass per acre. The 'Tangoio' variety has the highest protein content of willows, and is additionally very drought-tolerant.[29]
- *S. schwerinii* (sometimes listed as *S. kinuyanagi*), known as Japanese fodder willow, is reported to grow as much as 12 to 16 feet after coppice.

For silvopasture, willow not only offers wide site adaptability and fast growth but also some promising benefits as fodder for animals. There is a considerable body of research to support this, and in other temperate regions of the world, such as Bhutan, willow fodder can account for up to 20 percent of livestock diets.[30] This material is not only fed live during the growing season, but harvested and packed in plastic to ferment as willow silage, as discussed earlier in this chapter.

Not surprisingly, the different types of willow offer different nutritional benefits. In one study comparing voluntary animal intake of the hybrid versus the osier species, research found that sheep and goats consumed 22 percent less of the osier, which was attributed to the higher concentrations of lignin and condensed tannins.[31] This led the researchers to conclude that the hybrid offered a "better" food source, since the sheep would consume more.

But as discussed in chapter 3, tannins are an important secondary medicinal compound, one animals can use to regulate their needs. Sheep especially have proven they are able to select their diet to control parasites, and other research shows that willow specifically leads to a reduction in parasites and eggs in sheep.[32] We do not know the parasite loads of the sheep in the study, which found a preference for hybrid over osier willow, but chances are good that their intake would increase in relationship to their parasite load.[33]

What this means is that one willow isn't "better" than the other, and we again return to the diversity principle, encouraging you to plant a variety of willow in your pastures so that animals can select from the trees as they see fit.

A final word on willow is that it offers great promise as a tree for many other uses, many of which could complement its productive role in silvopasture.

Craft material. Willow has been valued all over the world for its flexible strength, which makes it great for baskets, living fences and structures, and even an ecofriendly material for coffins.[34]

Water purifier. Willow can be grown as a filter for cleaning toxins from water while also producing useful wood products for biomass energy or other uses.[35]

Charcoal and biochar. Willow is highly prized as a material for artist's charcoal[36] and is also a good material for soil-building biochar.[37]

Medicine. Willow bark is the original aspirin, and has been shown to be very effective in reducing inflammation and for pain relief.[38] The inner bark is harvested and can be preserved using a tincture or by consuming it in a tea infusion.

Sustainable energy. Biomass from willow, coupled with its carbon-sequestration effects, is a net energy gain. For every unit of fossil fuel used in cultivation, it can generate 18 to 43 units of equivalent biomass energy.[39]

From its high adaptability, many uses, and benefits to ruminant livestock, it's pretty clear from the preceding characteristics why one might argue that willow could be beneficially incorporated into many silvopasture systems.

Black Locust: High-Protein Fodder and Valuable Wood Products

Almost as easy to grow as willow, the black locust is a notorious outlaw that offers incredible benefits to silvopasture. This tree grows and spreads so well that it is illegal to plant it in Massachusetts and Minnesota, and is on the watch list for many other states, including New York and Connecticut. Its native range is complicated and in debate, but it is found throughout the US and into Canada.[40] Black locust spreads effectively and quickly both by seed and from root suckers. Once established, it can become the all-consuming species in a pasture. An unmanaged pasture, that is.

A tree that is "out of control" can definitely be problematic, unless of course it is a valuable food source for animals. Animals can control the spread of the tree, and benefit from its high nutritional content, which includes high protein content (around 20 percent), a nutrient profile very similar to that of alfalfa.[41] If planted at close spacing and cut when young, it could potentially be cut and baled like hay for later storage and use.[42]

Along the lines of densely planting high-nitrogen species comes the concept of intensive silvopasture, which is currently practiced mainly with *Leucaena leucocephala* on more that 500,000 acres in Australia, Colombia, and Mexico.[43] In this system woody trees are planted at extremely high density (up to 4,000 trees/acre) and then grazed wholesale by livestock. The benefits of intensive silvopasture are impressive and well documented,[44] from increased animal health and production (three to five times higher meat and milk production[45]) to improved water quality and the possibility of much greater than 30 times the carbon sequestration of organic agriculture. In cool temperate climates black locust would serve a similar role to the *Leucaena*, though only the most preliminary system has been proposed, and more needs to be done to address how it would work best in this climate. However, the potential is very much there.

In addition to its excellent food value for livestock, the black locust is one of the most rot-resistant woods available, and can be grown as a cash crop for landowners to sell as lumber or polewood, provided good improved stock is planted (see the sidebar on page 200). It has the potential to replace pressure-treated lumber, if only more effort were to be put into breeding and management.[46] This tree offers by far the best potential timber and polewood of the top four silvopasture trees recommended here. Black locust also offers one of the highest heat values of any wood; in fact, it can burn so hot it can melt the insides of a cast-iron stove. (Spoken from experience.) It's best to use it mixed with other woods in the firebox.

The challenge? Some states prohibit importing, selling, or trading black locust, including Massachusetts,

Figure 5.19. In addition to all the benefits of black locust, the tree sports a beautiful flush of fragrant flowers in late summer, which are edible and quite delicious. Photo by Wikimedia.

and it is restricted in Minnesota, Michigan, and New York. This is not necessarily a complete list—check with your state regulators before deciding how to proceed. Each state has its own specific regulations. States that restrict the planting, sale, or use of the tree often cite it being "non-native" and "invasive" in nature. I would argue (along with many established foresters and land managers) that the benefits of this tree outweigh the risks. If properly managed, it's a critical tree for a regenerative future.

In New York a regulated plant cannot be knowingly introduced into a location where it isn't already present. It's hard to believe there is even such a place in the state, and likely not in any location where farming has occurred, since the tree has a long history of value to both Native Americans and colonizer/settler farmers around the state. In any case, in New York the trees can be purchased, sold, propagated, and transported legally. Nurseries are required to attach a disclaimer to any material they sell.

Assuming you are clear to work with black locust, it's important to consider the genetic stock you source trees from, especially if your goal is to grow straight poles or trees that can be milled for lumber. Locust is

BLACK LOCUST AS A TIMBER CASH CROP
By Brett Chedzoy

"What tree can I plant on my land to make money?" and "What should I do with the field behind my house so that I don't have to keep mowing it?" are two questions frequently posed to foresters (especially this one). After breaking down the illusions of profitable fruit trees or Christmas trees, the conversation usually steers back around to black locust (*Robinia pseudoacacia*). This native of the Appalachian Range has many attributes that merit consideration for any tree-planting project.

Few if any trees in the Northeast match black locust's favorable attributes for silvopasturing. Consider the following:

- A legume, locust fixes significant amounts of atmospheric nitrogen, creating free fertilizer for companion plants.
- It is attractive and stable, and its highly decay-resistant wood is much in demand today as an alternative to pressure-treated lumber and posts.
- Due to a lack of shade tolerance, locust readily self-prunes in low-density plantings. This also results in a shallow and porous canopy through which higher-than-normal levels of sunlight can

infiltrate. Locust is also one of the last trees to leaf out in the spring. These characteristics are conducive to good growth by cool-season forages in the understory.
- Fast growth and "pioneer species" qualities make locust much easier to establish than most other hardwoods. Mature stands can be readily regenerated through coppicing, suckering, and the existing seed bank.
- Locust flowers are quality bee fodder, and the foliage can be very palatable and nutritious for livestock. However, locust foliage may have significant levels of condensed tannins, which could be potentially toxic to some livestock. If animals are not regularly exposed to locust foliage, they should be gradually acclimated and monitored.

But one of the best reasons to consider black locust is that it's probably the only tree we can grow profitably in this area. That isn't to say that other trees don't have their place in planting projects. But locust, thanks in large part to its rapid growth and suitability for smaller-diameter products like posts and poles, can be grown as a cash crop where

incredibly crooked in its "natural" form, and so seed selection, and sometimes pruning, is a critical factor for success. Ironically, the Hungarians identified the awesomeness of black locust a long time ago (in the 1700s), deciding to intentionally import seeds and engage in an intensive breeding program.[47] As a result, some of the best stock today comes from Eastern Europe, and nearly 70 percent of the forests in Hungary comprise black locust.[48]

Propagation of black locust is not as easy as that of willow, but it's not difficult, either. The seeds are wrapped in an extremely thick coating that allows them to persist in the soil for many years. To start from seed, you must scarify the coat by either soaking it in acid or hot water for 12 to 24 hours, or scraping it with a file.

Once this is complete, the seeds can be planted into pots or a prepared seedbed.

If you'd like a clone, especially from selected trees that exhibit improved characteristics, you can get root cuttings during the dormant season by seeking out a root flare from the base of an established tree and taking roots about the thickness of your thumb. These can be cut into 2- to 3-inch sections and planted.

Our experience at Wellspring Forest Farm is that black locust can be incorporated into paddocks after four or five years, once the bark is hard enough to resist stripping by sheep (or rubbing by cows). In our experimenting we had several incidences of bark strip, only to have almost 100 percent of the trees recover and persist despite the impact.

a dollar invested should yield more than a dollar in return.

Recently the New York State Department of Environmental Conservation added black locust to its list of restricted plants, meaning that it can potentially be invasive in some ecosystems like the pine barrens of Albany and Long Island—and therefore should not be introduced into new areas where it does not currently exist (though I struggle to think of any corner of New York State where locust is not already well established).

This bad reputation comes in part from locust's hardiness and survivability where other plants struggle (due to its ability to fix atmospheric nitrogen), and also its rapid spread through prolific seed production and suckering (root sprouting). Because of this tendency to spread into adjacent areas, it should not be planted near utility rights-of-way, roadways, or other locations where trees could become a nuisance.

With that disclaimer aside, locust is a near-ideal tree to grow on old-field sites where the choices are: (1) mow at least annually (expensive); (2) watch the field grow into multiflora rose and other noxious plants; or (3) invest in a locust plantation that can be profitable and pleasurable.

For locust to grow well and be profitable, diligent site selection and preparation are a must. Locust can tolerate a wide range of soils but does best in fertile, well-drained soils. Locust will struggle on poorly drained or seasonally wet sites. Pure plantations are also more susceptible to common pests like locust borer and locust leafminer, so mixing it up with some other tree species will help. Deer, mice, and rabbits can also cause considerable damage to young trees if left unprotected.

Once established, locust can be harvested for posts and firewood in as little as 10 years or gradually thinned to produce high-value sawtimber and poles within 30 years. And just how much money can be made from growing locust? Well, that depends on numerous variables—some controlled by people, and others by nature. One important variable, however, that will have a big influence on the economics is the genetic quality of the planting stock. Locust as a species can be highly variable in its growth habit, ranging from an almost "bonsai" form to super-straight "shipmast" varieties.

Cornell Cooperative Extension and the NRCS/ USDA Plant Materials Center (PMC) in Big Flats have been working on an "Improved Locust" project the past several years to identify and disseminate superior strains. The future potential to utilize this tree on farms around the Northeast is beyond great.

Mulberry: Good Protein, Minerals, and Fruit!

Holding steady in the top four is mulberry, a native tree that is well described in the sidebar (on page 202). Here the text will focus on what is known about its food value for animals, which is the highest of any of these species. This is due to a long and storied history of cultivation, breeding, and selection for silkworm production around the world.[49] The mulberry is one of the most researched and written-about tree crops for use by animals as a feed.

Mulberry has a global range and is utilized as a fodder crop for animals in India[50] and many other locales such as China and Afghanistan. It is considered highly digestible (over 80 percent) and can be a good food source for ruminants and even monogastric animals (pigs, chickens, horses, and more). The nutritional value is very high, with some of the most well-documented research putting its protein content at 15 to 28 percent in many countries around the world, with low amounts of fiber and very high mineral content.[51]

Unique to mulberry is its wider application beyond ruminants, making it a very important tree to have in pig and poultry production. Researchers in Japan found that egg size and production from chickens were similar to control feeds with up to 9 percent of commercial feed replaced by a mulberry leaf meal. They further discovered an improvement in the color of the yolks and a notable increase in vitamin K and beta-carotene.[52] Other studies suggest that mulberry leaf meal could be used for 30 to 50 percent of a pig's diet,[53] though a significant body of knowledge confirming this is lacking. In China it was found that silaged mulberry or sun-dried mulberry fruits could offer the same nutritional benefits to cattle as corn or cottonseed meal.[54]

Mulberry is unique in this group of trees because it offers a fruit crop, which is often overwhelming in its high productivity. This means plenty of fruit for humans, livestock, and wildlife, too. It's a fruit high in vitamin C, iron, antioxidants, and many other nutrients and minerals. They can be harvested and eaten fresh or dried. Any dropped fruits will be good food for livestock and wildlife.

Propagation is a bit more challenging with mulberry than with the previous two trees. As noted in the sidebar, seed is the most effective method, provided there are male and female trees growing together. Cuttings are possible, with some additional nursery infrastructure and skills. Mulberry is known to coppice and pollard easily, and also has a high heat value for firewood, though many say its high water content means it needs more time to dry before use.

A GIVING TREE: THE MULBERRY
by Akiva Silver

While some adults hate mulberry trees, virtually every songbird and kid loves them. They are such generous trees that in tropical regions are capable of bearing fruit 12 months of the year. Here in the Northeast, we can find individual trees that will ripen their delicious berries from June until the end of summer. These trees are capable of holding unfathomable amounts of fruit every year, and that is one reason I call them giving trees.

In some circles the mulberry has a reputation for being invasive. This is an unfortunate accusation, particularly here in New York State. White mulberries have been here for centuries, and red mulberries are native. They have had ample time to become invasive. However, I have never seen more than a few scattered trees in one place. In the Northeast they do not form dense thickets like Norway maples or ailanthus trees. A field left untouched is about a thousand times more likely to fill with honeysuckle, European buckthorn, autumn olive, and multiflora rose than with mulberry trees.

Mulberries can be anywhere from 15 to 70 feet tall, and they are often just as wide. They have similar shapes and growth habits to box elders. There is a lot of diversity among the trees in both form and fruit. Two species grow in the Northeast: *Morus alba*, the white mulberry, which is native to Eurasia; and *M. rubra*, the red mulberry native to the eastern United States. These two species freely hybridize with each other in the wild. The names *red* and *white* are misleading, as the fruits of white mulberries can be white, red, black, or lavender. Red mulberries are anywhere from red to black.

The berries vary in taste as much as in color. Some mulberries are just not that good, while others are outstanding. Don't judge the entire species based on a couple of taste tests; excellent-flavored trees are out there. The berries look like blackberries but are often much sweeter. It is almost impossible to find a fruiting mulberry tree that is not covered in songbirds. That is why they are sometimes called living bird feeders. Mulberries are true bird magnets—I have never seen a more reliable way to attract orioles.

These spreading trees often have low drooping branches for easy picking.

The dark fruits stain fingers and tongues, and purple bird poop stains cars and sidewalks. There are no poisonous look-alikes to the mulberry.

Mulberry trees can grow just about anywhere, and fast. They grow out of the cracks of sidewalks, in rich floodplains, in vacant lots, and in just about any open area. They can tolerate flooding, drought, and unreasonable pH levels at either end of the spectrum. They can be cut down again and again, and keep sprouting back. Milky sap oozes from their wounds,

Poplar: A Variety for All Places

The *Populus* genus is the last of the top silvopasture species. Research has found it similar in productivity and nutrition to willow, but with lower tannins and thus higher intake.[55] This is not surprising, since poplars, cottonwoods, aspens, and willows all share the same family, Salicaceae. They, along with willow, are highly encouraged as silvopasture plantings in New Zealand grazing farms, most notably because they offer water-quality benefits. Much of this usefulness emerged during drought times, when farmers turned to the trees originally planted as a conservation measure. Reports suggest that farmers have had good experience using both willow and poplar in coppice and pollard systems, and that ecosystem services persist even when managed for livestock feed.[56]

As with willow, poplars are great for wet and moist lands, but also tolerate more upland conditions. As we climb in elevation, where more and more species are unable to survive, these species quickly become critical in silvopasture. They are especially important in the high-elevation boreal forests found in many parts of North America.

just like on a fig tree. The roots are bright orange, and their intensely rot-resistant heartwood is bright yellow when first cut, darkening to a beautiful orange-brown with age. Mulberry wood is very hard and strong, with uses ranging from fence posts to cutting boards.

The leaves of mulberries can be all sorts of shapes, even on the same tree. They can have weird irregular lobes, or just be heart-shaped. The bark of mulberry stems strips easily and is a very strong fiber with multiple survival uses.

Figure 5.20. Mulberry fruit. Photo by Eric Schmuttenmaer/Wikimedia.

The mulberry has been domesticated for thousands of years as a food for silkworms in China. This was attempted in America during the early colonial days up until the mid-1800s. (The US silk industry never amounted to anything, primarily because of cheap overseas labor.) One tragic legacy of this episode was the accidental introduction of the gypsy moth, which was being bred as a possible silkworm substitute.

If you want to grow mulberry trees for yourself or for wildlife, it's important to know that some trees are male and some female. The female trees will bear fruit with very few seeds if no males are nearby, while the males will never make fruit. A seedling can take as long as 10 years to begin flowering, and then may change sex shortly afterward.

Grafted mulberries, or those grown from cuttings, can begin fruiting within a year or two, because they are skipping the juvenile phase. These clones can be guaranteed to be female, but of course they are evolutionary dead ends. The genetic diversity of seedlings is invaluable in a changing world. If you have the space, plant seedlings as well as one or a few named varieties.

If ever there was a tree to love, a tree so generous and fun that every neighborhood, chicken yard, hedgerow, animal pasture, and old field needs at least one, it is the mulberry.

For more articles on trees by Akiva Silver and information on his nursery, visit www.twisted-tree.net.

While it's similar to willow in many regards, poplar's lower tannin content means it can become a staple food for livestock, whereas willow intake will likely be more limited depending on the needs of an individual animal. Because we're used to relatively abundant rainfall in most cool temperate climates, drought can be especially surprising when it happens, and poplars could prove to be critical elements of a drought-resilient farm. One research study found poplar cuttings to be an important supplement that actually increased the reproductive rate in ewes that were grazing during both pre-mating and mating periods.[57] It would be great to have more research done comparing the fodder values of various species in this group.

We can think of possible cultivar selection most easily alongside the goals we might want for their incorporation:

Columnar poplars are beneficial in landscapes if you're seeking visual screens or windbreaks. Poplar offers superior structure for windbreaks, and columnar varieties maintain a narrow width of 10 to 15 feet, while growing 60 feet or more in height. Popular species include the Lombardy poplar (*Populus nigra* 'Italica') and the Bolleana poplar (*P. alba* 'Pyramidalis').

Aspen poplars do better on slopes and upland conditions. They are distinguished by more rounded leaves with a point. In the northeastern and western United States, quaking aspens (*P. tremuloides*) are most common, while warmer climates support bigtooth aspens (*P. grandidentata*) as the most common variety.

Cottonwood poplars are a set of trees most tolerant to wet and flood conditions, especially in hotter temperate climates. Varieties include eastern cottonwood (*P. deltoides*), narrowleaf cottonwood (*P. angustifolia*), Rio Grande cottonwood (*P. wislizeni*), and Fremont cottonwood (*P. fremontii*).

There are also a number of hybrid poplars out there, bred as crosses between one or more species for a variety of reasons, most often vigorous growth. All poplars are about as easy to propagate as willow is, from pencil- to

Figure 5.21. Poplar trees offer valuable feed and are aesthetically pleasing in the landscape. Pictured is the quaking aspen (*Populus tremuloides*), which was a volunteer in our pasture. Photo by Jen Gabriel.

thumb-thick cuttings that are taken during dormancy and established in rooting material.

THE RUNNERS-UP: TREES WITH SOME SILVOPASTURE POTENTIAL

Before anyone gets offended, the remaining trees we will profile in this text are not bad trees, but in the frame of silvopasture may have only one or two of the characteristics that we see with the previous choices. These species are also less researched, and we know very little about their potential interactions with animals. They are all great trees for different purposes; they just don't have all the traits that make them best for silvopasture.

In addition, all the preceding trees are quite tolerant of varying soils and climate types, from the coolest temperate to the warmest zones. They also have a number of subspecies to adapt to even the smallest variation in the landscape. The trees that follow are a bit more particular, so they may work well for some readers and less so for others.

What follows is a list of potential trees, in order of the amount of information we actually have in regard to their silvopasture potential.

Figure 5.22. Impressive pod production on a grafted honey locust 'Millwood' at the Virginia Tech Silvopasture Research Farm.

Honey Locust: Pod Production in Warmer Places

This tree (*Gleditsia triacanthos*) is not a relative of the black locust, but stands on its own as a species. Its most promising application to silvopasture is its sometimes prolific production of pods that make a very valuable food source for livestock, most notably the endangered 'Millwood' variety,[58] which is currently being researched extensively at Virginia Tech. The greatest potential for the pods is after they drop to the ground, as a late-summer and early-fall high-sugar feed for animals, comparable nutritionally to oats or barley.

While winter-hardy to -34 degrees F (-37 degrees C), honey locust isn't usually found in the cooler states of the northeastern United States; it prefers the Appalachian, warmer midwestern, and southeastern states. It tends not to grow well on heavy clay or gravel soils, where its taproot has trouble establishing itself. Where it does thrive, it can be relatively fast growing. While it is a member of the legume family, it doesn't appear by most accounts to fix nitrogen, though some research at the Yale School of Forestry found possible evidence of non-nodulating fixation happening.

(Nitrogen fixation tends to come from bacteria nodules formed on a tree's root system.)

There has been a fair amount of attention paid to honey locust, with one decent compilation coming from an (outdated) website set up by Andy Wilson of Virginia. This is an effort to compile a body of information from multiple universities that have conducted research over the past century, which was reported as 30 institutions within temperate zone countries, in 1994.[59]

While it's clear from the research that many varieties have potential, it is nearly impossible to source the material at the moment, as there are only a handful of nurseries with stock. As with the Hungarians taking on much of the low-level breeding work for black locust, the French recognized the potential of this US species earlier and have several varieties in the works, though yields, while decent, tend to be highly variable from year to year.[60]

Of all the second-tier trees we have listed here, this one has the most potential. What is needed is more energy and resources devoted to developing viable uses. Thus, it's not really ready for prime time.

Osage Orange

Actually a member of the mulberry family, the Osage orange (*Maclura pomifera*) is a tree that can grow in much of the United States, being hardy to zone 4. This tree isn't necessarily a go-to for silvopasture, but it's hard to pass up, with a relatively decent profile as an edible fodder with around 10 percent protein and a digestibility rating around 70 percent.[61]

Its pre-settlement range was more in central Texas and Oklahoma, but it has since naturalized to much of the United States save for the farthest northern reaches. Much of this spread was due to the Roosevelt-era windbreak planting mentioned in chapter 4.

Osage orange is a fast-growing tree that produces excellent wood, valued for tool handles, fence posts, and other craftwood needs. It is additionally a good windbreak species and burns even hotter than black locust. It was also traditionally used as living fence for livestock, and can be grown as part of a multi-functional hedgerow.

Mimosa

This tree (*Albizia julibrissin*) is not native to the United States but deserves a brief mention, as it offers some of the same properties as the black locust and in a side-by-side comparison offered more total nutrition.[62] The researchers noted that the leaves exceeded the nutritional nitrogen requirements of growing cattle and goats, whereas black locust would require additional supplemental feed.

Native to Southwest and East Asia (Iran, Korea, and Japan are well-known growers), the mimosa has been naturalized in the United States, mostly in southern regions, being hardy to zone 7, with some varieties reportedly winter-hardy to zone 6. As a nitrogen fixer it is considered moderately good, so it's a worthwhile tree to trial if you are building a silvopasture in a warmer temperate site.

Alder

Rarely mentioned is the alder genus (*Alnus* spp.), a group of highly adaptive nitrogen-fixing trees.[63] This would lend itself to being a high-protein food, and ruminants seem to readily consume it, in our observation. Still, there is little research or information available on this tree or its potential benefits in silvopasture systems.

Conifers and Silvopasture: Old Plantations and Timber Production

The role of conifers in silvopasture is arguably more straightforward than the multilayered possibilities of deciduous species. Conifers aren't going to be good feed sources, so the main goals are either to provide year-round shelter for animals, grow a decent timber/pulpwood species, or a little of both. Additionally, animals can have a possible place within Christmas tree production, but we will leave that discussion to the specific scenarios section, on page 215. The most prevalent conifer silvopasture in the United States is currently in the Southeast, and much of it involves conifer plantations grown for timber and pulpwood. Some growers also harvest pine needles for mulch, baling them and selling the material for a premium.[64]

Since broadleaf trees can take a long time to reach harvestable maturity, pines offer a more economical return, and nowhere is this more true than in the South. The pine species best adapted to this warmer temperate zone include loblolly/southern pine (*Pinus taeda*), longleaf pine (*P. palustris*), and slash pine (*P. elliottii*). For cooler temperate zones, suitable species include firs (*Abies* spp.), ponderosa pine (*Pinus ponderosa*),[65] and also larch (*Larix* spp.), which is gaining some interest as a niche local product among a growing number of gardeners looking for rot-resistant wood products. (Larch is best used for outdoor applications *not* in contact with the soil.) Up north it used to be common practice to plant Scots and red pine, but those markets are poor in many locations. Any of these trees alone or in combination could be suitable for creating a green or living barn (see the sidebar in chapter 3).

With any silvopasture system for timber, and especially with conifers, pruning is absolutely essential to achieve maximum value. Pruning lower branches as the tree grows results in less knotwood, a higher-quality log, and thus a better profit margin. It's recommended that pruning start when the trees reach 15 to 20 feet tall or the diameter reaches 5 inches. Pruning should remove all the branches where the trunk diameter is more than 4 inches, while retaining a live crown and never removing more than one-third of the total vegetation (see figure 5.23). Proper tools and technique are essential to avoid damage to the tree, and the best time to do the work is late in the dormant season.

In addition to newly planted conifers, old plantations and abandoned Christmas tree farms are all too common in many locales. Silvopasture offers a unique opportunity to reclaim these spaces, and improve both

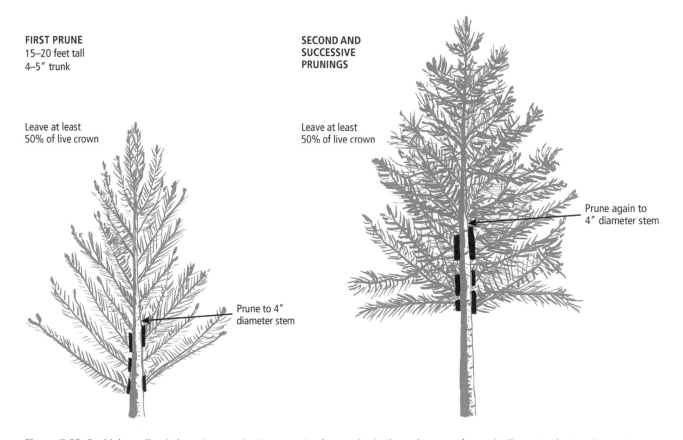

FIRST PRUNE
15–20 feet tall
4–5″ trunk

Leave at least
50% of live crown

Prune to 4″
diameter stem

SECOND AND SUCCESSIVE PRUNINGS

Leave at least
50% of live crown

Prune again to
4″ diameter stem

Figure 5.23. For high-quality timber, pines need to be pruned at least twice in the early years of growth. Illustration by Camilo Nascimento.

Figure 5.24. Goats grazing in a southern pine silvopasture at Atkins Agroforestry Research and Demonstration Site, Tuskegee University, Alabama. Photo by Uma Karki.

their productivity and the biodiversity they can offer. It's important to note when a conifer plantation is a mimic of the native forest type (such as in many locations down south and in the northwest United States) versus imposed (such as many locations in the northern US). We should always have an eye to the native forest types as a template for silvopasture patterning.

In most cases old plantations will demand a thinning to open up more light for forage establishment. This must be done carefully and not too quickly; tall, spindly conifers are especially susceptible to windthrow if they are opened up too much. Depending on the site, it may be necessary to leave a denser cluster of trees to protect against prevailing and seasonal winds. Soil testing is also critical prior to any attempt to seed a site.

Conifers offer another option for fast-growing trees and production of polewood, pulp, or timber where

local markets are good. The USDA has a comprehensive guide specific to silvopasture in pine plantations, called *Silvopasture: Establishment and Management Principles for Pine Forests in the Southeastern United States*. It is available for download, along with a free self-directed online course, at www.silvopasture.org. The guidebook from Tuskegee University discussed on page 259 is also an invaluable resource. Conifers can offer a valuable return, when the local and regional markets have the demand.

Integrating Silvopasture into Specific Tree Crop Systems

We began this discussion by looking at the types of trees you could add to a system assuming it was more or less livestock-centered. Now we will flip the focus and start with a specific type of production system, examining the nuances of applying silvopasture to fruit orchards, nut orchards, and Christmas tree operations.

It should be noted that this section differs greatly from the previous one in that we are entering another level of intensification in terms of the labor required for the crop. In many cases those already committed to the work of orcharding or maintaining Christmas trees will welcome the concept of integrating animals, as it often means a reduction in labor or input costs that are already being expended to maintain the trees. The bottom-line question is: Can animals be integrated in a way that supports the needs of both crop and animals, without doing irreversible damage to the crop?

Fruit Orchards and Vineyards

Animals and orchards have been together for a long time. Orchard grazing has been around for hundreds of years, and was quite common in North America until the 1950s.[66] There are many advantages to grazing orchards, but there are also challenges. The least fun part is the inconsistent and various regulatory issues you must master if you want to both graze and sell the tree crops for people to eat (see page 212).

Orchards, which are traditionally spaced widely, are great environments for grazing. It's just key to keep in mind that with orchards, needs are seasonal, and so the timing, number of animals, and patterns of grazing all need to remain flexible. While the primary goal might be to eliminate mowing and other maintenance tasks as a direct result of grazing, this might not happen fully for several years, when all the kinks have been worked out. By managing the understory with grazing animals, you minimize competition with the trees, better control pest and disease outbreaks, and reduce fuel and/ or chemical inputs. Additionally, you can get a better economic return and improve your farm's cash flow. Not to mention that grazing animals look great in the orchard, too!

Many examples of win-win scenarios can be found in orchard grazing, where animals are fed and the health of the trees supported. For instance, research at Michigan State University found that grazing pigs in orchards on the aborted apples that drop in June was significant in remediating plum curculio, a challenging orchard pest.[67] In Turkey one study found that fruit tree leaves in the fall offer significant nutrition to livestock, with plum, peach, and apricot among the best.[68] Animals can benefit from consuming these in autumn when available forage quality is low, and also help remove fungal pathogens from the orchard that might persist in the leaves if they were left on the ground.

Vineyards, too, can be great spaces for grazing animals, though in most cases there is more training involved—grape leaves can be a highly desirable forage. Operators in the western United States developed a very clear process for training sheep to avoid eating vines using lithium chloride, as mentioned back in chapter 3.[69] Others coordinate training with leaf shoot thinning needs in the spring, preferring Katahdin/ Dorper crosses as the breed for the job.[70] In New Zealand sheep are actually encouraged to harvest leaves, taking the role of machines and handwork in a portion of management known as leaf plucking, where some of the vegetation is removed to improve grape quality. A comprehensive guide describes the process, noting that sheep should graze only one variety at a time, and that timing is essential to avoid any crop loss.[71] One farmer-led research project found damage to be at most 10 to 15 percent, with many of the damaged vines returning.[72] This approach to vineyard management is not only effective for many, but a great promotional piece, too, as many vineyards often host visitors for tours and tastings.

Several successful case studies of French farmers grazing sheep (again, mostly Shropshires) in cider orchards have been documented in France, with many farmers reporting a drastic decrease in mowing needs, and some reducing fertilizer and pesticide use. Coordination of any spraying and grazing is, of course, essential, especially given that sheep are very sensitive to copper. Pruning may need to be adjusted to keep lower branches above browse height (one more reason sheep are likely preferred over other animals). In France farmers generally exclude animals for 30 days prior to harvest, while in England 56 days is more common.[73]

Figure 5.25. Turkeys (with guard geese) grazing an asparagus and apple silvopasture / alley cropping system at the Good Life Farm in New York. For orchards, birds are among the best animals, since they offer weed and pest control without damaging the crops.

Figure 5.26. Sheep grazing in a traditional cherry orchard in Sheldwich, England. This is a community orchard where sheep are grazed to manage the sward for biodiversity under the project Kent Orchards for Everyone (www.kentorchards.org.uk). Photo by Pippa Palmer.

In addition to sheep, the other common animals used in orchard management include pigs and weeder geese. For pigs, orchardists usually only rotate them through post-harvest to clean up the drops, while others employ pigs three times a season: in the spring to remove insects and provide light tillage; in midsummer to collect premature drops; and after harvest to clean up the leftovers.[74] Certain breeds are better than others, one favored one being the Gloucestershire Old Spot, known in the UK as the orchard pig. Some orchards even plant root crops in between the rows of fruit trees to provide supplemental feed as the pigs rotate through.[75]

There is little information specific to geese in orchards, other than the general management considerations outlined in chapter 3. Cows, goats, and other poultry are scarcely mentioned, and for the former two there is likely just too much risk given size and appetite. Sheep of certain breeds already mentioned appear to be the overwhelming preference.

NUT ORCHARDS

Much of the information gleaned from working with animals in fruit plantings can be applied to nut orchards as well. While there are some scattered examples and research findings, more work is needed to understand the dynamics at play. For one thing, unlike fruits that are mainly harvested off the tree or vine, nuts often fall to the ground before being swept up. This poses particular challenges to silvopasture, and might mean a longer gap between a grazing event and nut harvest, though at least in the case of pecan farmers, crops are

GRAZING WITH FOOD: WHAT ARE THE REGULATIONS?

When seeking to integrate livestock and food crops, we enter a complicated regulatory landscape, mostly stemming from some legitimate concerns about the interaction of manure and crops being harvested for human consumption. The concern from a regulatory standpoint is that manure increases human exposure to pathogens such as *E. coli* and salmonella. Occasional outbreaks and confusion around the issue create misunderstanding and sometimes hysteria, and farmers and practitioners are advised to read up and understand the different considerations of farming in ways that reduce risk.[76]

If the fruit or nuts are intended for personal consumption rather than for sale, then you are able to do as you see fit. Once they become products for sale, there are a number of possible legal hurdles to practicing silvopasture. Some of these are voluntary, and one is coming into play for almost all commercial farms in the United States over the coming years.

Current Good Agricultural Practices (GAP) standards along with National Organic Program (NOP) standards limit the application of raw manure to 90 days before harvest for crops not in contact with the ground and 120 days for crops that do have ground contact.[77] Additionally, GAP states that after application, no crop can be planted for at least two weeks. This standard, by default, can be extended to grazing/crop interaction (depending on how it is interpreted), which means that for many tree crops, animals can only graze in the early springtime, or just after harvest. The unfortunate part of this rule is the lack of recognition that it was intended for the bulk application of manure to fields from a spreader or sprayer, and not from the actions of grazing animals on rotation.

Both GAP and NOP standards are voluntary; that is, the producer chooses to participate in order to get a certification, which in some cases increases market value or is required if buyers are to purchase the crop (most often large grocery chains and institutions). But these regulations put a burden on farmers, greatly restricting their ability to be flexible with the timing of grazing, with many farmers noting that the three to four months before harvest is essentially the entire growing season in many areas.

Even if you don't have an interest in GAP or NOP certification, all commercial enterprises will be subject to a new set of standards over the coming years. The FDA's new set of rules governing food safety, known as the Food Safety Modernization Act (FSMA), passed under President Obama, offers the most up-to-date look at the role of domestic animals on the farm landscape. Note that farms with a three-year average gross revenue of less than $500,000 that sell the majority of their product to markets within 275 miles are "qualified exempt" from the regulations, though the FDA is allowed to revoke this exemption at their discretion and farms must still comply with some aspects of the regulations.

The standards originally mandated a ridiculous nine-month gap between manure applications and crop harvesting. After a rather large public outcry on a number of the standards led to 15,000 comments being submitted, the FDA was remarkably responsive, changing several standards, including their perspective on how domestic livestock fit into the farm landscape.

This excerpt is reprinted in full, a truly unique commentary from a regulatory agency in the matter of an integrated farm system, which was included in the final ruling of the FSMA[78]:

> *Although the final rule does not require establishing waiting periods between grazing and harvest, the FDA encourages farmers to voluntarily consider applying such intervals as appropriate for the farm's commodities and practices. The agency will consider providing guidance on this practice in the future, as needed.*

Further, responding to concerns expressed in public comments, the document notes that:

> *. . . currently available science does not allow us to identify a specific minimum time period between grazing and harvesting that is generally applicable across various commodities and farming practices. Rather, the appropriate minimum time period between grazing and harvesting would need to be determined based on the specific factors applicable*

to the conditions and practices associated with growing and harvesting the commodity [sic]. We are eliminating the proposed requirement for an adequate waiting period between grazing and harvesting in proposed §112.82(a) [sic!].

However, we encourage covered farms to voluntarily consider applying such waiting periods, as appropriate for the farm's commodities and operations. We will consider providing guidance on this practice in the future, as needed.

In such cases, you must assess the relevant areas used for a covered activity for evidence of potential contamination of covered produce as needed during the growing season (based on your covered produce; your practices and conditions; and your observations and experience) (§112.83(b)(1)).

If you find evidence of potential contamination during that assessment (such as observation of significant quantities of animals, significant amounts of animal excreta, or significant crop destruction), you must evaluate whether the covered produce can be harvested in accordance with the requirements of §112.112, and you must take measures reasonably necessary during growing to assist you later during harvest when under §112.112 you must identify, and not harvest, covered produce that is reasonably likely to be contaminated with a known or reasonably foreseeable hazard (§112.83(b)(2)).

FDA believes this suggestion goes beyond what is reasonably necessary to minimize the risk of serious adverse health consequences or death, to prevent the introduction of known or reasonably foreseeable hazards into or onto produce, and to provide reasonable assurances that produce is not adulterated under section 402 of the FD&C Act. We acknowledge the longstanding co-location of animals and plant food production in agriculture, and we do not believe it is necessary to prohibit grazing in areas where covered produce is grown to achieve the statutory purposes set forth in section 419 of the FD&C Act. We are requiring farms to assess relevant areas used for a covered activity as needed during the growing season for evidence of potential contamination, to evaluate whether produce can be safely harvested, and to take measures reasonably necessary during growing to assist the farm later during harvest when the farm must identify, and not harvest, covered produce that is reasonably likely to be contaminated with a known or reasonably foreseeable hazard when, under the circumstances, there is a reasonable probability that grazing animals, working animals, or animal intrusion will contaminate covered produce (§112.83). We believe this rule requires an appropriate level of public health protection while also appropriately providing sufficient flexibility considering the diversity of production and harvesting of produce (sections 419(a)(3)(A) and (c)(1)(B) of the FD&C Act).

It is a rare moment in regulatory history when an agency puts the power of good judgment into the hands of the farmers, acknowledging that historically this type of integrated practice has been done without ill effect. While it's good news that at least we don't suffer from regulatory oppression in regard to intermingling animals and tree crops, we should still be concerned with food safety. So what do we know about it?

Research has found that levels of *E. coli* and salmonella in grazing and orchard scenarios depend on the type of animal and environmental conditions. A study from the 1970s found that there were about six times more *E. coli*–contaminated pecan samples from cattle in grazed versus ungrazed orchards, though it is not clear if the cattle were given continual access to the orchard, or rotationally grazed.[79] Pigs grazing on a rotation through organic orchards in Canada resulted in little to no *E. coli* presence in samples of soil, leaves, and fruit.[80] And sheep in California grazing on rotation resulted in very low levels of harmful pathogens, with only 1.8 percent of fecal samples and 0.4 percent of soil samples containing *E. coli*, and only 0.8 percent fecal and 0.4 percent soil for salmonella.[81] These results would suggest that the risk level is low, at least in rotational systems.

The reality with much of regulation and fear around human pathogen contamination from livestock systems goes hand in hand with industrial feedlot systems that create the conditions to perpetuate *E. coli* and other bacteria.[82] While the risks might be higher, there is still evidence that *E. coli* strains can persist in grassfed animals at equal levels to those fed grain in confinement,[83] so we aren't out of the woods (or pasture) when it comes to minimizing risk.

The question at hand is about what types of precautions should be taken when mixing tree crops for human consumption and animals, and the answer is . . . we don't really know, and also "it depends." The best we can do is offer a short list of good practices, and continue to develop the conversation:

1. Exclude animals for at least a month before harvest.

Start by using 30 days as a minimum guideline for crops harvested from trees, and 45 days for those harvested off the ground, knowing that the longer the duration, the lower the risk. It may take as long as one to two months for a cow patty to fully decompose during the warmest summer months.[84]

2. Evaluate conditions.

Start taking observations of the deposition of manure by your animals, and how long it takes for it to dry up, then to fully disappear. While the previous studies mentioned seem to indicate a low presence of *E. coli* in rotationally grazed systems, knowing the timing for a dung deposit to dry up and disappear gives a range of time to apply to step 1.

3. Watch where you walk.

Be aware of, and train employees to be aware of, the potential for contamination from animal dung to a harvested crop. An example would be stepping in a cow patty, then climbing a ladder, which would potentially bring your hands into contact with the feces. Good education and awareness can go a long way toward reducing risk.

4. If you're seeking GAP or organic certification, use the FSMA standard as leverage and lobby for a change or clarification to the standard.

Existing certified farms and those that need one of these certifications to sell their products are, for now, potentially susceptible to the stricter 90/120-day rule.

5. Help determine more specific guidelines.

If you are committed to silvopasture and tree crop integration, get on board to help define and discover answers to the many questions that arise around this issue.

At the end of the day, with the issue of human pathogens and our food crops, the fear any silvopasture practitioner might have is not really about getting "busted" by a regulator, but in the importance of providing a safe product for your customer. No farmer who intends to stay in business wants to put the people buying food at any risk. This is why the FSMA decision is so heartening, as it actually puts dignity and trust in the ability of the farmer to operate in a way that supports the health of the general public. With around 2 million acres in fruit and 1 million in nut crops in the United States, there is ample potential to bring animals into these systems, with a lot of potential good to be had.

exempt from regulations[85] and the system has a long history of successful integration.

As with fruit, one big advantage of bringing animals into a nut planting would be the post-harvest cleanup, which can especially break pest cycles. Pigs are most suited for the task, though other animals may be able to help, too.

One of the older relationships in silvopasture within productive orchards is a history of grazing cattle with pecan trees in southern states.[86] The author of one article from the Noble Research Institute on the subject notes an interesting challenge in management: Farmers generally favor one crop over the other, and rarely are able to manage both to an optimal state. Not that this is a problem, as the benefits of mixing the two likely outweigh having a perfect crop of either. Trials conducted through the organization found returns of grazing to be $42.25 per acre in the first year and $60.75 in year

Figure 5.27. Cows grazing a mature pecan grove in Kansas. Grazing cattle offers two important benefits: additional income for the farmer, and a great reduction in the cost of mowing. Photo by William Reid.

two. The integration eliminated the need for three mowings in year one, and two in year two.[87]

In many cases the legacy of cows in the pecan orchards is merely an afterthought, and so researchers are working with farmers to improve grazing practices by incorporating the seeding of better-quality forage crops. One effort was able to produce 700 pounds of pecans and 250 pounds of beef per acre with no additional fertilizer needs.[88] These types of yields for nut trees are all in more southern states, though some nuts are being bred for productive potential in colder states, including hybrid hickories, a close relative of the pecan tree.[89]

The black walnut (*Juglans nigra*) is another commercial nut crop that offers some potential, with an income from livestock being a useful way to help justify the establishment of a planting, through a number of different potential scenarios.[90] Walnut has the unique distinction of being both valuable for nut production and high-value timber in the current (and likely future) market. One of the often cited concerns with walnut is toxicity from juglone, which is exuded from the roots of the tree and persists to some degree in the leaves and nuts, though only horses appear vulnerable to its toxicity.[91] And while some forages may be difficult to establish, there are plenty of good ones to incorporate that will thrive in the light shade canopy of the trees.[92]

Additional nut crops to consider include hybrid hazelnuts and chestnuts. Likely the best fit for hazels would be poultry, since the lower shrub form of the plant will leave it much more vulnerable to browse damage. Chestnuts offer an additional challenge of spiny outer husks that could inflict painful wounds on animals, though most are likely to adapt; it would not likely be an overwhelming problem. In nut crops it's important to distinguish "wild" or seedling nut trees from commercially improved varieties. In almost all cases the former won't produce the volume or consistency to justify harvesting a nut crop economically.

Nut trees could in fact be grown simply for their shade value, with the nuts being a food for foraging. This is a great use for existing walnut groves that won't likely produce a high yield or be worth the labor to harvest and market the nuts. This saves the cost and concerns with selecting or grafting high-production varieties in the orchard. There is considerable market value and potential for nut-finished pork, something many chefs are willing to pay a premium for. Feeding pigs a substantial amount of nuts before harvest is said to affect the flavor of the meat, though you must be careful to not oversell the concept to a relatively naive market.[93] While it's true that "nut-finished" could be technically any amount, it's not honest to market your meat as such unless the majority of the pigs' diet in the last few weeks is coming from nut resources.

CHRISTMAS TREES

There is a growing interest in using organic techniques to cultivate Christmas trees, and sheep and/or geese might be part of the solution. In England the practice has been promoted for many years, specifically with the Shropshire breed, which is said to be uniquely qualified for the task, though other resources mention several breeds being feasible, including Dorset, Hampshire, Rambouillet, Romney, Leicester, and Suffolk.[94]

Christmas trees are likely too sensitive for the demands of curious goats or clumsy cows. The other animals to consider would be poultry, and while any might be helpful, it is geese that are likely to do best with grass control. The tighter spacing of Christmas tree orchards means a tighter fit for many animals, notably grazing animals that like to flock or herd together.

Rather than regurgitate the words of others, if you're interested in this topic I would highly recommend a publication called *Two Crops from One Acre: A Comprehensive Guide to Using Shropshire Sheep for Grazing Tree Plantations*, which offers practical information based on a number of farmer experiences in the UK and other parts of Europe.[95] Some of the main suggestions for management include:

- Avoid browse early on by potentially waiting one or two seasons before integrating sheep into the system. Another research paper suggested that it's possible sooner, but close monitoring is necessary to avoid damage to terminal leaders and lateral branches.[96]
- The breed and even sub-breed or breed line are critical. There should be zero tolerance for sheep who nibble trees.

Figure 5.28. Sheep assist in maintaining a long-established Christmas tree farm in Michigan. Photo courtesy of NRCS/USDA.

- Sheep should not be introduced in early spring when tree growth is fresh. Allow it to harden off first. A good potential pattern is to introduce grazing in midsummer after hardening off and continue until late October (on a rotation), then resume from after Christmas until dormancy breaks.[97]
- Many say that older ewes are the safest to graze, and that all lambs, or at the very least lamb rams, should be kept out, as they can be too rambunctious for the trees.
- The most palatable species are fir trees. Pine is of little interest, and spruce is the least desirable from the animal browse perspective.
- Unlike other grazing scenarios, the goal is not optimal stocking but rather lower stocking, to focus the sheep on the task. Establishing lush pasture was little recommended; a better plan is to just use the sheep to graze the persistent weedy vegetation. A small number of sheep might get the job done just as well.
- Ongoing observation and decision making will be critical, especially in the first years.

Reading the guide, which also includes some information on fruit and vineyard systems, gives great optimism for the successful integration of sheep into this type of system. It also underscores the role the farmer must play as an active and engaged participant in the system, including a lot of focused energy on learning and fine-tuning the behavior of the animals involved. This might suggest that an alternative model is one where orchard or vineyard owners partner with livestock managers, so that two enterprises operate in concert on the same land, without overburdening any individual in the process.

Summary: Choosing Trees and Systems

Congratulations! You made it through the complex and often overwhelming considerations for choosing various trees and systems for silvopasture. In many ways it's much easier to start with an existing forest, as the species have already been decided; your choices

involve managing these trees over time. Yet with open lands and tree planting comes a great opportunity to shape a system in a way that can best meet your aims and support the land at the same time. Your personal goals and abilities will be the foundation for the decision concerning which trees to ultimately include in your silvopasture.

Silvopasture trees represent a balancing act between diversity and keeping things simple enough to manage. This all comes down to scale. A 1-acre backyard with a food forest and a grazing flock of poultry might have dozens of tree species, but only one or two of each. A 100-acre farm might have only two or three species that are replicated over the whole landscape. Regardless of size and scale, almost any design could benefit from the top four species we started with. From there you could go in almost any direction. Before investing too heavily in an idea, we recommend test-planting a handful of species you are interested in growing, to see how they perform on the site.

At our farm we initiated a test planting in our second year on the land, planting a wide range of species including black locust, honey locust, red alder, poplar, elderberry, birch, sycamore, willow, hazelnut, apples, peaches, cherries, and others. At the time we hadn't decided on grazing anything other than ducks. So our goal was to select species we were interested in and plant three to five of each, seeing which took with the least care. As we decided that sheep would become a part of the farm picture, our priorities shifted (more black locust, please!). We also learned about our site and its intense prevailing winds and hard clayey soils, which limited the growth of many species.

Over time we've pruned (pun intended) the list down, as we've clearly developed the farm enterprises we will focus on and learned that it's far easier to manage a few species than a larger number. From that original list, we've kept a few fruit trees, but only for personal consumption. Moving and observing our sheep along with the trees that do best, we've opted to focus on willow, black locust, poplar, and alder as our stable silvopasture trees. Research for the writing of this book has us eager to get some mulberry going, too.

To be clear, just because this book didn't profile a specific species does not mean it isn't a good idea for a silvopasture. Any species can be used, so long as it matches your goals and is suitable and adaptable to site conditions. This book focuses mostly on the tree species that have been most used specifically in silvopasture, have research and information behind them, or offer the most potential benefits to a silvopasture system. For instance, on our farm we are also planting out sycamore and birches, which we mostly enjoy for their benefits to wildlife and for aesthetics. We encourage you to explore the wide range of possibilities and combine tree species in the way that works best for you.

It's always best to start slow and small, whether in trees or livestock. Too many times we've seen others (and ourselves, too) fall victim to the fact that, as humans, many of us think bigger ideas than we can actually accomplish on the landscape. It's this, coupled with our enthusiasm, that leads us to purchase 800 trees at once, or start with a flock of 40 sheep. It's far better to start with 50 trees and four sheep. Get the basics down and check in to see if you actually *like* the work. Better to figure this out early on, and with low investment, than after you've dumped all you have into time and materials.

With trees, it's hard to keep a foot on the brake. You always feel like you're behind; as the Chinese proverb goes, "The best time to plant a tree is yesterday." If you are feeling impatient and eager for results, then plant fast-growing trees. It's amazing how they can truly transform the landscape in just a few years' time. Get something going, then cultivate patience. This work is for more than just us. It is the work of shifting agriculture into something longer-term and better thought out. It is the work of multiple generations. It is work for the community, and for the planet.

Planting Trees

When it comes time to plant trees, a few basic principles apply. Whether the tree is a $0.50 seedling or a $50 rare fruit cultivar, the same fundamentals apply. Don't rush

it, and remember that tree planting is a roll of the dice. Not all trees will survive, and that's okay.

CHALLENGES TO ESTABLISHMENT AND MAINTENANCE OF TREES

Trees take time to grow, and the first two to three years are the most critical. Getting past this hump of establishment means being able to step back and worry less, as the trees should mostly do fine. The work also transitions, from protecting young trees against potential damage and weed pressure, to thinning, pruning, and shaping the plantings you've installed. A sense of gratification also increases, as it's truly a wonderful thing to see trees become established and transform the feeling of a space.

Regardless of what you plant, and the pattern(s) you choose, the success or failure of the planting will depend on:

- The health and vigor of the stock you are planting.
- Your soil type(s) and efforts to improve quality.
- The process you use to dig an appropriate hole.
- Proper planting techniques.
- Weed suppression, mulching, and cover cropping.
- Efforts to protect your trees from the elements and predation.
- Availability of water and fertilizer during the establishment years (the first two or three).

There are four main questions to ask when developing a plan for planting:

1. Are the site conditions and soil suitable?
2. How will I protect against livestock and wildlife?
3. How will I reduce weed pressure?
4. What is my plan for dealing with low rainfall/ drought?

Each of these factors alone can equal tree mortality, but together they create a complex demand for management. This is, in fact, likely the largest deterrent for a landowner to plant trees in the first place, as no one wants to invest a bunch of time and money into something that only gets swallowed up by the grass.

Figure 5.29. Planting young black locust seedlings, which had an almost 100 percent survival rate and are nearly 20 feet tall after three years. The trick? Good stock, soil prep, and proper planting technique.

It's best to be realistic, even a bit pessimistic, as you approach a plan for management. Assume the lowest level of available time and energy to maintain the trees, and develop a strategy from there. It can be done. Most failures or even disappointing outcomes come from poor planning and forethought.

The classic pattern is when folks (I have done this!) start receiving catalogs in the mail for trees, in winter, and they randomly choose species that "look good," ordering a random number while being unsure where they will go. Suddenly one day in spring, the trees all arrive, and there is a mad scramble to not only plant them but figure out where they even belong. This all-too-common scenario should be avoided whenever possible.

Let's cruise through these four factors of tree planting management and cover some of the recommendations to increase success.

Site Conditions and Soil

Simply put, if your tree isn't happy where you plant it, it will die. Trees are surprisingly resilient, however, so this process might take a long time. It's almost worse when a tree looks pretty good for the first three years, then suddenly takes a turn for the worse. Remember: Accept tree mortality. It's part of the process. But learn from it, too. If a whole row of trees dies, question why. The answer usually isn't genetics; it's more likely that some aspect of the site caused the mass morality.

The most common site conditions that lead to trees struggling or dying are:

- Poorly drained soil or heavy compacted clay that roots can't grow through.
- Excessively wet or waterlogged soil, which some species can't handle.
- Exposure to wind and blowing, drifting snow.
- Drought, excessive rain, or high variation in hydrologic conditions.
- Inability to survive the coldest temperatures (aka lack of hardiness).
- Poor soil prep or improper planting procedure.

Figure 5.30. For this planting, we first tilled with a BCS rotary plow to loosen the soil, then added a few extra shovelfuls of topsoil when we planted the trees.

Note that for each of these conditions, there are tree species that are highly evolved to thrive in them. There is a right tree for *every* scenario. Often, within a given genus, different species are adapted to different ecotypes. For instance, with alder, the red alder (*Alnus rubra*) is more adaptable to higher, dry places, while the Italian or grey alder, and the speckled alder, prefer wetter conditions. It's important to sort out these differences in the planning process.

The first step to preparing soil for tree planting is a soil test. See the information on page 176 for resources to help with this. A soil test will help determine if the pH of the soil is off balance, and if there are any nutrient deficiencies. These would limit the ability of trees to grow at an acceptable rate. Follow your local extension office's recommendations for addressing any imbalances or deficiencies. In silvopasture you can use the same soil test you use for pasture as your preparation for tree planting.

After soil testing, the next stage is preparation. Establishing trees on pasture poses three particular challenges. The first is that pasture soils are overwhelmingly bacteria-dominant, whereas forests need a healthy population of fungi. The second is that in almost all cases, decades or even centuries of tillage, plowing, and driving over soils with heavy machinery have compacted the soil. The third factor is that dense grasses and forbs can compete with young trees as they seek to establish their roots.

It follows, then, that the approach to preparing soil for trees is not simply digging a hole, tossing in a tree, and walking away with fingers crossed. Instead, the approach is to:

1. Test your soil and address any deficiencies.
2. Remove/disturb the sod layer.
3. Take measures to decompact the soil.
4. Increase fungal presence in the soil.

Options for removing or disturbing the sod layer include tilling or plowing, mulching, or spraying to kill grass. Keep in mind that disturbing the soil doesn't prevent the grass from coming back. More on reducing weed pressure on page 224.

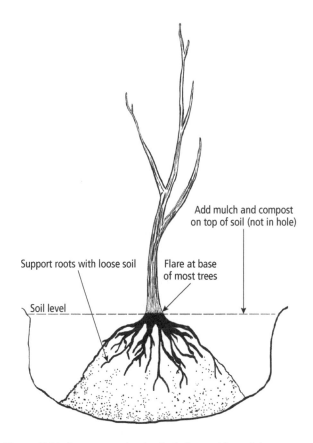

Add mulch and compost
on top of soil (not in hole)

Support roots with loose soil

Flare at base
of most trees

Soil level

Figure 5.31. Proper tree planting includes a wide and deep hole, support for the root system, and planting at the proper depth. Careful attention to these details can mean the difference between high mortality of planted trees and them thriving. Illustration by Camilo Nascimento.

Before planting, create a zone of decompacted soil at least two to three times the size of the tree's root spread. For row plantings, this might mean conditioning the whole row with tillage or plowing. You could also use a broadfork or other tools to open and decompact the area around the base of the trees.

It's critical to plant trees at the proper depth. Locate the flare of the tree at the base just above the termination of the surface roots. Plant the tree so that the flare is partially visible at the soil surface—not higher, not lower. Be sure to plant the tree a little lower than grade, so that water will find its way to the base (see figure 5.31).

When planting, have some extra topsoil on hand, and create a mound in the hole to set the roots on. It's helpful to plant in pairs, as one person can position the

tree and the other shovel soil back in. The soil should be stepped on and compressed more than you might think. Too much air from loose soil is actually a bad thing. Step around the base a few times, with the natural weight of your body pressing down.

Protection from Livestock and Wildlife

No matter what the site conditions, your trees will need protection from critters large and small, and—since livestock is a given with silvopasture—protection from your animals as well. This all comes down to excluding the tree from being eaten, rubbed, or trampled until it is either too big to succumb and/or resilient enough to recover. Depending on the species, and the amount of impact, this can range from two to five years.

The tree essentially needs two things: A significant portion of its leafy mass needs to be above a given animal's browse line, and the bark needs to be thick enough that it can't be stripped by pigs, goats, or sheep (some cows might go for this, as well). It's hard to know exactly when this will occur, but our experience with fast-growing trees seems to indicate around four to six years, depending on the species. Slower-growing trees will, naturally, likely take longer.

The strategies for protecting trees from both livestock and wildlife have different pros and cons, and depend very specifically on the animals you are bringing into the silvopasture, as well as the level of wildlife pressure your site experiences. The one animal everyone worries about is, of course, deer. And if you ask a hundred people about the best deterrent strategy, you will get a hundred different answers. Rather than repeat the considerable literature out there about deer deterrent strategies, we'll cover some of the tips we've learned along the way.

What we recommend is to consider what you know of your animals and site, and gauge the various options against your budget. Start with test plantings and see how the results stack up. It's all a cost-benefit analysis. Be honest about your ability to tend and maintain trees. And start with smaller test plantings to determine what scale you can reasonably keep up with.

The amount of loss you can tolerate also depends on the amount you've invested in trees. Many of the species favorable to silvopasture can handle some impact and still recover. The name of the game is not to completely eliminate damage, but to reduce its impact long enough for trees to get large enough. In areas prone to high deer damage, the cost of purchasing larger stock to begin with might outweigh constant monitoring or the cost of deterrents.

A mix of strategies is always going to work better than just one. For example, on our farm we use a combination of our dogs, tree tubes, electric netting, and a bit of acceptable damage, which has worked well for us. The shape of our land and the larger vegetation patterns of our landscape mean that deer tend to move through our site and not linger too long. Ironically, we might

Figure 5.32. This tree enjoys deluxe protection: a welded wire cage, a tube to protect the base from rodents, and a weed barrier. Also, it is weed-whacked weekly. This setup might cost $5 per tree or more.

see this pattern change as we fill in the open pasture on our lands. Fortunately, by that time the trees will be out of harm's way.

Fencing out trees. You'll need fencing to exclude livestock from your young trees, with the exception perhaps of some poultry. As discussed in chapter 3, the more wire and wood you use for your fencing, the more likely it is to exclude animals as a physical barrier. It's likely not practical, though, to invest in this material if you are doing extensive plantings. This creates yet another incentive, as previously discussed, to train your animals to a hot fence, which is the cheapest option. We simply sacrifice a fence to surround our trees with a net; we don't bother electrifying it, and the sheep leave the trees and shrubs alone.

While it's possible and might make sense to fence out individual trees, the most economical way to protect young trees from animals is in many cases to fence out rows or clusters with the portable, temporary fencing now widely available. Ultimately your goal is for trees and pasture to mingle together in a paddock. If you stagger your planned tree establishment, you can reuse a few sets of materials, moving them from one location to the next after the first area is established.

This approach is also beneficial because surrounding individual trees with fencing or tubes is costly and a poor investment when the future of one individual is uncertain. The exception is if you're planting grafted trees and selected cultivars—trees into which you've invested more money because they have valued qualities and a higher likelihood of success. Spacing for such trees will also be different, because you'll be assuming that more will survive.

If deer browse is heavy on your site, single-strand fencing can be an affordable deterrent, whether you electrify it or not. Another strategy is to install a 3-D fence on the perimeter of the paddock you plant trees in, which can keep deer out and grazing animals in (see figure 3.37 on page 109 for an example).

Tree tubes. One of the more hotly debated strategies out there is the use of tree tubes. These are relatively straightforward to install and usually quite effective,

Figure 5.33. We have found that simply wrapping a net fence around trees (or, in this case, elderberry shrubs) keeps the sheep from eating them—no electricity needed!

at least for protection from deer browse and some rodents. Tubes can sometimes be problematic, however, as they're desirable habitats for mice and voles, bird nests, and even wasp and hornet nests. Some place a light plastic mesh cap over the top to prevent these intrusions, but then it's important to remember to remove the cap once the tree emerges from the tube.

Almost all sources agree that vented and translucent tubes perform best, which is indeed what we've found on the farm. These can actually help accelerate the growth of the tree, as they transmit UV light and create a slightly warmer microclimate, which is a boon for trees in colder climates especially. It's also nice to purchase tubes that have a slit and so can be removed without having to pull the tube up and over the tree once it emerges.

Tubes can be purchased as short as 30 inches and as long as 72 inches. We have saved some money by purchasing the 72-inch tubes and cutting them in half.

A different approach involves using wraps made of plastic or mesh that protect the trunks of older

Figure 5.34. Trees planted at a Virginia Tech Research Farm were given a loosely tilled soil on contour, which harvests water runoff. Each has a tree tube to protect the young seedling until it can handle some minor deer browse.

trees without creating a cavity for critters to nest in. Also useful are bud caps, which are placed over the terminal bud of trees in the winter—often the time when deer pressure is highest, since there is less other browse in the landscape. Bud caps can be purchased commercially, or simply made from notecards or cardboard and slipped over the tree until springtime.

Reducing Weed Pressure

Next on the list is protecting trees against competition from other grasses and forbs in the pasture. Note again that the faster-growing pioneer species are better adapted to outcompete other plants. Many government publications speak of the absolute necessity of all-out war on surrounding species, with many promoting the use of herbicides to kill the grasses around the base of trees. While this might be warranted in some scenarios, we've never needed more than a bit of cover crop, mulch, mowing, and occasionally a load of wood chips to keep our trees content.

Our strategy at the farm has been to do our best and employ resources when they're available and easy, aiming to cut or mulch the surrounding vegetation

Figure 5.35. This planting benefited from the duck rotation: We cleaned out the duck house and mulched the trees with the phosphorus-rich straw used to line their shelter.

two to three times a season, at best. The materials and techniques we've used include:

- Mulching with poopy duck straw when we clean out the duck house three to four times a year. We move the material to the closest row of trees, which benefit from the mulch and get a bump of phosphorus from the duck poo.
- Filling the pickup with a load of wood chips (free) from the town municipal pile on the way home from work. We can drive right down the row and mulch in 20 or 30 minutes.
- Actually letting the weeds grow tall, then using a scythe (and sometimes a weed whacker) to clean and mow the tall vegetation between the trees. The scythe is better than the weed whacker, as it's more pleasant to use and because it cuts the long grasses and forbs well. We can then effectively mulch with this material, slowing down the next phase of regrowth.
- While the sheep are kept out of the trees with fencing, when we move the ducks/geese in they happily intermingle with the new trees, weeding some grasses and trampling the others.
- When inevitably we don't mow a paddock at the optimal time, we end up with clumps and clusters of tall grasses once we do get to it, which impedes the regrowth of pasture. Taking some time to rake this material and add it to rows of trees nearby feels like a double win, since we both mulch the trees and enable better pasture regrowth.
- When we establish trees in areas where we perform soil disturbance, we follow with a cover crop such as annual rye, tillage radish, field peas, clover, or buckwheat to establish more tree-friendly plants before the weeds come in.

We are also considering trying to establish ground cover and herbaceous plants between trees, which would exclude weeds while also eventually providing a more diverse fodder for the animals to eat. The more we venture down the road of planting communities, not just individual trees, the more excited we get by the range of possibilities.

It's key to treat these maintenance activities as a pleasure, not a chore. Part of this involves setting reasonable expectations. Manage patterns and dynamics, rather than worrying about what plants you do and don't want. Stack your efforts and see a maintenance task as a "harvest." Take time to observe and design your system so that maintaining trees works with the other elements in your landscape, and with your work preferences. And above all, don't feel guilty. Do your best.

Watering and Dealing with Drought

The fourth major consideration for trees is adequate water to feed the roots, and ultimately the shoots. Fortunately, the cool temperate climate is defined by adequate rainfall in most seasons to support tree growth, and so supplementing isn't really necessary. That is, in *many* circumstances. With water, it's good to have a plan,

Figure 5.36. Whenever feasible, give young trees ample water; this is a necessity if the site isn't getting regular natural rainfall. A soaker hose conserves water, putting it where it is needed, at the roots of the tree.

then a backup plan if the first plan fails. Adequate water is most critical in the first year of establishment, and more explicitly in the first four to eight weeks after planting.

Water can, of course, be given manually from a hose under pressure, if a supply is close by. For trees planted farther afield, using a tank on the back of a trailer with gravity feed or a small pump is doable but very time consuming. Employing a sprinkler is okay, but tends to waste water. If this is absolutely necessary, it's best to water in the early morning or later in the day, when less water is lost to evaporation.

A better option is to set up drip tape or soaker hose when you plant the trees, which uses only a fraction of the water, and puts it right at the base and root zone of the tree, where it is most needed (soil moisture versus watering). In the event of a severe drought, consider waiting to plant trees at all until a better season presents itself. During the severe drought of 2016, we heeled our trees in a good bed of soil in the woods, opting instead to plant them out the following season.

All this said, we've established most of the trees on our farm without supplemental watering by planting mostly in the spring (April/May) and sometimes in the fall (October/November). In several cases soaker hose was installed, but barely used.

Review

In this chapter we assessed our existing pasture patches, identified our goals and appropriate tree species to meet those goals and site conditions, and then planned for stock selection, patterned planting, and successful establishment of trees. All these steps are necessary for success, but often they get sidelined by the farmer's emotions. The familiar story goes like this: It's late winter, the tree catalog arrives in the mail, and the order goes in. In spring a bunch of trees show up, which are placed in the field a little later and perhaps without as much preparation as the farmer had hoped. Some survive, many others die, and the result is a few trees for considerable time, money, and stress.

While we'd love to think that all farmers will plan their hearts out, carefully documenting, measuring, keeping spreadsheets, and ordering trees well in advance of planting, it's simply unrealistic. Hopefully, however, laying out the process in this chapter has at least illuminated the important considerations for planning and encouraged you to make a few scratches on the page and a few lists made well in advance of the coming season.

From my experience as a farmer, I know that each time I don't plan things out, I regret it, while every time I take that extra time for planning I save time, money, and frustration. The best we can do is get a little better each season. We may always be imperfect, but that means we can always improve.

Equipped with a full sense of how the animals, trees, and forages all fit together in both existing woody ecosystems and open pasture, we are ready to bring all these elements together into a more coherent silvopasture grazing plan in our final chapter.

KEYLINE DESIGN AND PLOWING FOR TREE ESTABLISHMENT IN PASTURES
By Connor Stedman

Keyline design is a water and soil management strategy developed by P. A. Yeomans in Australia in the 1940s and '50s. It has been applied to the creation of silvopastures as a site improvement and layout technique for new tree plantings. The keyline system has been applied around the world and further developed by the work of Darren Doherty of Regrarians, Ltd. (www.regrarians.org), among many other practitioners. While a full exploration of keyline design processes and considerations is beyond the scope of this book, we are including this brief overview to highlight its potential applications and encourage silvopasturists to learn more and experiment.

The full suite of keyline design strategies includes:

- Small agricultural ponds placed at inflection points (keypoints) in a site's topography.
- Irrigation ditches moving pond-collected water to and across larger land areas.
- Specialized off-contour subsoiling (keyline plowing).
- Integration of soil-building forages, tree crops, and rotational grazing patterns.

Here we focus on keyline plowing (subsoiling) and its use as a site improvement technique for establishing silvopastures. The major benefit of keyline plowing on pastures is an increase in water infiltration and soil water retention capacity. This supports both pastures and silvopastures through greater water availability to forages and trees, and therefore greater forage production and faster tree establishment.

In heavier soils keyline plowing can also move significant quantities of water horizontally across the landscape, and either concentrate it or disperse (spread) it depending on the layout utilized. Keyline design and plowing is often best suited for landscapes with one or more of the following characteristics:

- Gently to moderately sloping.
- Soils containing at least some silt or clay component (loams, silts, clays).

- Current vegetation is pasture, coarse herbaceous, and/or low, shallow-rooted woody.
- Compacted soils.

Keyline subsoiling is not generally well suited for significantly rocky, hydric (wet), steeply sloped, or forested sites, although keyline layout and design approaches can still be utilized on many of these sites using earthworks rather than subsoiling. Soil moisture conditions are also significant for the timing of keyline plowing. Subsoiling should generally be avoided in soils that are either in exceedingly dry conditions or that are much wetter than a wrung-out sponge.

Keyline design and plowing can be utilized on many sites as a site preparation strategy for the conversion of pasture to silvopasture. Typical steps include:

1. Survey representative contour lines for each distinct topographic land unit of the pasture or field in question. These are often lines running through "keypoints," or inflection points, in the landscape's topography. Consult a keyline design resource or specialist to determine how to select these representative contours appropriately for your site.
2. Convert the surveyed contours into one or more "keylines" and mark these in the landscape. This conversion of contours into keylines is where much of the art and science of keyline design lies. Keylines are typically slightly off contour and generally run off the above-mentioned keypoints, aiming to move surface water flows from areas of concentration (valleys) into areas of dispersal (ridges). Again, we recommend consulting a dedicated keyline design resource or specialist for this step. Many landscapes require multiple keylines to account for significant changes in topography.
3. Plow with a Yeomans (keyline) plow or similar subsoiler in parallel strips along the marked keyline(s). The subsoiler, rather than turning soil as in conventional tillage, cuts a narrow rip through the use of a rotating coulter and a

Figure 5.37. Plowing with a keyline plow specially designed for the task (*left*), and the residual line patterning after plowing (*right*). Photos by AppleSeed Permaculture, at Q Farms in Sharon, Connecticut.

following shank. Each rip increases localized infiltration and encourages root growth while also acting as a "micro-swale" across the landscape, collecting and moving water laterally during heavy rain events or snowmelt. It thus can decrease erosion risk on sloping land while encouraging plant establishment and growth.

4. Lay out tree establishment locations in these subsoiled lines based on your chosen spacing and planting pattern. The subsoil rips provide an ideal, partially prepared soil environment for tree planting. The even spacing between rips allows for easy establishment of curving rows for alley or row silvopasture patterns. Not all ripped rows are planted—rows of trees are laid out at the appropriate row distance for the silvopasture system being utilized.

5. Plant silvopasture trees as laid out!

6. Plow one or two more times in future years in between planted tree rows, at greater soil depth to further increase infiltration and root growth.

Over time, observed effects and benefits of keylining plowing have included:

- Breakup of agriculture hardpans (plow pans, cowpans), allowing for deeper root penetration and increased plant growth.
- Greater water infiltration and storage capacity.
- Redistribution of water across the landscape, turning wet areas drier and dry areas wetter.
- Increased earthworm populations in the soil.
- A pattern (via plowing) for tree layout and planting.

Keyline plowing can also have potentially detrimental effects, such as creating a bumpy pasture texture for the first one to two years after plowing. This effect can be mitigated by the use of an attached crumble roller on the plow, and will also naturally recede with time. Keyline plowing can also temporarily increase frost heaving and rock surfacing in stonier soils.

It should also be noted that in certain pastures and grazing systems, a diversity of moisture levels can be beneficial for early- and late-season forages. Some keyline layouts in heavier soils can moderate these moisture extremes, which may not be beneficial for some grazing plans. Alternative keyline layouts can avoid this moisture redistribution if it is determined to be undesirable. Keyline subsoiling can also be utilized as a drainage strategy on some sites, requiring less intensive landscape modification than tile. We recommend contacting an experienced keyline designer to learn more and discuss how these considerations may relate to your particular site conditions and pastoral or silvopastoral system.

Recommended resources to learn more about keyline design:

- *Water for Every Farm* by P. A. Yeomans.
- *Making Small Farms Work* by Richard Perkins.
- Regrarians Platform, Darren Doherty: www.regrarians.org.
- AppleSeed Permaculture, Connor Stedman: www.appleseedpermaculture.com.
- Keyline Vermont, Mark Krawczyk: www.keylinevermont.com.

Figure 5.38. Ripped lines with cover crop growing, measuring for trees. Photo by AppleSeed Permaculture, at Shellbark Farm in Accord, New York.

Figure 5.39. Trees planted and tubed. Photo by AppleSeed Permaculture, at Shellbark Farm in Accord, New York.

6

Putting It All Together

Having covered the relevant definitions, history, ecology, and considerations for working with both existing forests and integrating silvopasture into pasture, at last we arrive at the point of putting together all the pieces. From the outset, it's important to know that the work in silvopasture is never finished; nor is it going to look the same year after year. Our goal is to set up a process to pull together a plan that is coherent and sets the course for the activities you will undertake.

In this chapter we first summarize the key points of this text, and outline a road map for silvopasture design, using our farm plan as an example. We also cover some calculations and planning tools, including one to help you determine the stocking rate of land for a given number of animals, grazing charts to track your

Figure 6.1. Developing silvopasture takes patience. This photo was taken the first time sheep were given access to a willow planting, four seasons after planting. The initial wait is hard, but this will provide fodder for decades to come.

rotational grazing practice, and a budgeting tool. This chapter, and subsequently the book, ends with some big-picture concepts: silvopasture economics, getting help, marketing the image of silvopasture, land tenure and the next generation.

Key Silvopasture Points

Let's begin by summarizing the key points that have been mentioned throughout this text:

1. Start integrating silvopasture with the most marginal parts of the land.

Skip the most lush pasture, best soils, and mature forest when getting started. Begin silvopasture in those spots that are overgrown, wet, and have poor soils. Reclaim scrublands, hedgerows, and forgotten edges of the landscape. You might be surprised at what you find, and how trees can add value to these landscapes.

2. Work with what you've got; invest in improvements later.

There is much to be said for starting by improving existing pasture, rather than spending money to condition soil or renovate and reseed. A lot of improvement can be made by simply improving your grazing management practices. Almost any pasture can be improved in a few seasons. So rather than rip it all up and start over, focus on grazing management, and see what shows up.

3. Start small, and build slowly. Don't have too much diversity.

Avoid purchasing hundreds of trees of dozens of different species. Start with a small test planting to see what does well on the land. Figure out your methods for planting trees, moving fence, and providing shelter and water. Confirm you actually like the work before scaling up.

4. Take a lot of time selecting the animal and breed, and know you might need to train them.

While you can train many breeds, starting out with one that is best suited for your plans will save considerable time and energy. Ask questions of the people you source animals from, and make sure they are used to browsing a lot of diverse forages. Don't buy animals that have mainly been fed hay or been in confinement. Make sure the animals you buy are familiar with the fencing system you plan on using. Focus on breeds that have a proven track record. Start small with the number of animals.

5. Start with animals well under the stocking rate.

Don't max out your pasture right away. Allow for flexibility by understocking, but not so much that you can't have a functioning rotational grazing system. This means, arguably, at a minimum, four to eight sheep or goats, two to four cows and pigs, and a dozen or so poultry. Rotating too small a herd or flock won't make sense. Know your capacity, but don't stretch to it—grow into it.

6. Honor the animals' body wisdom.

Recognize that animals know what they need, if only given the opportunity. Make time to build a relationship with the animals. Identify the leaders of the group. Provide opportunities for animals to be curious and engaged with the landscape you design.

7. Ruminants are most silvopasture-ready.

Of all the animals, ruminants are the best choice for low-input silvopasture. They are the only livestock that don't need supplemental grain. The majority of silvopasture research has centered on ruminants. For orchards, vineyards, and Christmas tree operations, sheep seem to be the standard.

8. Poultry are low-risk and low-impact.

Chickens, turkeys, ducks, and geese are a good option for smaller spaces, and can be integrated with more confidence. They just won't have the same impact on the ecology as a herd or flock of ruminants. Poultry are well suited for existing farm needs around pests and disease. They will, however, most likely need supplemental grains.

9. EXERCISE EXTREME CAUTION WITH PIGS.

Pigs have a bad history with woodlots and trees. It's not the fault of the pigs, but rather of the humans managing them. Spend a lot of time researching breeds, visiting farms, and thinking long and hard. Know that the norm in pig-rearing is destruction. It's extremely hard to avoid this without moving them frequently, so it's better to have a plan to do so. Start very small. They will need a lot of high protein inputs (grain), and a hot fence.

10. ACCEPT AND PLAN FOR TREE MORTALITY.

In the progression of the forest, many trees won't make it to the canopy. It's the same with the trees you plant. Increase the likelihood of success by testing soil, prepping planting space, and using proper techniques to plant. Don't beat yourself up when trees die. Celebrate when many live. Ask questions about why certain trees do or don't survive. Constantly learn from the experience.

11. VALUE THE SHADE, SHELTER, AND FODDER FUNCTIONS OF TREES.

Among the greatest benefits animals receive from trees are these three. Other yields are a bonus, and it's good to aim for as many as possible. But do the math such that these yields specifically are worth the effort, if that meets your silvopasture goals.

12. KEEP SOME PASTURE AS PASTURE, SOME WOODS AS WOODS.

Diversity means also having some spaces not in silvopasture. Leave some pasture open, and keep animals out of the best woods you have.

13. PLAN ON MOWING, PRUNING, CUTTING, AND MORE FOR SEVERAL YEARS.

Don't throw out the mower, brush cutter, or chain saw just yet. Over time you should be able to do these things less. Or differently. But in the beginning you may have decades of unchecked growth to contend with. Go at it incrementally, and consider how to set yourself up so that your energy inputs decrease over time.

14. REMEMBER THAT WHAT WE KNOW IS LESS THAN WHAT WE DON'T.

Remain humble, curious, and honest. We know enough to get started, but we aren't exactly sure where we are going. Make time for observations, be your own critic, talk with others, and constantly ask questions. Strive to improve the system each season. Be kind to yourself and others in the learning curve. Make sure you are enjoying it.

Designer's Checklist

At this point it's beneficial if you've completed the activities outlined in chapters 4 and 5. This would include stand maps of your forest and open lands, along with observations of their current status and some goals for each, plus some initial concepts for tree patterning you might want to incorporate. If you haven't completed these activities, we recommend that you revisit the appropriate sections of the book and do so before you read on.

As we go through the following checklist, we will describe the process we undertook for Wellspring Forest Farm, which is neither complete nor perfect. We're detailing it not because it's exemplary, but because it is what we know best and are working each and every day to improve.

CHECKLIST SUMMARY

❑ Articulate overall goals for your silvopasture.
❑ Complete stand maps for forest stands and pasture patches.
❑ Decide on animal(s) and breed(s).
❑ Determine paddocks and estimate stocking rate.
❑ Finalize and prioritize tree planting patterns and species selection.
❑ Develop a grazing map and chart and keep records.

GOALS AT WELLSPRING FOREST FARM & SCHOOL

Our farm's goals document is long, and breaks down all of our enterprises. Here are some excerpts specific to our silvopasture:

Life Goal

Our life is a balance of work and recreation, where we build relationships of mutual support, challenge ourselves and learn new things, support capacity-building in our community, recognize our privilege, and share our wealth.

Farm Goal

Our farm is developed in the image of a forest. Our practices are regenerative and we leave a forest in our footsteps as we produce maple syrup, mushrooms, duck eggs, lamb, and elderberry. It serves as a place to teach others about the forest, and ways they can cultivate a working relationship to it, through the apprentice program, workshops, events, and yurt rentals.

Silvopasture Goals

Our silvopasture systems support our efforts to reforest the landscape by providing shade, shelter, and habitat for our animals. The trees are mainly selected to be eventually browsed by the sheep as a food source, but some material will be developed into wood products.

We reclaim the abandoned edges and hedgerows to support the healthiest remaining trees and provide cool summer microclimates.

We maximize the available forage per acre, are able to graze from April 1 to December 15, and have a profitable small-scale sheep enterprise.

Overall, our system develops into a mature forest, with our management shifting from planting to harvesting woody material.

We will know we have succeeded when no paddock on the farm needs the portable shade structure and there is an established forest canopy in the pasture.

ARTICULATE OVERALL GOALS FOR YOUR SILVOPASTURE

While you were asked previously to state goals for a particular stand or patch, it's time now to zoom out to the big picture and ask yourself some tough questions. The more you know what motivates you and others involved to do the work, the better you can design your system to meet those criteria.

Farming and homesteading are hard work. Unlike many people before us (and many people around the world currently), many of us get to choose if and how we interact with land and grow food and other products. That is a privilege, the ability to choose. In many ways it's easier (and cheaper) to buy food, firewood, lumber, and other goods from the store. So why do it?

Some reading this book may not have a choice. Making the land work better might mean the difference in being able to keep or lose it. Available jobs and access to good food are not a given, and are scarce in many places. And the way the system works means that some people have greater access to resources than others.

A goal statement can be written out however you choose. It might help to jot down:

- Your needs.
- Values that are important to you.
- Ways you will meet your needs and values.

You are encouraged to think beyond the farm or homestead, to your life as a whole, since these things are not separate.

Some questions that might help along the process:

- How expansive do you want shade access to be on your pastureland?
- What plants and locations have species you'd like to control?

- Are there areas of the land you'd like to clear for future use? (Say, for a road or annual tillage.)
- What crops/yields do you want to co-produce alongside grazing?
- What additional acreage do you want to develop for grazing?
- What is your interest in incorporating woody fodder into the silvopasture?
- What financial constraints do you have to implementing silvopasture?

COMPLETE STAND MAPS FOR FOREST STANDS AND PASTURE PATCHES

After spending some time with overall goals, bring together your forest stand map, paddock patch map, and stand/patch description table for the entirety of your farmscape. (You may have already done this.) If you want to refer back to the sections for guidance, you can find the forest stand assessment process starting on page 122 and the pasture patch on page 172.

The goal of bringing together the two assessment maps is to develop an overall perspective and recognize the areas that are highest priority for developing silvopasture. Perhaps a scrubby edge and low-quality pasture are next to one another, giving a clear directive to a place worth taking action. Spend some time having a look over the maps.

It's helpful to have a written description of each area as well as a goal or two, and likely in the process of assessment you discovered some patterns and areas that you most want to deal with first. Looking at the map, propose some short, medium, and long-term areas to address. This is an important time to bring in anyone who is a participant or stakeholder in the farm decision-making process, to get their input as well.

Poor Quality   Good Quality   Excellent Quality

Figure 6.2. A completed map combining the forest stands (letters A through I) and pasture patches from previous assessment chapters. This level of detail is adequate to make a number of management decisions over the lifetime of the farm.

DECIDE ON ANIMAL(S) AND BREED(S)

This aspect of the planning process was discussed extensively in chapter 3 as well as in our examinations of the applications of silvopasture to orchards, vineyards, and Christmas tree plantings in chapter 5, where some species are said to be preferred. Before you can determine paddock size and complete grazing materials, this aspect of the decision making has to be complete. While you might read up and decide on a breed based on the literature, often a number of realities come into play. The availability of the breed and breeding stock is critical. Local knowledge, experience, and community connections are also useful.

For example, we settled on Katahdin sheep as one of our finalists, but knew that we needed to make sure there were enough around that others were raising so that we could get access to a new ram every few years. (This is true for all breeding operations, unless you go with artificial insemination.) We found about half a dozen others nearby raising the same breed, so we can swap tips and stories. And we also paid attention to what the animals were used to eating, and of the "culture" of the flock. We ended up finding great stock just down the road, and even share the ram with our neighbor.

DETERMINE CARRYING CAPACITY, PADDOCK SIZE, AND REST PERIOD

This is a new element of the planning process, one deliberately left until now because often it distracts from doing good site assessment. Paddocks are defined boundaries where animals will be given access to graze for a given duration. Ideal rotational grazing practices involve moving animals from one paddock to the next daily, although for many it's probably more realistic to move them every two or three days. After three days, grass regrowth could be compromised.

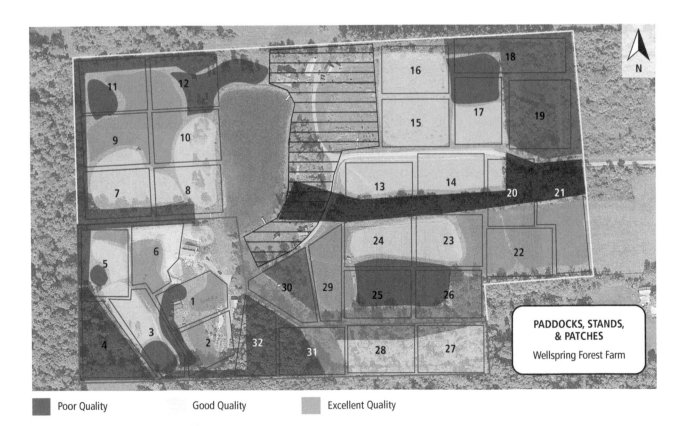

Poor Quality   Good Quality   Excellent Quality

Figure 6.3. Eventually, your actual paddocks will likely include multiple forest stands and pasture patches. We have roughly ½-acre paddocks on our farm, which we calculated using information presented in this chapter.

There are a number of different ways to answer the question, "How many animals can my land support?" It might be one of the best questions to answer with "It depends." The quality of forage, season, weather, animal size, and time of year all come into play. Finding the balance between the number of animals and the available acreage to graze means ensuring no harm is done, production is high, and extra work is low. In other words, too few animals means you will have to mow and maintain weedy vegetation, and too many means you will degrade the landscape unsustainably and have to purchase feed.

For pigs and poultry, the following calculations are not as relevant, since much of their diet will come from grain. As mentioned in chapter 3, claims on the appropriate stocking of pigs/acre range wildly, from as many as 25 per acre (!) to 10 to just 1 per acre. As discussed in the pig section, researchers studying the dehesa system have stated that extremely low densities were sustainable, as low as 0.5 to 1 pig per acre. This is congruent with other pig forage systems in Europe. In any case, the stocking rate should definitely be lower than with ruminant animals, so 1 to 4 pigs/acre would seem to be a logical maximum. Observation and being ready to move them before it's too late will be key.

For poultry, it also depends. You might want to rely most on observation if you are attempting to maximize production. Often mentioned are numbers of around 100 to 200 birds to the acre. Again, it's harder to pin

down an exact figure. Reading through the ways calculations are done for ruminants will provide some good information on the variables to consider.

For ruminants, we can calculate the maximum number of animals by first collecting the following information. As we walk through these factors, we will use Wellspring Forest Farm as an example, grazing about 50 sheep on 23 acres, 15 of which offer substantial forage at ground level.

Length of grazing season in days. This will depend on where you are. For our farm in Central New York, we assume April 1 to November 1, which is 214 days. If you aren't sure where to pin this down, consider looking up the average frost-free days for your location, and add on a month (30 days), since pasture still grows a few weeks before and after the frost season sets in. Also consider that the length of day does affect grass growth. For our latitude (41 degrees) we find that growth is restricted before the spring equinox (March 15) and after the fall equinox (September 15), though we can stockpile and save more pasture in the latter part of the year.

Average weight of animals. Our sheep average 90 pounds based on our records. If you're unsure of weights, use 100 pounds per sheep/goat and use 1,000 pounds per cow.

Daily utilization (4 percent of weight). NRCS/USDA recommends 4 percent for all animals because

Table 6.1. Average yields of Dry Matter (DM)/acre and monthly distribution

		Annual Yield Lb./Acre DM	May	June	July	August	September	October
Cool-Season Perennial Grasses	Good	6,430	24%	31%	15%	10%	15%	5%
Average of six species	Poor	2,430	15%	37%	12%	13%	16%	7%
Warm-Season Grasses	Good	5,166	0%	10%	35%	33%	20%	2%
Average of three species	Poor	2,673	0%	10%	40%	40%	10%	0%
Legumes	Good	5,480	18%	38%	28%	13%	3%	0%
Average of three species	Poor	2,750	18%	38%	28%	13%	3%	0%

Note: The percentages are a rough estimate of the allocation of total yield of DM for a cool temperate farm (USDA/NRCS data).
Source: Data from Undersander, D. J., et al. *Pastures for Profit: A Guide to Rotational Grazing.* Cooperative Extension Publications, University of Wisconsin, 2002. Available at https://www.nrcs.usda.gov/Internet/FSE_DOCUMENTS/stelprdb1097378.pdf.

Table 6.2. Estimated calculations of available forage for a given month at Wellspring Forest Farm

	Lb./Acre/Year	May	June	July	August	September	October
Cool-Season Grasses	2,437	585	755	366	244	366	121
Legumes	650	117	247	182	85	19	0
per acre	3,087	702	1,002	548	329	385	121
× 15 acres	46,305	10,530	15,030	8,220	4,935	5,775	1,815

Note: These numbers were based on assuming a 75% grass/25% legume mix, and further buffered by reducing the yields to 65% of the total.

ruminants range from 2.5 to 3 percent intake, with 0.5 percent trampling loss; 0.5 to 1 percent is added as a buffer.

Average yield of pasture per acre. Table 6.1 shows average annual yields for forages, as well as their seasonal distribution. We will need to utilize both annual and monthly data for our calculations.

Since our pastures are usually a mix, we are going to use 75 percent cool-season grasses and 25 percent legumes, at a less-than-good rating of 5,000 lb./acre/year for grasses and 4,000 lb./acre/year for legumes, based on table 6.2. This equals 3,750 for grasses and 1,000 for legumes, giving us a total of 4,750 lb./acre/year. This would be for a moderately good pasture, with uniform distribution of the forage. I know my choice pastures are about 65 percent of the total, so I am going to knock that off the estimate, leaving us with an average of 3,087 lb./acre/year.

Depending on your situation, you might want to know how many animals you can fit on your acreage, or you might want to know how many acres you need for a given number of animals.[1] In either case the following calculations will help you play with capacity estimates coming from these two main angles.

Calculate Total Carrying Capacity of Land

If you have a set number of acres, then you can determine what they might be able to handle in terms of number of animals. For this example, we are using 15 acres because, though we graze closer to 23, some of these acres are in transition and don't offer sufficient forage at this time.

$$\frac{\text{number of acres} \times \text{available forage per acre}}{.04 \text{ intake} \times \text{average animal weight} \times \text{grazing days}} = \begin{array}{c}\text{total}\\ \text{number}\\ \text{of}\\ \text{animals}\end{array}$$

$$\frac{15 \times 3{,}087}{.04 \times 90 \times 214} = \begin{array}{c}60\\ \text{animals}\end{array}$$

Calculate Total Land Needed for a Given Number of Animals

If you have a set number of animals already or are thinking of a specific group size, you can use this formula to determine how many acres they need:

$$\frac{(\text{number of animals}) \times (\text{average weight}) \times (.04) \times (\text{grazing days})}{\text{average yield per acre/year}} = \begin{array}{c}\text{total}\\ \text{number of}\\ \text{acres}\end{array}$$

$$\frac{50 \times 90 \times 0.04 \times 214}{3{,}087} = \begin{array}{c}12.5 \text{ acres}\\ \text{needed}\end{array}$$

Looks like we have plenty of land, or could have many more animals. We have to remember, however, that forage doesn't grow the same year-round, as table 6.2.

We can determine the amount of feed needed each month by taking our utilizing number (4%) and multiplying it by 30 days (= 1.2). Then the formula is:

Amount of food needed each month =
(1.2) × (number of nimals) × (average weight)

1.2 × 60 animals × 90 lb. = 6,480 per month
1.2 × 50 animals × 90 lb. = 5,400 per month

Looking at table 6.1, we can see that if we had 60 animals, we'd be in a deficit for the months of August – October. Keeping us at 50 means we are pretty on target for most months, knowing that we also have the additional transitional acreage to work with. Once we get more acres into reliable forage production, our stocking can go further up.

Calculate Paddock Size

Finally, with some sense of the number of animals we can handle, we can look at paddock sizing using the following formula:

$$\frac{(\text{average weight}) \times (0.04) \times (\text{number of animals})}{\text{available dry matter/acre}} = \text{paddock size, acres/day}$$

We will use our 50 sheep, averaging 90 lb., and use 500 lb./acre as the available forage, when we hit those slumpy times of summer. The calculation looks like:

$$\frac{90 \times 0.04 \times 50}{500} = \text{0.36 acre needed each day}$$

Since we want to move our animals ideally every day or two, we decided to make them around ½ acre, figuring we would be good. This has been working out, though it becomes clear in the field that this type of standardizing isn't how it plays out in real life. Pasture paddocks are highly variable; some are much better in quality, others much worse. We end up adjusting things a bit, and always move the animals when we observe in the field that it is time. But these calculations can help us chew on the numbers, think about the elements, and ultimately give us some sense of what to aim for.

If I wanted to get very precise, I could actually measure available forage in different parts of pasture (see the sidebar), as well as track these over the course of the

---

RESOURCES TO MEASURE AVAILABLE FORAGE IN PASTURE

Calculating Available Forage, Utah State University Extension: extension.usu.edu /rangelands/ou-files/Calculating_available _forage_NR_RM_03.pdf.

Estimating Pasture Forage Mass from Pasture Height, West Virginia Extension: extension.wvu.edu/files/d/230d97c5 -7ced-40af-92f1-bfdb90fdb83d/estimating -pasture-forage-mass-from-pasture-height -2016.pdf.

Estimating Available Pasture Forage, Iowa State Extension: store.extension.iastate.edu /Product/Estimating-Available-Pasture -Forage-PDF.

---

year. Most of us won't likely do that more than once, so we can take some of these as educated guesses, start out smaller, and see what the land has to tell us.

Once you've determined the best paddock size, you can begin a new layer on your map to determine the size and shape of your paddocks. Aim to keep the boundaries of a paddock within a defined stand, though a single stand might contain several paddocks. Other factors that determine boundaries include areas you want to exclude from grazing, places where trees are planted (if they're on a boundary, they can more easily be excluded while getting established), and inclusion of both pasture and hedgerows. Just because a boundary is drawn on a map does not mean it is fixed. We adjust and change the boundaries all the time, as needed, in the field. Google Earth is a wonderful tool to help you organize paddocks.

FINALIZE AND PRIORITIZE TREE PLANTING PATTERNS AND SPECIES SELECTION

If you need to, refer back to the discussion on page 185 and refine any concept sketches of placements and patterns of trees. This may change as you figure out paddock sizes and lay them out next. It's also wise to prioritize areas for planting, as you will likely establish trees over several years.

DEVELOP A GRAZING MAP AND CHART AND KEEP RECORDS

Once you have your paddocks determined, you can utilize one of the most useful tools for record keeping: a grazing chart. This chart tracks your activity in various paddocks throughout the growing season and provides priceless information on your operation, which will allow you to accurately adjust the numbers for stocking rates and paddock sizes, and also make for a more accurate enterprise budget over time as you collect more real data. The trick is to make collecting the data simple.

We used to print out the grazing chart, which has been offered as a printable document by Troy Bishopp, aka "the Grass Whisperer," for many years, on the wonderful blog *On Pasture*.[2] The chart can be sent down to a print shop and blown up; simply hang it on the wall of your shop or barn, and shade in the days that your animals occupy a given paddock. Use different colors for multiple types or groups of animals, and be sure to also mark when a paddock was mowed, seeded, or otherwise disturbed (see table 6.3). Some go so far as to make weekly rounds and mark on the map the heights of forage for a selection of paddocks, to get a sense of forage regrowth dynamics over the course of the season.

After a few years with the large chart on the wall, we switched to a binder and a sheet of paper, along with a handy copy of the map with paddocks labeled. We simply make a note of the activity that occurred in writing, then transfer the data to the computer once per month (at best), or at the end of the season (at worst), shading in the data for each block of time.

Other data you might want to track:

Figure 6.4. The grazing map is simply a map of the paddocks on a site, used to number paddocks and help with record keeping. If this is done in Google Earth, each "pin" can hold a lot of notes and information, and serve as an easy way to document activities in each paddock.

- The labor associated with moving fence, moving animals, mowing, planting trees, maintaining trees, and other tasks.
- Receipts for purchases, including feed, vet care, fencing, equipment, tree stock, and other material goods.

Keeping good data doesn't come easy for many people (myself included), but every time we collect, the feedback and information are well worth it. It's been critical to make this system work, whether that involves putting up a chalkboard and transferring the items to the computer later, carrying around a small notebook in your pocket, or creating a system for multiple people to access and help in the effort. Data allows you to make decisions based on fact, and not emotion. Try it.

The Big Picture: Economics and Marketing

The question every farming activity seems to come down to is, "Will it pay?"

The prefix *eco* is rooted in the Greek *oikonomia*, which means "home management." And frankly, we farmers aren't always the best at keeping up with the business side of things, or, much worse, we don't see the farm as a business in the first place. Many farmers (myself included, past tense) have gotten deep into enterprises without really knowing if they make any money or not. So a discussion of economics in silvopasture is really two parts. The first is to look at what others have documented, and the second is to discuss the microeconomic system that is specific to the farm enterprise.

There are a number of examples of economic research into silvopasture, though to be fair most research has focused on the biophysical aspects of management, and there is ample room for more financial analysis. There is really little doubt that conceptually having both grazing animals and trees together should yield more dollars per unit of land. What is less certain, however, is: (1) whether the additional management is worth the extra yield, and (2) whether either system (trees or animals) hinders the productivity of the other.

Let's start by looking at some of the research around the economics of silvopasture. We encourage you to study these citations further if they're relevant to your situation.

1. THERE ARE MANY STUDIES LOOKING AT THE COST BENEFITS OF PROVIDING SHADE FOR ANIMALS, WITH DRAMATIC INCREASES IN YIELD.

- The University of Florida found that artificial shade from permanent structures resulted in a 10 to 19 percent increase in milk production, and Texas Tech

Table 6.3. Excerpt from our grazing chart records

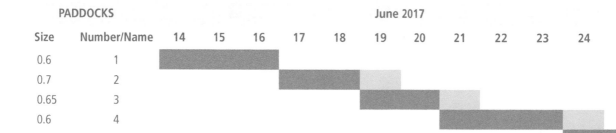

Size	Number/Name	14	15	16	17	18	19	20	21	22	23	24	25
0.6	1												
0.7	2												
0.65	3												
0.6	4												
0.6	5												
0.65	6												
0.7	7												
0.8	8												

PADDOCKS — June 2017

Sheep   Ducks   Mowing

A PLAN FOR SILVOPASTURE
By Jono Neiger

Starting silvopasture operations is an investment in the future and a long-term endeavor. As farmers gain interest in silvopasture, they need support in identifying the best trees and varieties for their fields, defining the spacing, and working out the details of installation and maintenance. Mick and Louise Huppert came to Regenerative Design Group with a desire to change their grazing system on their small acreage in Petersham, Massachusetts. As an older couple they needed a fairly low-effort system and were interested in incorporating tree crops into existing pastures for silvopasture management. We focused on design and planning of the tree crops and silvopasture system, and then installation and care.

The Hupperts own a small farm made up of two properties. Like many in the Northeast, the property has relatively small fields on marginal soils not suitable for tillage agriculture. The home property with the farmhouse and barn is almost 8 acres; a larger 35-acre field is a few miles away. The home 8 acres has several pastures, which were planned for a conversion to silvopasture in 2016. Windbreak and water diversion swales were planned at that time as well, and the windbreak was planted in 2016. Initial planting of the silvopasture trees was in spring 2016 with widely spaced chestnut, honey locust, alder, willow, and mulberry.

An initial assessment of the land identified fairly good soil drainage and fertility, open sun conditions across most of the fields with the exception of several areas near the southern and eastern tree lines, and a septic system somewhat centrally located, which can't have trees planted too close. There is water coming into the property from above, and at the bottom of the slope is a forested wetland.

The fields have plentiful water coming in from an adjacent farm above the land. But the water is

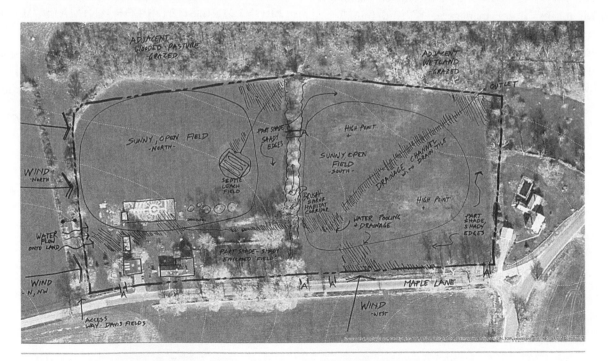

Figure 6.5. Assessment map for the Huppert Homestead. Design planning by Regenerative Design Group.

concentrated through a channel that goes by the house and garage, where it causes minor flooding and makes access difficult. Beginning as high up on the property as possible, the plan is to intercept the water as it enters the land; direct it across slope through an off-contour swale; and then spread it through an on-contour swale across the fields.

We explored various silvopasture options to give a range of options of varying levels of grazing intensity, cost of establishment, and skill and knowledge required to maintain the system. Each option also included a windbreak and fodder crop planting on the north edge of the property. The first option, Simple Silvopasture, consists of widely spaced trees that function to provide: (1) high-value nut crops, and (2) supplemental fodder for the livestock. A permanent perimeter fence outlines the two fields, and movable interior fences define the paddocks.

The second option, Habitat and Alley Crops, is more oriented to building up the ecological and habitat capital of the fields with rows of perennial crops such as blueberry, hazelnut, and elderberry and alleys maintained with small livestock such as ducks and geese. Early successional woodland edges and pollinator support plantings surround the alley crops. This option doesn't meet the goal of utilizing livestock in a silvopasture system, but highlights the ability to produce suitable crops while still integrating some (smaller) livestock grazing opportunities.

The third option, Silvopasture Trials (pictured), recognizes that we don't know enough about which systems are best in this specific area of the Northeast; nor do we know the costs and benefits of establishment and maintenance. We outlined a range of systems from a simple open savanna type, to alley crop silvopasture, a pollarded fodder crop and coppiced hedgerow areas, a mixed-tree-crop silvopasture, and an intensive silvopasture with intensively grazed fodder shrubs. This experimental mix would give more information and add to our understanding of temperate zone silvopasture.

The first phase of the project, beginning in 2016 at the nearest pasture to the house, started with the simple addition of high-value nut trees and support trees

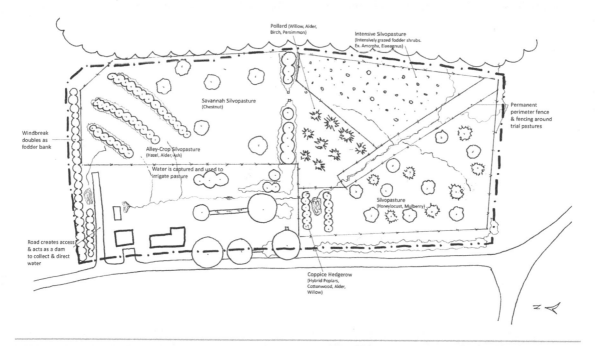

Figure 6.6. Design of several trial silvopasture systems for the Huppert Homestead. Design planning by Regenerative Design Group.

at wide spacing. It is a blending of the high-value nut grove silvopasture and the mixed-species silvopasture. Hybrid chestnuts ('Colossal', 'Dunstan', 'Tanzawa', 'Tsukuba') with Chinese and American genetics were planted at 70-by-70-foot spacing—enough to give plenty of light once the trees reach their full height and width of 40 feet.

Trees are protected from browsing and rub damage from the livestock with 5-foot-high rigid welded wire fencing, supported by three heavy-duty T-posts. Siberian peashrub (*Caragana arborescens*) were planted by each chestnut to give additional support through their nitrogen fixation. Other fertility support plants were added such as Italian alder (*Alnus cordata*) and honey locust (*Gleditsia triacanthos*). Additional fodder and crop trees are included in the planting, such as mulberry (*Morus alba tatarica*) and willow (*Salix* spp.).

A simple fence system consisting of a perimeter fence and internal, movable paddock fences creates the ability to manage the livestock. An alley for moving the livestock down the field into each paddock was established as well. The approximately 2-acre initial pasture has been grazed in four separate paddocks. Initial grazing within the pasture has been with a small group of two to three Devon cows. They have been easy to work with around the tree cages, and the plantings have added a minor difference in the grazing management.

In loving memory of Michael "Mick" Huppert.

Figure 6.7. One of the strategies employed to protect young trees from cattle. Photo by Jono Neiger.

University saw an increase of 17 pounds per animal for shaded heifers versus those unshaded in a feedlot situation. Note that these are in confinement scenarios, where animal stress is higher than on pasture.[3]

- Research at the University of Kentucky Animal Research Center found that shade in pasture results in weight gain increases during late spring and early summer of 1.25 pounds per day for cows, 0.41 pound for calves, and 0.89 pound for steers.[4]

- In a silvopasture analysis done by the University of Missouri, a farmer converted just under 3 acres of pasture edge to silvopasture and saw from June 15 to August 15 an average daily gain of 1.6 to 2.1 pounds per head per day. Normally, this farm saw gains of 0 to 1 pound per head per day. Projected over the season, this would result in a $130 to $170 increase in gross income per animal.

2. OTHER ANALYSES HAVE COMPARED THE ECONOMICS OF SILVOPASTURE MANAGEMENT TO OTHER FORMS OF LAND USE.

- Joe Orefice conducted a financial analysis of a thinning project at North Branch Farm in the Adirondack region of New York, comparing returns from open pasture, silvopasture, woodlot, and no management (eventual timber harvest). He found that the silvopasture and no management scenarios were the highest, with the silvopasture treatment surpassing the pasture because of the future value of a timber harvest at 30 years.[5]

- In Mississippi a study of silvopasture treatments demonstrated the ability of cattle rearing to increase cash flow in the short term, finding a 100 percent increase in cash flow on a per-hectare basis for silvopasture versus woodlots and traditional pine plantations.[6]

- In Argentina, it was estimated that silvopasture yields a higher annual income than alternative agriculture, cattle ranching, or forestry management.[7]
- In Florida silvopasture was more profitable than either traditional forestry or pasture management, and was found to be a good strategy to support restoration of native longleaf pine forests, at least in some parts of the state.[8]
- A research project in Louisiana found that silvopasture generated a greater internal rate of return than either managed timber or open pasture.[9]
- Research models from the Coastal Plain found that loblolly-pine-forage-cattle systems may have up to 70 percent greater net present value than a stand-alone forestry operation.[10]

3. LIVESTOCK IS OFTEN THE SHORT-TERM DRIVER, AND RESEARCH HAS BEEN DONE ON THE ECONOMICS OF LIVESTOCK PRODUCTION IN ROTATIONAL GRAZING SYSTEMS.

- An analysis from the University of Kentucky showed a $7,000 increase in returns from a rotational versus a continuous grazing system.[11]
- In Wisconsin graziers grossed on average about $200 more per cow than did confinement operations in data collected from 1995 to 1999, according to the University of Wisconsin Center for Dairy Profitability.[12]
- Research on a Pennsylvania dairy farm found a higher gross margin of $121 per cow on the grazing farm than on continuous grazing or grain feed.[13]

It does seem evident from the above research that silvopasture has promise in many locations when compared economically with other forms of land use, which is good news. This means that, while timber offers a longer-term income, the focus of silvopasture for most people will be on profitable livestock production and perhaps some niche wood products. Others might focus on ways to incorporate animals to help offset costs in maintaining orchards or vineyards and

produce more yields from the same acreage. In all cases, the benefits of shade and shelter, along with grazing, appear to potentially improve the bottom line.

Each of the preceding scenarios analyzed is in a very specific context, making the case for a particular set of species, in a particular location, with specific market conditions. As these variables change, so do the economics. Also important is that these studies, and the paradigm in general, focus on trees grown for timber and animals sold on the commodity wholesale market. Both of these industries have thin profit margins, at best.

No comprehensive analysis has been done for other specialty wood products (such as black locust posts) or animals sold in direct markets, or even local or regional wholesale markets, which tend to fetch much higher prices. And while more direct sales mean more labor and marketing on the part of the farmer, the profit margins can be better. More research and economic modeling would be beneficial for specific silvopasture systems, but ultimately good business planning on the part of the individual farmer is the most likely way to achieve a profitable enterprise.

On the farm scale, economics has to do with choices you and the others managing the operation make. One joke we like to make at the Cornell Small Farms Program is around a disease that infects many farmers known as "shiny equipment disorder." This is when farmers use getting into a new enterprise as justification for buying that new tool or piece of equipment. These large purchases can literally make or break a given enterprise. Trying to minimize expenses is critical to realizing a profitable farm.

Since shade and shelter increase gains in animals' weight or milk production, this alone might justify tree planting. This would be a valuable study for a researcher to perform. Better still if those trees also have potential secondary markets that provide additional income. This requires some creative thinking and specialization, but could take many forms, such as:

Black locust posts selected for fence and hop posts. Fence posts are generally 8 to 12 feet long, depending on how deep you want to bury them. Hop posts

Figure 6.8. One example of an auxiliary enterprise that could fit within a silvopasture system is making charcoal—either briquettes for grilling or high-quality artist's charcoal—from willow.

are long, at 20 to 22 feet, and need to be relatively straight. Prices often start at $1.00 to $2.00 per linear foot, at least in New York.

Larch for fence and hop posts, or milled for rot-resistant lumber.

Biochar and charcoal. Making these materials takes some skill, but the returns can be worth it. Artist's charcoal made from willow can fetch $1 per piece, and hardwood charcoal can be sold at a premium.

Nursery stock, where potted trees sell for $10 to $20 apiece.

Craft materials, woven fences, living willow structures, and so on.

Make no mistake, none of these is a turnkey market, where a grower can readily sell the materials, though locust and larch posts are currently in such high demand (at least in the Northeast) that good sales are almost a guarantee.

In addition to the of sale tree products, animals can of course be a source of income as well. Good breeding stock are selected for their competence in browsing brush or working with trees. If you are willing to travel, solar farms are seeking animals for clearing brush; you could start a business providing sheep or goats for hire.

All this isn't to say that the livestock alone can't be profitable. In fact, with the large growth in demand for grassfed meats, a small enterprise selling to niche markets can do quite well. That is, if the enterprise is managed like a business.

The Microeconomy of Small Farms

In addition to the sales of livestock, their products, timber, and other wood materials, farmers often don't pursue all the advantages that are available to them to save costs associated with land and even the efforts of reforestation and erosion control. Getting to know these available programs may very well mean the difference between being able to implement silvopasture versus feeling it is out of your reach. All operations that are in the business of selling farm products *at any scale* should explore the following:

Create a business structure. Many farms sign up as a DBA (which stands for "doing business as") with their county clerk's office, which gives them title to the name and allows them to open a bank account and track expenses separate from their personal account. When selling products, an LLC (limited liability corporation) is a wise structure to pursue.

File a Schedule F. Once you begin pursuing sales of farm products at any volume, begin filing the "Profit or Loss from Farming" form with your taxes, which is Schedule F of Form 1040. This allows you to claim expenses from farming activities against any income. Often, especially in the early years, claiming a loss is likely, which can help offset wage income from another job. Note that this is not available to homesteads or hobby farms, but for operations intending to (eventually) make a profit. Consult with an accountant if you are unsure. Also refer to Publication 225 for complete information on tax benefits.[14]

Save on sales tax. Once you begin filing a Schedule F, you are able to also avoid paying sales tax on goods purchased for farming activities in many states. In New York, for example, the simple process involves filing a form (ST-125) with places where you purchase goods; taxes are then automatically taken off at the point of sale. If outside of New York, check with your state agency to determine the process.

Apply for ag assessment. Depending on where you live, your state sets a threshold for the amount of sales you need to make to potentially qualify for reductions in your property taxes when the land is in agricultural production. Check with your extension office or Farm Service Agency (FSA) to see what is available where you live.

Explore IRS tax deductions. We highly recommend you consult the document from the University of Missouri about additional tax deductions related to agroforestry and forestry that may be relevant to your situation.[15] Deductions can be made for materials and expenses related to reforestation, provided the trees are for timber production. Farmers can also get deductions for efforts to reduce soil erosion and plant windbreaks. The pdf of this document is available at www.silvopasturebook.com as well as from the University of Missouri.

Check out cost-share programs. The Natural Resources Conservation Service (NRCS) has a number of programs that can potentially pay for part or all of some improvements for silvopasture. See the "Getting Help" section on page 255 for more information.

Each of these programs has its pros and cons, and it will take some time to wade through the paperwork. Consider seeking support from the resources mentioned to help you navigate this. The benefits of doing it right can be tremendous, and significantly contribute to the health of your farm economy. Ultimately, though, the most important thing you can do to support a viable farm enterprise is to understand and work with budgets and cash flow sheets, to which we now turn.

BUDGETS

Initially, a budget will be a projection of your estimated income and expenses; it will become more accurate over time as you (hopefully) collect real data to support it. We've developed a simple Excel budget for silvopasture, available at www.silvopasturebook.com.

Note that this is a sample budget, and by no means a comprehensive financial analysis. It's good to start small and call on others for help if you need to dive deeper into the finances.

A budget includes several aspects of the business:

Fixed costs, the "known" expenses that occur regardless of the productivity of your enterprise. Examples could include facilities and infrastructure, supplies, insurance, taxes, utilities, and depreciation.

Variable costs, those that are in relation to production and all the unknowns that might arise. Feed, labor, vet bills, and slaughter and sales costs often fit into this category.

Revenues, including the sales of primary and secondary products from the enterprise.

With our silvopasture, we are building a livestock enterprise, with secondary potential yields from timber, firewood, and other products. Given the confounding number of variables, it's challenging to budget for wood products, especially timber.[16] With animals, it's much easier to calculate so many pounds of meat or gallons of milk, and project from there. Wood products are much more volatile, and the price is determined heavily by the current market forces.

You can see from the sample budget for our sheep enterprise in table 6.4 that we focus on the tree aspect of silvopasture as a cost of raising sheep. In other words, we are seeing planting trees and clearing silvopasture as the cost of raising healthy, productive sheep, since the main functions we are aiming for are shade, shelter, and fodder. The budget includes some projections for possible secondary products, but none is a focus of our personal operation.

Budgeting was incredibly helpful as we looked at how we wanted and needed to scale our enterprise. We originally planned on retaining about 15 ewes each year, which would result in an average of 22 lambs a year raised and sold to market. When we crunched the numbers, this scale was essentially break-even for our time and labor. Deciding to keep 20 ewes for breeding each season meant we could turn a small profit on the system and meet our overall farm goals.

Table 6.4. Sample budget for a 20 ewe/30 lamb silvopasture sheep operation

FLOCK COMPOSITION

Number of Ewes	20
Number of Lambs	15
Number of Ram Lambs	15
Total	50

INCOME CALCULATION	No. Head	Lbs./Hd.	Net Price	Unit	Total	Per Head
Market Lambs	25	50	$6.75	lb.	$8,438	$337.50
Cull Ewes	5	65	$6.00	lb.	$1,950	$390.00
Lamb Pelt	18		$80.00	pelt	$1,440	$80.00
TOTAL INCOME					**$11,828**	

OPERATING COSTS				Unit	Cost	Total	

FEED COSTS	No. Hd.	Lbs./Hd.	Days	Per Lb.			Per Head
Hay	20	5	150	$0.12	$1,800		$90.00
Minerals	50				$504		$10.08
Grain Supplement	50				$240		$4.80
Slaughterhouse Fees	30			$65.00	$1,950		
Hide Tanning	20			$30.00	$600		

LABOR	Weeks	Hrs./Week	Hrs./Year	Wage		
Moving Fence	28	5	140	$15	$2,100	
Winter Feeding	24	4	96	$15	$1,440	
Health Support			12	$15	$180	
Maintenance			20	$15	$300	
Tree Planting/Maintenance			40	$15	$600	
Marketing/Sales			30	$15	$450	

HEALTH

Vet Bills	$250
Vaccines	$80
De-Worming	$0
Hoof Trimming	$120
Shearing	$0
Materials & Supplies	$600
New Trees	$300

TOTAL OPERATING COSTS	**$11,514**	**$230.28**
RETURN TO FARMERS FOR LABOR	$5,070	
RETURN TO FARM	**$314**	
COST PER POUND LIVE WEIGHT		**$6.80**

Note: For an annotated version of this budget visit www.SilvopastureBook.com.

A WORD ON PRICING PRODUCTS

Many beginning farmers set prices based on: (1) what others are selling the product for, and (2) what they think customers are willing to pay. While both of these drivers are important to consider, using them as a starting point for pricing can lead to a loss down the road. Prices are not fixed; they all depend on what someone is willing to pay. This is highly variable, given the geography and current market of the region you are seeking to sell in.

Rather than allowing the market to control you, it's far better to start by figuring out what price point sustains your operation: in other words, what you need to sell your products for in order to make it work. If that price is much higher than the current market average, you can then look for ways to reduce costs, or you can seek a marketing strategy to justify the higher price.

We looked at auction pricing first for our sheep, which averages around $2.50 to $3.00 per pound live weight. Our budget (see table 6.4) was handy for telling us that we needed a higher price per live weight to be profitable. So we looked at retail markets, including selling directly to customers through the freezer trade (half and whole lambs), and restaurants. We had already begun developing these outlets with our sales of maple syrup and mushrooms, so bringing another product to the table was much easier.

Direct marketing and freezer trade take a lot of education. Consumers don't know what a "half" or "whole" is, much less if it will fit in their freezer. Pictures that showed the sizes, along with information on how to prepare individual cuts, has turned a scary proposition into one that our customers have become excited about.

was improving our technique for moving fence, by far the most intensive part of the operation. Eventually we were able to cut the time we spent doing this chore in half.

It took about four years to get to this scale. We started with just four pregnant ewes and have not paid for a sheep since, save a rental fee for the ram, who visits each season. This means that we don't have the cost of investing in new animals, which other operations will have. Additionally, since we've timed the operation in accordance with the season, the influx of lambs in March and April and culling in October and November mean that the majority of the flock is fed on pasture resources, which we can always improve, yet there is no per-day feed cost to account for. We just need to purchase hay to tide our ewes over for the winter, until the cycle begins the next year.

Constructing this budget was a very different experience from when we did the same for meat ducks and chickens. Since grain feed is such a high-cost input, the question was always how we could reduce the amount of feed while still getting birds to marketable weight in the fastest time possible. This, on top of labor efficiency as a concern, means that the margins for poultry and pigs are probably thinner, but doable. Budgeting and strategy would be even more critical in these enterprises.

Many farmers avoid budgeting because they fear it will tell them what they don't want to hear: that their enterprise isn't financially feasible. While this was true for many of the first iterations we developed, what changed was our relief and security in using these tools to follow a plan toward profitability. While at this scale our operation is not enough to pay one person full-time, it meets our goal of being a profitable enterprise within our diversified income streams (maple syrup, mushrooms, duck eggs, and so on).

CASH FLOW

While budgeting is a useful exercise, it only projects or summarizes the patterns of a given season. The real challenge in farming is often in the seasonal fluctuations that mean a lack of cash on hand during certain times of the year, while you're flush at other times. Often these flows of scarcity and abundance don't match up well with the fluctuations in costs. Mapping cash flow

It's important to note that our budget accounts for paying for labor, whether it be our own, or paid out to a laborer. We estimated this initially, then updated the guesses with actual numbers from our notes. Labor accounts for well over half the cost, so it's important to assess ways to save time. One area we focused on heavily

Table 6.5. Sample cash flow for a silvopasture sheep operation

CASH FLOW	Jan	Feb	March	April	May	June
INCOME						
Market Lambs						
Cull Ewes						
Lamb Pelt	$400.00	$440.00				
TOTAL INCOME	**$400.00**	**$440.00**	**$0.00**	**$0.00**	**$0.00**	**$0.00**
EXPENSES						
FEED COSTS						
Hay						
Minerals	$42.00	$42.00	$42.00	$42.00	$42.00	$42.00
Grain Supplement	$20.00	$20.00	$20.00	$20.00	$20.00	$20.00
Slaughterhouse Fees						
Hide Tanning						
LABOR						
Moving Fence				$300.00	$300.00	$300.00
Winter Feeding	$240.00	$240.00	$240.00	$120.00		
Health Support	$15.00	$15.00	$15.00	$15.00	$15.00	$15.00
Maintenance	$30.00	$30.00	$30.00	$30.00	$30.00	$30.00
Tree Planting/Maintenance			$200.00	$200.00		
HEALTH						
Vet Bills			$100.00	$150.00		
Vaccines				$80.00		
De-Worming						
Hoof Trimming				$60.00		
Shearing						
Materials & Supplies	$50.00	$50.00	$50.00	$50.00	$50.00	$50.00
New Trees			$300.00			
TOTAL EXPENSES	**$347.00**	**$347.00**	**$647.00**	**$957.00**	**$407.00**	**$407.00**
TOTAL INCOME	$400.00	$440.00	$0.00	$0.00	$0.00	$0.00
TOTAL EXPENSES	$347.00	$347.00	$647.00	$957.00	$407.00	$407.00
DIFFERENCE	**$53.00**	**$93.00**	**-$647.00**	**-$956.00**	**-$407.00**	**-$407.00**

July	Aug	Sept	Oct	Nov	Dec	TOTAL
			$2,812.00	$2,812.00	$2,814.00	$8,438.00
			$1,950.00			$1,950.00
					$600.00	$1,440.00
$0.00	**$0.00**	**$0.00**	**$4,762.00**	**$2,812.00**	**$3,414.00**	**$11,828.00**
			$900.00	$900.00		$1,800.00
$42.00	$42.00	$42.00	$42.00	$42.00	$42.00	$504.00
$20.00	$20.00	$20.00	$20.00	$20.00	$20.00	$240.00
			$1,950.00			$1,950.00
			$600.00			$600.00
$300.00	$300.00	$300.00	$300.00			$2,100.00
			$120.00	$240.00	$240.00	$1,440.00
$15.00	$15.00	$15.00	$15.00	$15.00	$15.00	$180.00
$30.00	$30.00	$30.00	$10.00	$10.00	$10.00	$300.00
				$200.00		$600.00
						$250.00
						$80.00
						$0.00
			$60.00			$120.00
						$0.00
$50.00	$50.00	$50.00	$50.00	$50.00	$50.00	$600.00
						$300.00
$407.00	**$407.00**	**$407.00**	**$4,107.00**	**$1,577.00**	**$477.00**	**$11,514.00**
$0.00	$0.00	$0.00	$4,762.00	$2,812.50	$3,414.50	$11,828.00
$407.00	$407.00	$407.00	$4,107.00	$1,577.00	$477.00	$11,514.00
-$407.00	**-$407.00**	**-$407.00**	**$655.00**	**$1,235.50**	**$2,937.50**	**$314.00**

over the course of the seasons, along with seeking ways to better balance it, is a critical aspect of any solid farming enterprise.

A cash flow sheet, then, stretches the budget lines of a year into monthly columns to help complete the analysis and determine any gaps you need to plan for. Livestock animals are often harvested prior to winter, since costs go up in the winter with purchased or stored feeds (hay and/or grain). This can lead to a large expense in slaughterhouse fees, which might not be paid back if the product is then sold over the next several months. On the other hand, animals sold by the half or whole through a custom butchering situation are often reserved well in advance, which means you recoup the slaughter fees almost immediately through the deposit and sale of the entire animal at slaughter time. If done in the fall or early winter, this revenue can go directly into feed costs for the remaining animals in the herd or flock.

There are many ways to arrange your cash flow. It may be useful to look at the whole farm if you have a diversity of products. For instance, at our farm we balance cash flow curves by producing maple syrup, which provides springtime revenue to balance a lack of income from the sheep. Often revenue from the sheep enterprise helps pay for maple sugaring supplies in late winter, when we haven't yet made any new maple syrup. Mushrooms offer stable income during the summer months, though dried product is often something we sell in the fall and winter each year. All told, a farm should strive to even out gaps and develop a positive cash flow, where there is always some money in the bank for both expected and unexpected expenses.

The spreadsheet (see a sample in table 6.5) is also available for digital download at silvopasturebook.com and shows another example of how annual budgeting feeds a cash flow analysis for the farm. As with the

Figure 6.9. Sharing good photos of your grazing system is a helpful marketing strategy, since the visual appeal of silvopasture is very high. You can also promote the climate and forest preservation benefits supported through purchasing silvopasture products. Photo by Brett Chedzoy.

budget, this starts out as a projection, with real-time data helping to fill out the real picture over time. Working through these documents is never a useless exercise; it always provides insight and understanding.

MARKETING THE
IMAGE OF SILVOPASTURE

For most small-scale producers, effective marketing is going to be a critical aspect of a farm enterprise. It's often not the reason anyone gets into farming, but it's a necessary aspect of the work, at least if you want to *keep* farming. Challenge yourself to seek out the most ideal market, price, and niche for your product. Understand that you are both in community, and sometimes in competition, with other farmers. And most important, realize that the story of your farm is absolutely critical. The farm narrative is what gets customers to support you and your products, and is easily the most important aspect of the sales end of the business. Take lots of photos, write articles, and share the story of silvopasture with your customers.

As mentioned previously, many economic analyses of silvopasture crunch their numbers in the context of commodity prices. Especially if you are sold on practicing silvopasture, pursuing a niche market will be key. Research has found that many customers are willing to pay more for meats that are lower in fat and better for the environment; the chances of customers paying more also increase when they are first educated to understand the differences among meat options.[17] Food safety, health benefits, and environmental concerns are all cited as reasons that consumers are willing to pay more, though taste, tenderness, and quality are incredibly important,[18] often overshadowing the impacts of how and where meat is raised, though those rank as very important, too.[19] All in all, market a multitude of attributes for best success.[20]

With tremendous growth and interest in organic and naturally grown foods over the past decade, there is likely room to expand on consumer awareness, marketing silvopasture products in an entirely different way. They offer many of the same attributes as grassfed animals, and we can further promote the narrative of "forest-grown" or "climate-smart" as attributes

TOWARD
SILVOPASTURE STANDARDS

The best set of standards would be developed by a group of practitioners who mutually agree on a reasonable set of practices that would qualify someone as a silvopasture practitioner. The following is merely a list of my proposed thoughts for a checklist to help others determine if they are practicing it well.

Ideally a group of practitioners, researchers, and those with expertise in standards programs would help flesh this out more.

1. Are animals merely being put in the woods, or is each layer (trees, animals, forages) being actively and intentionally managed to support the others?
2. Are there measured increases to percent organic matter (soil) or rapid tree growth (tree girth) to support the theoretical benefits?
3. Is there any bare soil on the site? What is the action plan for addressing this condition when it arises?
4. Are animals rotated at least once a week at a minimum, more ideally every one or two days?
5. Are highly sensitive or healthy areas of the landscape being protected from animal impacts?
6. What are the calculated rates of sequestration based on the number of trees planted?

Of course, silvopasture standards could also link up with grassfed grazing and animal welfare standards, drawing on those as part of a certification process. As more practitioners emerge, hopefully a conversation can be started.

uniquely assigned to silvopasture. Indeed, many environmental advocates are quick to name livestock as a major problem, arguing that if everyone just stopped eating meat all the world's problems would be solved (see chapter 1). As practitioners of silvopasture, we can tell a different story, where meat and dairy consumption in tree-based systems can actually sequester more carbon than any other form of farming. The trees, as we've discussed, are the critical component of this, and pictures of silvopasture tell a thousand words.

On that note, if we are going to leverage the potential power of silvopasture as a narrative for marketing, it's important first that we get it right. For the sake of the larger community, please do silvopasture well, or *don't call it silvopasture*. There are far too many people out there using the word without having a system that backs it up. As the sidebar mentions, it may be time to determine common standards so that this practice can be promoted to benefit all. It only takes one bad apple to ruin the perception of many people. If you aren't sure you've got it, then hold off on using *silvopasture* as a term until you are sure.

ASSESSING MARKETING CHANNELS

There are many options for where to sell your products, and they can be thought of in two main categories: direct and indirect/wholesale. Direct markets include outlets where the farmer interacts directly with the customer for a transaction. Examples include farmers markets, community-supported agriculture, farm stores, and the like. Indirect or wholesale markets are where the farmer sells to someone who will in turn sell to an end consumer, such as grocery stores, restaurants, and retail stores.

Your size and scale, your goals, and how much energy you want to expend in sales all help determine your best marketing channel(s). Prices tend to be higher in direct markets, but in many cases more labor on the part of the farmer or hired help is required, which may end up meaning the price is actually lower per given unit. It all depends. Properly assessing the best channel for your operation takes time, some trial and error, and the ability to be critical toward a channel that may not be working well.

With livestock, small-scale operations generally choose one of two main routes. In the first, animals are slaughtered and sold via freezer trade direct to consumers, by the quarter, half, or whole animal. This option means a quick turnaround for getting meat to the consumer, though your base is limited to those who both can afford a higher volume of meat and have the freezer space for it.

Other operations opt to sell by the individual cut, whether to individuals, to restaurants, or at markets.

This can drastically increase the price per pound for many cuts, but considerable time can be involved making the sales happen, and you'll need on-farm freezer storage between slaughter and sale. Animals can also be sold at auction, often live, which is the simplest method from a time standpoint, assuming an auction site is close by. This is also the lowest price option, though selling live animals means you save slaughtering fees and time spent marketing.

Ultimately, a market channel assessment depends largely on volume. If you have a smaller number of animals, you can likely afford to spend more time and space storing, marketing, and selling them at a higher price. Higher-volume operations need to move more product, and faster. This factor of scale along with your preference for how you want to spend your time are the most important considerations. For instance, some farmers love spending evenings or weekend days at farmers market, valuing the social experience and time they can interact with customers. Others would prefer to spend their time on the farm, and aren't as interested in customer interaction. Highly recommended is a complete market channel assessment guide from the Cornell Cooperative Extension.[21]

FINAL THOUGHTS ON THE FARM ECONOMY

The combination of budgeting, cash flow, and market channel assessment helps you project a picture of how your farm enterprise will perform into the future. These planning tools, along with efforts to take maximum advantage of ways to reduce costs, are the foundation of a farm-specific economy. Still, the work is not just in the analysis but also is in compelling marketing that tells the unique story of your farm and draws in supportive customers. These aspects of a silvopasture are as important as the technical parts we've spent the majority of the book discussing.

Neither of these actions is immediate, as building the small farm economy takes considerable time. This is a good reason to start small, and scale up, since starting with just a few animals means less pressure selling in large volume, allowing for markets to be built slowly. New farms should be thinking in the scale of 5 to 10 years from start-up to viability. Patience is perhaps the hardest aspect of good farming to cultivate.

Getting Help: Support from Government, Industry, and Private Consultants

Very few silvopasture practitioners possess the unique set of skills that silvopasture demands all by themselves. This is, in part, because for so long forestry, grazing, farming, and pasture management have all been maintained as separate tasks. As we've discussed in depth, bringing them together has many promising benefits, but also increases the complexity of the system. It's best to build relationships with local resources and individuals, many of them free, as you go about building a silvopasture system. The more eyes on the system, the more it's likely to succeed.

As a reminder, it's important to know that many professionals you approach, especially foresters, may be skeptical at first, to say the least. This is not personal, but likely due to their experience in the field or what they were taught in forestry school. Remember that we unfortunately have a long history of woodland abuse by livestock. Do your homework, acknowledge this history, and express your desire to do it right. This will go a long way toward getting you good assistance from many individuals. This book should help you support your reasons for doing it in the first place.

GOVERNMENT AGENCIES

The first place to start is by getting to know the agencies and institutions locally and regionally that can help. These include extension services, land grant colleges, USDA programs, and your state environmental agencies.

Extension and land grant colleges. Each state has a college or university that is considered a "land grant" and has special access to land and resources, with the goal of serving the development of communities in the state. Some states have a county-based extension system (such as New York), while others are more regional or statewide (such as Vermont). The original concept of the land grant and extension affiliates is to engage in a two-way conversation, where community needs are shared with extensions, who relay them to institutions for research. The research results are then, ideally, shared back with the communities. Extension services are mainly educational, often offering low-cost classes and assistance with things such as soil or forage testing. Some have the capacity to do site visits and help with various assessment activities. Contact your local office to see what is available.

USDA programs. The United States Department of Agriculture (USDA) has a number of offices and programs to support farming practices:

Farm Service Agency is often the first stop for farmers to register and get "into the system." This provides access to several opportunities, including NRCS cost-share programs, low-interest loans for land and equipment, and disaster relief payments for crop losses due to droughts, floods, or other unexpected setbacks (www.fsa.usda.gov).

Natural Resources Conservation Service (NRCS) is focused on conserving natural resources and offers a number of cost-share programs. A visit to the county or regional office is well worth the effort, as most agents are very willing to help out. The office can provide aerial maps with parcel boundaries, soil types, and topographic lines. The cost-share programs are for a wide range of practices from planting trees to protect waterways to actually helping establish silvopasture systems specifically (code 381)—a standard approved for funding on a state-by-state basis. Keep in mind that sometimes the funding isn't directly for silvopasture, but perhaps enhancements that are in alignment, such as clearing invasive brush or developing more habitat for wildlife and pollinators. Creative thinking goes a long way in securing good funding for establishing systems, and it can be done. Sometimes it's most helpful to contact local Technical Service Providers (TSPs), who are private contractors registered with NRCS to carry out the work. They are often some of the best resources to help you navigate the bureaucracy (www.nrcs.usda.gov).

Other projects, such as **Sustainable Agriculture Research & Education (SARE)**, offer opportunities for grants where farmers can conduct their own research and receive support through farmer grants. These programs pay not for routine supplies and equipment, but for the time and materials necessary to conduct on-farm research (www.sare.org).

In all these cases the process and protocol are not always the clearest. We recommend that you meet the people running these programs, and seek their guidance to match the opportunities to your situation. Building a long-term relationship with these folks will help your farm do well.

State Agencies

The big advantage of working with state agencies is that many of the staff are locals and invested in the land and in supporting those who want to steward it. While the best way to get there isn't always agreeable, it's important on the part of the farmer to get to know their local resource professionals and build a mutually beneficial relationship, over time.

Environmental. Each state has some type of entity tasked with supporting environmental health and quality. In relation to silvopasture, the most relevant assistance it can provide is often around forest management. In many states, a state forester can visit your land and help develop a basic management plan, sometimes also a prerequisite for the programs mentioned previously. While often not comprehensive, the management plan can help you identify the best and least valuable lands for growing good trees.

Soil and water conservation districts are county or regional organizations that are tasked with administering natural resource laws and management programs. Local offices vary, but can help with planning for drainage and pond construction, as well as often offering low-cost trees for conservation plantings (with many suitable species for silvopasture). The district office also may be involved in completing the steps necessary for agricultural tax exemption.

Figure 6.10. Agroforestry trainees involved in pruning, September 2015, Atkins Agroforestry Research and Demonstration Site, Tuskegee University, Alabama. Research sites and active farms are some of the best places to learn by seeing silvopasture in action. Photo by Uma Karki.

FINDING A FORESTER

In addition to your state environmental agency and extension foresters who can potentially offer limited free services to you, finding a good consulting forester is a long-term investment that often makes sense for landowners. The state agency can usually provide a list of consulting foresters, but is not able to recommend one over another. The Society of American Foresters membership also offers a good pool to choose from.

It's critical to shop around and find a forester who aligns with and understands your goals and ethics for your woods. You want to hear him or her talk about the health of the forest as the bottom line for any management. If they talk about making you lots of money as the main purpose of their work, be advised to stay away. Good forestry, at best, makes moderate gains in terms of income. A consulting forester is also going to be most helpful if familiar with silvopasture and sympathetic to its goals. Before starting any work, ask for some references, and view examples of their work. Once a tree is removed from the stump, it can't be put back.

A Logger Is Not *a Forester*

Whether you're hiring folks or doing the work yourself, it's important to separate the functions of forester and logger. The job of the forester is to assess the woodlot and weigh the consequences of decisions for managing it. Ideally, this is done with an eye for income, but not at the expense of the health and integrity of the woodlot. A good forester is there to be a voice for the woods. A logger, on the other hand, is a person who focuses on the safe felling and removal of trees, all while doing minimal harm to the woods. In many cases loggers and their employees don't have a background in ecology and management, and so shouldn't be making decisions about which trees should stay or go.

LOCAL, STATE, AND NATIONAL ORGANIZATIONS

An additional resource to consider utilizing is the range of local woodland, conservation, land trust, and farming organizations in your area. Often a woodland owners association gathers for educational programing and woods walks, helping to learn together the best strategies for management and even forming a network to help one another do some of the work. Many states have their own private woodland owners organizations.

The **National Woodland Owners Association** is a good place to start in the United States. The group offers resources and is a voice for the concerns of private woodland owners (woodlandowners.org).

American Tree Farm System offers a certification for woodlands from 10 to 10,000 acres, where good stewardship practices are recognized (www.tree farmsystem.org).

Women Owning Woodlands supports leadership women to manage their own woodlands (www .womenowningwoodlands.net).

Minority Landowner magazine offers insight into the stories and practices of minority landowners around the United States (www.minoritylandowner.com).

Livestock breed associations can be invaluable for you to connect with other farmers and homesteaders who are managing the same breeds you are. They can be a great resource for animal stock, tips and tricks, and efforts to preserve genetic diversity in livestock.

And finally, seek out **organic and sustainable farming organizations**, which are often state-based and can offer a good network of growers who gather and share information through field days, conferences, and technical assistance.

Looking Ahead: Tenure, Ownership, and the Next Generation

Zooming out from the details of your own farm and forest situation, all this discussion of silvopasture is threaded with a much bigger question. While the work now may be to get systems up and running, what are the long-term future prospects of sustaining silvopasture for multiple generations? As we've discussed, any management system involving trees and forests is long-term work; most of us will only begin to see the real fruits of our labor in our lifetimes.

This question brings us back to the challenging notions of landownership, tenure, and a shift in the priorities and practices of land transfer that were discussed briefly in chapter 2. We are each embedded in a system of private landownership that is a notion unique to civilization, and one that isn't likely to change anytime soon, at least not system-wide. Long ago, land was never owned, but managed by a community. More recently, privatized ownership was transferred from one generation to the next, a less common desire for many families today. With the more common sale of land from one party to another, priorities change, and trees that we painstakingly planted or managed might be cut in an instant. So how can silvopasture systems thrive, and survive?

At the root of this, land and agricultural production will need to continue to rise in value for people. If we understand and value a silvopasture system, and if it provides wealth to individuals and communities (not just monetarily), it will be preserved. We are at a critical juncture for this notion, where many people are just starting to see and reinvest in the landscape. The first step, then,

AGROFORESTRY RESEARCH AND EXTENSION EDUCATION
BENEFITS FARMERS AND LANDOWNERS
By Uma Karki, PhD

Well-managed agroforestry systems provide economic viability through regular, short-term incomes from crop and/or livestock components, and long-term incomes from trees. Most of the southeastern forest consists of pine trees, which require 20 to 30 years to mature. Landowners with sole pine plantations have to manage the tree stands (thinning, pruning, and burning) several times before trees are harvested, and pay property tax annually; a similar scenario is true with non-pine woodlands. Although hunting leases on woodlands can provide income, adoption of agroforestry practices can benefit landowners with additional regular income opportunities.

The southeastern region has a great potential for the development of various agroforestry practices because of its suitable environment. However, the adoption is currently negligible because of inadequate research and extension education. To serve this need, an agroforestry extension education program was developed at Tuskegee University, starting with silvopasture training in 2010.

The training event was conducted annually until 2013. During this time, 80 professionals, farmers, and landowners were trained. In 2014, with the collaboration of 1890 Agroforestry Consortium (1890 AC) member institutions (Alabama A&M University, Alcorn State University, Florida A&M University,

Figure 6.11. Longleaf and loblolly pine silvopasture, Atkins Agroforestry Research and Demonstration Site. Photo by Uma Karki.

and North Carolina A&T State University) and the funding support of Southern SARE, this program was expanded to incorporate other aspects of agroforestry—alley cropping, forest farming, windbreaks, riparian buffers, ecosystem services, and economics of agroforestry systems. With this collaborative effort, an agroforestry handbook was developed and used to conduct curriculum-based, hands-on training sessions in the Southeast.

There are 11 chapters in the agroforestry training handbook: Silvopasture Introduction, Establishment and Management of Trees in Silvopasture Systems, Forage Selection and Establishment in a Silvopasture System, Suitable Animal Species and Facility Requirements for Grazing in a Silvopasture System, Sustainable Grazing Management in a Silvopasture System, Non-Timber Forest Products: Forest Farming, Alley Cropping, Riparian Buffers, Windbreaks, Ecosystem Services, and Economics of Agroforestry

Systems. Chapters were reviewed by experts from various land-grant universities, National Agroforestry Center, and USDA Forest Service.

This handbook is now available online (www .tuskegee.edu/Content/Uploads/Tuskegee/files /CAENS/Others/agroforestry/Agroforestry _Handbook.pdf) and also at www.silvopasturebook .com to the public as a free download.

Five curriculum-based regional training sessions were conducted in 2014 and 2015 in different states of the Southeast (three in Alabama, one in North Carolina, and one in Florida). Agroforestry research and demonstration sites present at the facility of or near the host institutions/sites (Tuskegee University, Alabama A&M University, Florida A&M University, and North Carolina A&T State University) were used for hands-on activities, demonstration, and site tours. A total of 181 professionals, farmers, and landowners from Alabama,

Figure 6.12. Agroforestry trainees involved in collecting soil samples, October 2014, Atkins Agroforestry Research and Demonstration Site. Photo by Uma Karki.

Florida, Mississippi, North Carolina, South Carolina, Tennessee, and Missouri participated in the training events.

Participants learned different aspects of agroforestry practices: silvopasture, forest farming, alley cropping, tree management, soil management, windbreaks, riparian buffers, ecosystem services, grazing management with proper animal care, economics of agroforestry systems, beekeeping, and mushroom production on tree logs. Trainees increased their knowledge of various aspects of agroforestry by 23 percent.

Two landowner trainees used the learned skills and knowledge to improve an existing silvopasture or develop a new silvopasture system. Several other landowners expressed that they were considering adopting some type of agroforestry practices in the near future. Trained professionals were found educating their clientele about agroforestry practices.

Tuskegee University and other 1890 AC partners are working together, and independently, to continue the efforts promoting agroforestry research and extension education in the Southeast. Currently experts from Tuskegee University and Alabama A&M University are collaborating on a USDA-AFRI-funded project: Agroforestry-Based Cropping Systems for Sustaining Small- and Medium-Sized Landowners in the Southeastern US.

Under this project Tuskegee University, with partial support from McIntire Stennis Forestry Research Program, is evaluating the use of small ruminants for sustainable management of silvopastures and woodlands on two sites.

Site 1 is located near the Tuskegee campus. This silvopasture plot consisted of longleaf-loblolly mixed pine in sandy loam soil. In total there are three 1-acre plots at Site 1 with 11- to 12-year-old trees. Eight different cool-season forages—annual ryegrass (Marshall), arrowleaf clover, chicory, crimson clover, hairy vetch, MaxQ tall fescue, rye, and white clover—were cultivated in separate strips within each plot in fall 2014 and 2015. When forages were well established and attained the recommended grazing height, Kiko wethers goats were rotationally stocked in the plots based on the available forage biomass. Grazing started in early spring 2015 and early winter

Figure 6.13. Silvopasture demonstration site, Plantersville, Alabama. Photo by Uma Karki.

Figure 6.14. Kiko wethers debarking longleaf pine, Atkins Agroforestry Research and Demonstration Site (Site 1). Photo by Uma Karki.

2016, and ended around mid-May. Goats grazed all forages well from the start, with the exception of white clover, which was grazed moderately toward the latter portion of the grazing season.

Site 2 is located in Plantersville, Alabama. This silvopasture plot consisted of 14 acres of 18- to 19-year-old loblolly pines in fine sandy loam soil. The plot was subdivided into four paddocks after a second thinning. In fall 2015 a combination of Marshall ryegrass and crimson clover was grown in two plots, while MaxQ tall fescue and arrowleaf clover were grown in the remaining two plots. Grazing began in January 2016 with a mixed breed of goats, and continued until the end of June. Goats readily grazed all of the forages from the start. Additional work is continuing to evaluate warm-season forages in these sites.

Goats showed some debarking behavior while grazing in Site 1, especially on longleaf pines, resulting in the death of the heavily debarked trees. Studies to identify alternative grazing animals in order to minimize tree damage at this site are under way. Goats in Site 2 did not show any debarking behavior. These studies showed the potential of expanding grazing opportunities for animals by developing silvopasture systems after thinning southern pine plantations. Any of the cool-season forages evaluated in these studies or

a combination of these forages suitable to the soil type and climatic condition can be grown for developing cool-season grazing. However, the grazing manager needs to be watchful for damage the grazing animals may inflict on trees, and take appropriate action to avoid it.

Similarly, Alabama A&M University is developing cover-crop-based alley cropping of specialty vegetables with loblolly pine and pecan trees. Other ongoing research is assessing the soil quality changes associated with different agroforestry practices, and monitoring the economic benefits of silvopasture with small ruminants and cover-crop-based alley cropping of specialty vegetables.

Research findings are disseminated to professionals, students, farmers, and landowners through publications and presentations at the training events, conferences, and meetings. An agroforestry training event is conducted annually at the collaborative institutions, where the research and demonstration sites are used for site tours, demonstration, and hands-on activities.

Uma Karki is associate professor and state extension livestock specialist at Tuskegee University. Contact her at articles.extension.org/pages/71114/uma-karki-tuskegee-university.

---

is to articulate and share the value that trees and forests have to individuals and the larger community, and get people back into a working relationship with the woods.

A second task is to break away from our tendency to be individualistic in our pattern of landownership and management, instead looking at ways to engage on a community level. Embedded in this idea is that more need to acknowledge the incredible privilege of owning land, and in that recognition seek ways to advocate and support more equity and access to land for more people. This extends not only to private lands, but public ones as well.

In other parts of the world, community forestry programs train and develop a workforce of forest

stewards, giving them skills and access to the tools needed to be active participants. There are hundreds of thousands of acres of woodlands that offer space to mutually benefit people and the environment, if only we can better design those connections. This notion of community or social forestry was something that used to be embedded in cultures, much like agrarian life as a whole.

The younger generation needs to be better educated about the benefits, promises, and importance of farms and forests to both community and global health. There are very few opportunities for youth to even witness, much less engage with, land stewardship. The good food and farming movement is changing this, but it

Figure 6.15. A more whole relationship to animals, land, and people is at the heart of silvopasture. The invitation for a larger farm and forest culture beckons people from all walks of life to a different way to farm.

rarely includes forests and agroforestry in its dialogue. Cute animals, and the desire to be around them, might be one way into this conversation.

Another area to pay attention to is the policy and government arena. There are opportunities to get involved and build networks around common interests. For example, the maple industry is a coalition of producers, industry members, educators, and advocates that have banded together to advocate for their own needs and interests. These associations and collective efforts will be critical and have a storied history of success.

This ecological approach to farming promises a responsible and caring approach to animals, the reforestation of farmland, and the raising of viable products in ways that positively affect the climate. It's complex, but interesting.

Above all, the narrative and storytelling of silvopasture are key. Very few people simply know anything about it, and the more we tell compelling stories, the more interest, demand, curiosity, and implementation of the practice will grow. Whether or not you are an outspoken advocate, remember that in your interactions you are helping define silvopasture for everyone. It's okay, and necessary, to proceed without knowing everything about it, but walk in a cautious way and be adaptive and responsive, always learning so that you can improve. This is a basic tenet of good farming, and it is even more critical for the emerging and exciting practice of silvopasture.

Welcome.

Epilogue

This is only the beginning. While the basic concepts and tools outlined in this book allow us to get started with silvopasture, only a dedicated effort from a robust network will give it the momentum to flourish over the next several decades and generations. While the focus in many farming practices is technical, solutions tend to be compelled socially.

When we look at the landscape of grazing today, there is both great opportunity and several challenges. Most promising is that more and more farmers want to get animals back onto the landscape, and consumers seem willing to support this by paying for grassfed and pastured animal products. Yet in selling a product and developing viable farm enterprises, we can sometimes oversell, and sometimes look out over an idyllic pasture while forgetting that our "sustainable" or "regenerative" grazing practices may, in fact, be far from it.

The first hurdle to good silvopasture is the willingness to rotationally graze animals, and to do it well. Many graziers or potential graziers aren't willing to go there. I often consult on and see poorly managed grazing systems, and encourage clients to get their grazing plan down before bringing animals to the woods, or trees to the pasture.

Another complicated question, as we have discussed, revolves around the regenerative nature of grazing, and silvopasture, in regard to climate change and carbon sequestration. The conversation continues. What has been made clear from recent analysis is that rotationally grazing alone doesn't mean any of us are saving the climate. Further, cutting trees and opening up forested land to silvopasture is, at least in the short term, a net carbon loss. The only clear "winner" for the climate at this point is to plant trees. We could all do well by planting as many trees as we can in our lifetimes.

All of this doesn't mean that good grazing isn't good. It certainly uses fewer resources, can be a tool to restore healthy soil and landscapes, and can contribute to a more financially viable system for the farmer. We are on a path of discovery, and what's most important is to stay open, critical, and reflective. This is how we can collectively learn, grow, and change, so that our farming continues to get better for our family, our community, and our planet.

So consider this a callout to students, researchers, beginning farmers, policy makers, and anyone compelled by these concepts. The need is to first get clear on our common language and understanding of silvopasture, then to coordinate efforts to enact and further develop the process as a network. In working on this book and talking with many others during workshops, conferences, site visits, and conversations, I've started a laundry list of "needs" for silvopasture moving forward:

Education. More workshops, field days, and opportunities to expose people both to the overall concept of silvopasture and to more specific technical skills. These events need to target end users as well as technical service providers (state and federal agencies, extension, and so forth) who already interact with farmers and landowners routinely. Many of these support specialists are asked about silvopasture, but aren't able to articulate the concept well.

A curriculum, developed with multiple stakeholders, would help lay out the common points of understanding and agreement and support a consistent dissemination of the information. It would also help identify what is known with confidence, versus areas of less certainty that need to be further explored.

Demonstration sites. People can envision change when they see it firsthand. When you're developing a novel practice, consider sharing your work. It's important to document the process and be able to have an array of before, during, and after photos. A map of sites and information is being maintained at www.silvopasturebook.com.

Research. There is ample space for research at university farms and experiment stations, as well as on farms with differing variables in soil, microclimate, and the like. Some of the top priorities for research include:

- Fodder value and nutritional composition of trees and woody plants, and changes over the course of a season and between years.
- Specific tree/forage/animal studies to determine the success and possible drawbacks of interactions.
- Potential of tree fodders as a major source of nutrition for ruminants as well as monogastric animals.
- Methods of tree planting that define the most effective establishment techniques.
- Approaches to determining sustainable stocking rates and rotations for pigs and poultry.
- Economic analysis, especially for silvopasture systems that yield non-timber wood products and sell to retail markets (versus wholesale commodity).

And there are many more. We are maintaining a list of research questions in the Silvopasture in Practice listserv. More information on that on page 265.

Policy. It's important to promote agroforestry as a whole, and silvopasture specifically, to government, regulators, and farm policy organizations that craft the future of farming from many angles. These can lead to important considerations for regulations and encourage the implementation of incentives through existing means like grants, NRCS cost-share programs, and agriculture investment to support development. Papers that articulate needs and benefits in concise ways help convey the message to officials. We can also work collectively to think creatively about ways silvopasture can be funded through existing incentive programs (like wildlife and pollinator habitat) since the system has a multitude of benefits for those priorities as well.

Social justice. A big missing link in the conversation is the recognition that the history of our society and systemic barriers mean that not all people have equitable access to land, capital, and education for any type of farming, let alone silvopasture. Anyone involved in developing and advocating for silvopasture needs to recognize who is not at the table, and work on ways to increase access. Those of us with privilege have a great opportunity to leverage it for the wider good. Our farm and organizational work recognizes that we farm and work on land originally taken from native people, and that our nation's wealth and food system evolved in a way that is inequitable.

There are many central issues to consider, and there is sometimes a tension between small steps that help individuals and efforts that effect system-wide change. Some key areas of focus relevant to silvopasture include land access, access to training and education, and systemic barriers that make it harder for people of color, immigrants, and refugees to enter farming as a livelihood.

There are solutions, of course, but only if we find the will and the people to organize them in our communities. Collective management strategies might prove to be some of the most successful ways out of the mess, where individuals with varying expertise partner to manage land for the public good while still maintaining personal livelihoods, harvesting firewood, polewood, craftwood, and using the space for grazing. Some coppice groups in England and Europe still operate in these ways.

Networks. Many of the items on this list are long-term, big-picture endeavors that take multiple generations to shift. In order to get there, robust networks are important. Informal networks offer a place to connect with others, discuss ideas, and gauge interest in research and development topics. If silvopasture is to grow and do so successfully, it needs to be with many honest conversations, often about challenging and complex subjects. Only on a community level will we best articulate what good grazing is, and which silvopasture techniques offer the most benefits to farms, and to a changing climate.

Three good places to stay connected online:

- **Silvopasture Ning** (silvopasture.ning.com) is an informal network where resources, photos, and conversation are shared.
- **Facebook** (www.facebook.com/groups/silvo pasture) hosts an active group of those interested in silvopasture.
- **Silvopasture in Practice listserv** (groups.google .com/d/forum/silvopasture-in-practice). In addition to an email discussion thread, this group is co-managing a series of collective documents, including the research topics mentioned on page 264.

It is my hope that this book becomes just one resource among many others that are developed as we all continue to learn and grow. Silvopasture is at once old and new, traditional and modern, simple yet complex. Its great attribute is that it is flexible to many landscapes—yet this also challenges us not to take what has been researched and practiced out of context.

Start your silvopasture at home, getting to know the ecology of your region and considering all the ways this affects grazing and vegetation growth. Start small, and head in the best direction for your unique situation. Take photos and notes, and share them with the larger group. Ask plenty of questions. Together we might be able to see this practice in North America blossom and develop into a common way to value animals, trees, and forages and create a truly regenerative form of farming.

Notes

CHAPTER 1

1. Nash, Roderick. "The American Wilderness in Historical Perspective." *Forest History Newsletter* 6, no. 4 (1963): 2–13. doi:10.2307/3983142.

2. Mt. Pleasant, Jane. *Traditional Iroquois Corn: Its History, Cultivation, and Use.* Ithaca, NY: Natural Resource, Agriculture, and Engineering Service, Cooperative Extension, 2010.

3. Kerrigan, William. "Apples on the Border: Orchards and the Contest for the Great Lakes." *Michigan Historical Review* 34, no. 1 (2008): 25–41. doi:10.2307/20174256.

4. Oates, Gary. "Potential Carbon Sequestration and Forage Gains with Management-Intensive Rotational Grazing (Research Brief #95)." Center for Integrated Agricultural Systems. Accessed May 10, 2017. http://www.cias.wisc.edu/potential-carbon-sequestration-and-forage-gains-with-management-intensive-rotational-grazing-research-brief-95.

5. Garnett, T., et al. "Grazed and Confused? Ruminating on Cattle, Grazing Systems, Methane, Nitrous Oxide, the Soil Carbon Sequestration Question—and What It All Means for Greenhouse Gas Emissions." Food Climate Research Network, University of Oxford, 2017.

6. Toensmeier, Eric. *The Carbon Farming Solution: A Global Toolkit of Perennial Crops and Regenerative Agriculture Practices for Climate Change Mitigation and Food Security.* White River Junction, VT: Chelsea Green Publishing, 2016.

7. Orefice, Joseph, John Carroll, Drew Conroy, and Leanne Ketner. "Silvopasture Practices and Perspectives in the Northeastern United States." *Agroforestry Systems* 91, no. 1 (February 2017): 149–60. doi:10.1007/s10457-016-9916-0.

8. Sharrow, Steve. "Trees in Pastures: Do Cattle Benefit from Shade?" *Association for Temperate Agroforestry* 8 (2000). Accessed May 1, 2017. http://www.aftaweb.org/latest-newsletter/temporate-agroforester/51-2000-vol-8/july-no-3/27-cattle-benefit-shade.html.

9. Blackshaw, Judith K., and A. W. Blackshaw. "Heat Stress in Cattle and the Effect of Shade on Production and Behaviour: A Review." *Australian Journal of Experimental Agriculture* 34, no. 2 (1994): 285–95.

10. "Pasture-Based Farming Enhances Animal Welfare." *Eat Wild.* Accessed July 18, 2017. http://www.eatwild.com/animals.html.

11. Villalba, J. J., and F. D. Provenza. "Self-Medication and Homeostatic Behaviour in Herbivores: Learning About the Benefits of Nature's Pharmacy." *Animal* 1, no. 9 (October 2007). doi:10.1017/S1751731107000134.

12. Voth, Kathy. "How to Teach Livestock to Eat Weeds." *On Pasture*, April 9, 2013. http://onpasture.com/2013/04/09/how-to-teach-livestock-to-eat-weeds.

13. Smallidge, Peter, Brett Chedzoy, and Tatiana Stanton. *Enhancing Meat Goat Production Through Controlled Woodland Browsing.* Report. Natural Resources, Cornell University. Ithaca, NY: NESARE, 2003.

14. Fair, Barbara. "Coping with Drought: A Guide to Understanding Plant Response to Drought." North Carolina State Extension Publications, 2009. Accessed May 4, 2017. https://content.ces.ncsu.edu/coping-with-drought-a-guide-to-understanding-plant-response-to-drought.pdf.

15. "Here's Where Heavy Rain Is Increasing the Most in US." Accessed August 19, 2017. http://www.climatecentral.org/gallery/maps/heres-where-heavy-rain-is-increasing-the-most-in-us.

16. Crain, Rhiannon. "An Introduction to Habitat Connectivity." *Habitat Network.* Accessed May 9, 2017. http://content.yardmap.org/learn/habitat-connection.

17. Kelber, Almut, et al. "Light Intensity Limits Foraging Activity in Nocturnal and Crepuscular Bees." *Behavioral Ecology* 17 no. 1 (2005): 63–72.

18. López-Uribe, Margarita M., et al. "Crop Domestication Facilitated Rapid Geographical Expansion of a Specialist Pollinator, the Squash Bee *Peponapis pruinosa.*" *Proceedings of the Royal Society B: Biological Sciences* 283, no. 1833 (2016).

19. Winfree, Rachael, et al. "Native Bees Provide Insurance Against Ongoing Honey Bee Losses." *Ecology Letters* 10, no. 11 (2007a): 1105–13; Garibaldi, Lucas A., et al. "Wild Pollinators Enhance Fruit Set of Crops Regardless of Honey Bee Abundance." *Science* 339, no. 6127 (2013): 1608–11; Park, Mia G., et al. "Per-Visit Pollinator Performance and Regional Importance of Wild *Bombus* and *Andrena* (Melandrena) Compared to the Managed Honey Bee in New York Apple Orchards." *Apidologie* 47, no. 2 (2016): 145–60.

20. Winfree, Rachael, Terry Griswold, and Claire Kremen. "Effect of Human Disturbance on Bee Communities in a Forested Ecosystem." *Conservation Biology* 21, no. 1 (2007b): 213–23; Romey, W. L., et al. "Impacts of Logging on Midsummer Diversity of Native Bees (*Apoidea*) in a Northern Hardwood Forest." *Journal of the Kansas Entomological Society* 80, no. 4 (2007): 327–38; Hanula, James L., Michael D. Ulyshen, and Scott Horn. "Conserving Pollinators in North American Forests: A Review." *Natural Areas Journal* 36, no. 4 (2016): 427–39.

21. Hall, Damon M., et al. "The City as a Refuge for Insect Pollinators." *Conservation Biology* 31, no. 1 (2017): 24–29.

22. Batra, S. W. T. "Red Maple (*Acer rubrum* L.), an Important Early Spring Food Resource for Honey Bees and Other Insects." *Journal of the Kansas Entomological Society* (1985): 169–72.

23. Roulston, T. H., and James H. Cane. "The Effect of Pollen Protein Concentration on Body Size in the Sweat Bee *Lasioglossum zephyrum* (Hymenoptera: Apiformes)." *Evolutionary Ecology* 16, no. 1 (2002): 49–65.

24. Cane, James H., and Stephen L. Buchmann. "What Governs Protein Content of Pollen: Pollinator Preferences, Pollen–Pistil Interactions, or Phylogeny?" *Ecological monographs* 70, no. 4 (2000): 617–43.

25. Simone-Finstrom, Michael D., and Marla Spivak. "Increased Resin Collection After Parasite Challenge: A Case of Self-Medication in Honey Bees?" *PLoS One* 7, no. 3 (2012): e34601.

26. Gherman, Bogdan I., et al. "Pathogen-Associated Self-Medication Behavior in the Honeybee *Apis mellifera*." *Behavioral Ecology and Sociobiology* 68, no. 11 (2014): 1777–84; Manson, Jessamyn S., Michael C. Otterstatter, and James D. Thomson. "Consumption of a Nectar Alkaloid Reduces Pathogen Load in Bumble Bees." *Oecologia* 162, no. 1 (2010): 81–89.

27. Hawken, Paul, ed. *Drawdown: The Most Comprehensive Plan Ever Proposed to Reverse Global Warming.* New York: Penguin Books, 2017.

28. "GMO Crops Mean More Herbicide, Not Less." Accessed May 8, 2017. https://www.forbes.com/sites /bethhoffman/2013/07/02/gmo-crops-mean-more -herbicide-not-less/#2d5ca6f3cd53.

29. "We Already Grow Enough Food for 10 Billion People . . . and Still Can't End Hunger." *ResearchGate.* Accessed May 4, 2017. https://www.researchgate.net/publication /241746569_We_Already_Grow_Enough_Food_for _10_Billion_People_and_Still_Can't_End_Hunger.

30. "Animal Feed." GRACE Communications Foundation. Accessed May 4, 2017. http://www.sustainabletable .org/260/animal-feed.

31. "Regenerative Agriculture: The Definition of Regenerative Agriculture." Accessed May 8, 2017. http://www.regenerativeagriculturedefinition.com.

32. "Farms, Feedlots, Forests & Climate Change Issues." FEW Resources.org. Accessed May 8, 2017. http:// www.fewresources.org/farms-feedlots-forests--climate -change-issues.html.

33. Hawken, Paul, ed. *Drawdown: The Most Comprehensive Plan Ever Proposed to Reverse Global Warming.* New York: Penguin Books, 2017.

34. Teague, W. R., et al. "The Role of Ruminants in Reducing Agriculture's Carbon Footprint in North America." *Journal of Soil and Water Conservation* 71, no. 2 (2016): 156–64.

35. Hawken, Paul, ed. *Drawdown: The Most Comprehensive Plan Ever Proposed to Reverse Global Warming.* New York: Penguin Books, 2017.

36. Toensmeier, Eric. *The Carbon Farming Solution: A Global Toolkit of Perennial Crops and Regenerative Agriculture Practices for Climate Change Mitigation and Food Security.* White River Junction, VT: Chelsea Green Publishing, 2016.

37. Montagnini, F. "Función de los Sistemas Agroforestales en la Adaptación y Mitigación del Cambio Climático." In *Sistemas Agroforestales: Funciones Productivas, Socioeconómicas y Ambientales.* Turrialba, Costa Rica: CATIE, 2015.

38. Jones, Christine. "Ruminants and Methane." *The Natural Farmer*, Summer 2014, B-21; Nair, P. K. Ramachandran, and Vimala D. Nair. "'Solid–Fluid–Gas': The State of Knowledge on Carbon-Sequestration Potential of Agroforestry Systems in Africa." *Current Opinion in Environmental Sustainability* 6 (2014): 22–27.

39. Nordberg, Maria. *Holistic Management: A Critical Review of Allan Savory's Grazing Method.* Uppsala, Sweden: SLU/EPOK, Centre for Organic Food and Farming, and Chalmers, 2016.

40. "Why Eat Less Meat." Accessed May 8, 2017. http://www.whyeatlessmeat.com.

41. Logsdon, Gene. *All Flesh Is Grass: The Pleasures and Promises of Pasture Farming.* Athens: Ohio University Press, 2004.

42. Daley, Cynthia A., et al. "A Review of Fatty Acid Profiles and Antioxidant Content in Grass-Fed and Grain-Fed Beef." *Nutrition Journal* 9, no. 1 (March 10, 2010): 10. doi:10.1186/1475-2891-9-10.

43. Tansawat, Rossarin, et al. "Chemical Characterisation of Pasture- and Grain-Fed Beef Related to Meat Quality and Flavour Attributes." *International Journal of Food Science & Technology* 48, no. 3 (March 1, 2013): 484–95. doi:10.1111/j.1365-2621.2012.03209.x.

44. Yeager, Davis. "Grass-Fed vs. Conventional Beef." *Today's Dietitian* 17, no. 11 (November 2015). Accessed May 9, 2017. http://www.todaysdietitian.com/newarchives/1115p26.shtml.

45. "The Privilege and Classism of Organic Food." *Words*Things*Stuff* (blog), February 8, 2015. https://wordsthingsstuff.wordpress.com/2015/02/08/the-privilege-and-classism-of-organic-food.

Chapter 2

1. Smallidge, Peter. "Forest Succession and Management." *Cornell University Cooperative Extension ForestConnect*, November 2015. https://blogs.cornell.edu/cceforestconnect/files/2015/12/Forest-Succession-118hj3w.pdf.

2. Perry, David A., Ram Oren, and Stephen C. Hart. *Forest Ecosystems.* Baltimore: Johns Hopkins University Press, 2008.

3. Blair, John, Jesse Nippert, and John Briggs. "Grassland Ecology." In *Ecology and the Environment*, edited by Russell K. Monson, 389–423. New York: Springer New York, 2014. http://link.springer.com/10.1007/978-1-4614-7501-9_14.

4. Williams, Gerald. "References on the American Indian Use of Fire in Ecosystems." USDA Forest Service, 2001. Accessed August 21, 2017. http://www.wildlandfire.com/docs/biblio_indianfire.htm.

5. Prasad, V., C. A. Stromberg, H. Alimohammadian, and A. Sahni. "Dinosaur Coprolites and the Early Evolution of Grasses and Grazers." *Science* 310 (2005): 1177–90.

6. "Describe the Major Differences Between the Plant Families Used as Forages." Oregon State University, *Forage Information System*, May 28, 2009. http://forages.oregonstate.edu/nfgc/eo/onlineforagecurriculum/instructormaterials/availabletopics/plantid/differences.

7. Smeal, Daniel, and Bonnie Hopkins. "Turfgrasses for Northern New Mexico." New Mexico State University, Guide H-511. Accessed August 21, 2017. http://aces.nmsu.edu/pubs/_h/H511/welcome.html.

8. David Abel and the Globe Staff. "Northeast Will Experience Faster Warming from Climate Change, New Study Finds." *Boston Globe*, January 2017. Accessed August 21, 2017. https://www.bostonglobe.com/metro/2017/01/12/northeast-will-experience-faster-warming-from-climate-change-new-study-finds/nitce6eK8zqQN2LXZXgvwK/story.html.

9. Knapp, Alan K., et al. "The Keystone Role of Bison in North American Tallgrass Prairie: Bison Increase Habitat Heterogeneity and Alter a Broad Array of Plant, Community, and Ecosystem Processes." *BioScience* 49, no. 1 (1999): 39–50. doi:10.1525/bisi.1999.49.1.39.

10. Borer, Elizabeth T., et al. "Herbivores and Nutrients Control Grassland Plant Diversity via Light Limitation." *Nature* 508, no. 7497 (April 24, 2014): 517–20. doi:10.1038/nature13144.

11. McNaughton, S. J. "Ecology of a Grazing Ecosystem: The Serengeti." *Ecological Monographs* 55, no. 3 (1985): 259–94.

12. Dick, A. C., and Vern Baron. "Nutrient Management on Intensively Managed Pastures." *Agri-Facts*, August 2009. Accessed August 7, 2017. http://www1.agric.gov.ab.ca/$department/deptdocs.nsf/all/agdex12813/$file/130_538-1.pdf?OpenElement.

13. Yoshihara, Yu, Miya Okada, Takehiro Sasaki, and Shusuke Sato. "Plant Species Diversity and Forage Quality as Affected by Pasture Management and Simulated Cattle Activities." *Population Ecology* 56, no. 4 (October 1, 2014): 633–44. doi:10.1007/s10144-014-0443-4.

14. Chedzoy, Brett. "Bale Grazing: Feed the Cattle, Feed the Pasture." *On Pasture*, December 3, 2013. http://onpasture.com/2013/12/02/bale-grazing-feed-the-cattle-feed-the-pasture.

15. Dumont, B., A. J. Rook, Ch. Coran, and K.-U. Röver. "Effects of Livestock Breed and Grazing Intensity on Biodiversity and Production in Grazing Systems. 2. Diet Selection." *Grass and Forage Science* 62, no. 2 (June 1, 2007): 159–71. doi:10.1111/j.1365-2494.2007.00572.x.

16. Savanna Oak Foundation. "Oak Savannas: Characteristics, Restoration and Long-Term Management." Accessed November 19, 2017. http://oaksavannas.org.

17. Curtis, John T. *The Vegetation of Wisconsin.* Madison: University of Wisconsin Press, 1959.

18. Ford, P. L. "Grasslands and Savannas." *Earth System: History and Natural Variability*, vol. 3 (2009): 252.

19. Dovčiak, Martin, Richard Hrivnák, Karol Ujházy, and Dušan Gömöry. "Patterns of Grassland Invasions by Trees: Insights from Demographic and Genetic Spatial Analyses." *Journal of Plant Ecology* 8, no. 5 (October 1, 2015): 468–79. doi:10.1093/jpe/rtu038.

20. McPherson, Guy Randall. *Ecology and Management of North American Savannas.* Tucson: University of Arizona Press, 1997.

21. Pearce, Fred. *The New Wild: Why Invasive Species Will Be Nature's Salvation.* Boston, MA. Icon Books, 2015.

22. Baudena, M., et al. "Forests, Savannas, and Grasslands: Bridging the Knowledge Gap Between Ecology and Dynamic Global Vegetation Models." *Biogeosciences* 12, no. 6 (March 20, 2015): 1833–48. doi:10.5194/bg-12-1833-2015.

23. Nippert, J. B., and A. K. Knapp. "Linking Water Uptake with Rooting Patterns in Grassland Species." *Oecologia* 153 (2007): 261–72.

24. Harrison, Jeff. "Hydraulic Lift: A Review." Accessed May 16, 2017. http://prizedwriting.ucdavis.edu/past/1995-1996/hydraulic-lift-a-review.

25. Yu, Kailiang, and Paolo D'Odorico. "Hydraulic Lift as a Determinant of Tree-Grass Coexistence on Savannas." *New Phytologist* 207, no. 4 (September 2015): 1038–51. doi:10.1111/nph.13431.

26. Shepard, Mark. *Restoration Agriculture: Real-World Permaculture for Farmers.* Austin, TX: Acres USA, 2013.

27. "Great Lakes Ecosystem: 1994 Proceedings of the Midwest Oak Savanna Conferences [Nuzzo]." Accessed August 22, 2017. https://archive.epa.gov/ecopage/web/html/nuzzo.html.

28. "Geography of Oak Savannas." Accessed May 16, 2017. http://www.oaksavannas.org/geography.html#Anderson.

29. Frelich, Lee E., et al. *Fire in Upper Midwestern Oak Forest Ecosystems: An Oak Forest Restoration and Management Handbook.* USDA Forest Service, Pacific Northwest Research Station, 2015. https://www.researchgate.net/profile/David_Peterson8/publication/282295001_Fire_in_upper_Midwestern_oak_forest_ecosystems_an_oak_forest_restoration_and_management_handbook/links/560b182608ae840a08d68816.pdf.

30. "Longleaf Pine Silviculture." USDA Forest Service, Southern Research Station, Restoring and Managing Longleaf Pine Ecosystems. Accessed August 22, 2017. https://www.srs.fs.usda.gov/longleaf/research/silviculture.

31. Olea, L., R. J. López-Bellido, and M. J. Poblaciones. "European Types of Silvopastoral Systems in the Mediterranean Area: Dehesa." In *Silvopastoralism and Sustainable Land Management.* Oxfordshire, UK: CABI, 2005: 30–35.

32. Jacobson, Mike. "Dehesa Agroforesty Systems." *AFTA Newsletter* 13 (2005). http://www.aftaweb.org/latest-newsletter/temperate-agroforester/92-2005-vol-13/october-no-4/97-dehesa-agroforestry-systems.html.

33. Joffre, Richard, and Serge Rambal. "How Tree Cover Influences the Water Balance of Mediterranean Rangelands." *Ecology* 74, no. 2 (1993): 570–82.

34. Joffre, Richard, Serge Rambal, and Jean-Pierre Ratte. "The Dehesa System of Southern Spain and Portugal as a Natural Ecosystem Mimic." *Agroforestry Systems* 45, no. 1 (1999): 57–79.

35. Scholes, R. J., and S. R. Archer. "Tree-Grass Interactions in Savannas." *Annual Review of Ecology and Systematics* 28 (1997): 517–44.

36. Joffre, R., and S. Rambal. "Soil Water Improvement by Trees in the Rangelands of Southern Spain." *Oecologia Plantarum* 9 (1988a): 405–22.

37. Bernhard-Reversat, F. "Biogeochemical Cycle of Nitrogen in a Semi-Arid Savanna." *Oikos* 38 (1982): 321–32.

38. Frayer, Lauren. "This Spanish Pig-Slaughtering Tradition Is Rooted in Sustainability." NPR.org, March 2015. Accessed May 19, 2017. http://www.npr.org/sections/thesalt/2015/03/18/392177526/this-spanish-pig-slaughtering-tradition-is-rooted-in-sustainability.

39. Core, J. R. "Men in Trees: A Look at the Annual Portuguese Cork Harvest." *Core77*, October 2011. Accessed May 19, 2017. http://www.core77.com//posts/20839/Men-In-Trees-A-Look-at-the-Annual-Portuguese-Cork-Harvest.

40. "Montado, Cultural Landscape." UNESCO Centre du patrimoine mondial. Accessed May 19, 2017. http://whc.unesco.org/fr/listesindicatives/6210.

41. "Put a Cork in It." National Wildlife Federation. Accessed May 19, 2017. https://www.nwf.org/News-and-Magazines/National-Wildlife/News-and-Views/Archives/2006/Put-a-Cork-in-It.aspx.

42. González Alonso, Clara. "Analysis of the Oak Decline in Spain: La Seca," 2009. http://stud.epsilon.slu.se/55/1/Oak_Decline_Clara_Gonz%C3%A1lez_Alonso.pdf.

43. Adams, S. N. "Sheep and Cattle Grazing in Forests: A Review." *Journal of Applied Ecology* (1975): 143–52.

44. Ogawa, Yasuo. "Combined Forestry and Livestock Production in the Uplands of Kyushu, Japan." ASPAC Food & Fertilizer Technology Center, 1995. http://www.fftc.agnet.org/htmlarea_file/library /20110729153539/eb404.pdf.

45. Matsumoto, Mitsuo, Kenjiro Honda, and Juuro Kurogi. "Management and Yield Prediction of Kunugi (*Quercus acutissima*) Grazing Forests." *Journal of Forest Research* 4, no. 2 (1999): 61–66.

46. Wealleans, Alexandra L. "Such as Pigs Eat: The Rise and Fall of the Pannage Pig in the UK." *Journal of the Science of Food and Agriculture* 93, no. 9 (July 1, 2013): 2076–83. doi:10.1002/jsfa.6145.

47. Koenig, Walter, and Johannes Knops. "The Mystery of Masting in Trees: Some Trees Reproduce Synchronously Over Large Areas, with Widespread Ecological Effects, but How and Why?" *American Scientist* 93, no. 4 (2005): 340–47.

48. Szabó, P. "Rethinking Pannage: Historical Interactions Between Oak and Swine." In *Trees, Forested Landscapes and Grazing Animals,* ed. Ian D. Rotherham. Abingdon, UK: Routledge, 2013, 62–70.

49. Rotherham, Ian D., ed. *Trees, Forested Landscapes and Grazing Animals: A European Perspective on Woodlands and Grazed Treescapes.* Abingdon, UK: Routledge, 2013.

50. Smith, Jo. "The History of Temperate Agroforestry," 2010. http://orgprints.org/18173.

51. Adams, S. N. "Sheep and Cattle Grazing in Forests: A Review." *Journal of Applied Ecology* 12, no. 1 (1975): 143–52. doi:10.2307/2401724.

52. Oppermann, Rainer, Guy Beaufoy, and Gwyn Jones, eds. *High Nature Value Farming in Europe.* Ubstadt-Weiher, Germany: verlag regionalkultur, 2012.

53. Boutsikaris, Costa. "Episode 12—Swedish Fäbod Forest Culture." Woodlanders.com video 13:21. Accessed August 23, 2017. http://www.woodlanders.com/blog /2017/7/21/episode-12-swedish-fbod-forest-culture.

54. Herzog, F. "Streuobst: A Traditional Agroforestry System as a Model for Agroforestry Development in Temperate Europe." *Agroforestry Systems* 42, no. 1 (1998): 61–80.

55. Read, H. "A Brief Review of Pollards and Pollarding in Europe." Premier colloque européen sur les trognes, October 26–28, 2006.

56. Nilsson, Greta. "The Eastern Forests Endangered Species Handbook." Accessed May 17, 2017.

http://www.endangeredspecieshandbook.org/dinos _eastern.php.

57. Spiegel, Jan Ellen. "How Far East? Not Too Far." *New York Times*, November 4, 2007. http://www .nytimes.com/2007/11/04/nyregion/nyregion special2/04Rb3bison.html.

58. Oxford Dictionaries. "Definition of *Colonization* in English." Accessed May 22, 2017. https:// en.oxforddictionaries.com/definition/colonization.

59. Orefice, Joseph N., and John Carroll. "Silvopasture—It's Not a Load of Manure: Differentiating Between Silvopasture and Wooded Livestock Paddocks in the Northeastern United States." *Journal of Forestry; Bethesda* 115, no. 1 (January 2017): 71–72. doi:http://dx.doi.org .proxy.library.cornell.edu/10.5849/jof.16-016.

CHAPTER 3

1. Anderson, John D., Malcolm L. Broome, and Randall D. Little. "Rotational Grazing: Will It Pay?" Publication P2299. Accessed August 26, 2017. https:// georgiaforages.caes.uga.edu/events/GS15/11Econ /MS%20State%20-%20Rotational%20grazing.pdf.

2. McDonough, William, and Michael Braungart. "Why Being 'Less Bad' Is No Good." September 14, 2013. Accessed November 25, 2017. https://www.theglobalist .com/why-being-less-bad-is-no-good.

3. Lane, Woody. "Forage Growth and Intensive Grazing Basics." *Alberta Sheep & Goat Grazing Manual.* Accessed August 27, 2017. http://www1.agric.gov .ab.ca/$department/deptdocs.nsf/all/sg14664/$file /pfm-grazing-1-growth-final-press.pdf?OpenElement.

4. Undersander, D. J., et al. *Pastures for Profit: A Guide to Rotational Grazing.* Cooperative Extension Publications, University of Wisconsin, 2002.

5. Flack, Sarah. *The Art and Science of Grazing: How Grass Farmers Can Create Sustainable Systems for Healthy Animals and Farm Ecosystems.* White River Junction, VT: Chelsea Green Publishing, 2016.

6. Emmick, Darrell L. "Managing Pasture as a Crop." Middlebury: University of Vermont Extension, 2012.

7. www.behave.net. Accessed August 27, 2017.

8. Originally from a podcast "Animal Behavior with Dr. Fred Provenza." Accessed August 27, 2017. http://agriculturalinsights.com/episode-060-dr-fred -provenza-on-animal-behavior-and-phytochemical-rich -foods-as-it-relates-to-health. This webpage has since been deleted, but I confirmed this quote with Fred Provenza.

9. Ramírez-Restrepo, C. A., and T. N. Barry. "Alternative Temperate Forages Containing Secondary Compounds for Improving Sustainable Productivity in Grazing Ruminants." *Animal Feed Science and Technology* 120, no. 3 (2005): 179–201.

10. MacAdam, Jennifer W., Joe Brummer, Anowarul Islam, and Glenn Shewmaker. "The Benefits of Tannin-Containing Forages," 2013. http://digitalcommons.usu .edu/cgi/viewcontent.cgi?article=1354&context =extension_curall.

11. Jerónimo, Eliana, et al. "Tannins in Ruminant Nutrition: Impact on Animal Performance and Quality of Edible Products," 2016. https://dspace.uevora.pt/ rdpc/handle/10174/19651.

12. Puchala, R., B. R. Min, A. L. Goetsch, and T. Sahlu. "The Effect of a Condensed Tannin-Containing Forage on Methane Emission by Goats." *Journal of Animal Science* 83, no. 1 (January 1, 2005): 182–86. doi:10.2527/2005.831182x.

13. Lisonbee, Larry D., Juan J. Villalba, and Fred D. Provenza. "Effects of Tannin on Selection by Sheep of Forages Containing Alkaloids, Tannins and Saponins." *Journal of the Science of Food and Agriculture* 89, no. 15 (December 1, 2009): 2668–77. doi:10.1002/jsfa.3772.

14. Lyons, Robert K., et al. "Interpreting Grazing Behavior." *Texas Farmer Collection*, 2000. http://oaktrust.library. tamu.edu/handle/1969.1/86955.

15. Nikkhah, A. "Chronophysiology of Ruminant Feeding Behavior and Metabolism: An Evolutionary Review." *Biological Rhythm Research* 44, no. 2 (April 1, 2013): 197–218. doi:10.1080/09291016.2012.656437.

16. Burritt, Elizabeth A. "Mother Knows Best," 2013. http://digitalcommons.usu.edu/cgi/viewcontent.cgi ?article=1484&context=extension_curall.

17. Kennedy, Mark. "Animal Behavior: Impacts on Grazing," 2006. http://uknowledge.uky.edu/ky _grazing/2006-Jan/Session/8.

18. Meuret, Michel, and Frederick D. Provenza. "When Art and Science Meet: Integrating Knowledge of French Herders with Science of Foraging Behavior." *Rangeland Ecology & Management* 68, no. 1 (2015): 1–17.

19. Lyman, T. D., F. D. Provenza, J. J. Villalba, and R. D. Wiedmeier. "Cattle Preferences Differ When Endophyte-Infected Tall Fescue, Birdsfoot Trefoil, and Alfalfa Are Grazed in Different Sequences." *Journal of Animal Science* 89 (2011): 1131–37.

20. Voth, Kathy. "How to Teach Livestock to Eat Weeds." *On Pasture*, April 9, 2013. http://onpasture.com /2013/04/09/how-to-teach-livestock-to-eat-weeds.

21. Burritt, Beth, Morgan Doran, and Matt Stevenson. "Training Livestock to Avoid Specific Forage," 2013. http://digitalcommons.usu.edu/cgi/viewcontent.cgi ?article=1367&context=extension_curall.

22. Orefice, Joe. "Pigs 'n Trees." *Small Farm Quarterly* (Ithaca, NY), Fall 2016. Accessed November 1, 2017. http://smallfarms.cornell.edu/2016/01/11/pigs-n-trees.

23. Stedman, Connor. "Pigs in the Woods." *Renewing the Commons.* Accessed March 4, 2017. https://renewing thecommons.wordpress.com/2010/12/04/pigs -in-the-woods.

24. Shattuck, Kathryn. "Preaching the Gospel of the Forest-Fed Pig." *New York Times*, Food section, December 30, 2013. https://www.nytimes.com/2014/01/01/dining /preaching-the-gospel-of-the-forest-fed-pig.html.

25. Brownlow, M. J. C. "Towards a Framework of Understanding for the Integration of Forestry with Domestic Pig (*Sus Scrofa Domestica*) and European Wild Boar (*Sus Scrofa Scrofa*) Husbandry in the United Kingdom." *Forestry: An International Journal of Forest Research* 67, no. 3 (1994): 189–218.

26. Rodríguez-Estévez, Vicente, et al. "Consumption of Acorns by Finishing Iberian Pigs and Their Function in the Conservation of the Dehesa Agroecosystem." In *Agroforestry for Biodiversity and Ecosystem Services— Science and Practice.* InTech, 2012. https://www .intechopen.com/download/pdf/34866.

27. Spencer, Terrell. "Pastured Poultry Nutrition and Forages," 2013. Available at http://www.sare.org /content/download/73280/1060790/Pastured_Poultry _Nutrition_and_Forages.pdf.

28. Turner, James. "Leader Follower Grazing System." *Permaculture Research Institute*, May 8, 2015. Accessed August 30, 2017. https://permaculturenews .org/2015/05/08/leader-follower-grazing-system.

29. Grimes, Jesse. "Pasturing Turkeys." Accessed August 30, 2017. http://www.albc-usa.org/documents/turkey manual/ALBCturkey-3.pdf.

30. Winkler, Hannah. "Guinea Fowl: Your Overlooked Backyard Buddy." *Modern Farmer*, October 15, 2014. Accessed July 12, 2017. http://modernfarmer .com/2014/10/get-watch-bird.

31. Damerow, Gail. "Raising Guinea Fowl: A Low-Maintenance Flock." *Mother Earth News.* Accessed July 7, 2017. http://www.motherearthnews.com /homesteading-and-livestock/raising-guinea-fowl -zmaz92aszshe.

32. Clark, M. Sean, and Stuart H. Gage. "Effects of Free-Range Chickens and Geese on Insect Pests and

Weeds in an Agroecosystem." *American Journal of Alternative Agriculture* 11, no. 1 (March 1996): 39–47. doi:10.1017/S0889189300006718.

33. Bowen, Richard. "Rumen Physiology and Rumination." Accessed July 11, 2017. http://www.vivo.colostate.edu /hbooks/pathphys/digestion/herbivores/rumination .html.

34. Rinehart, Lee. "Pasture, Rangeland and Grazing Management." ATTRA, National Center for Appropriate Technology. Accessed July 11, 2017. https://attra.ncat.org/attra-pub/viewhtml.php?id=246.

35. Coffey, Linda, and Margo Hale. "Tools for Managing Internal Parasites in Small Ruminants: Pasture Management." Accessed August 31, 2017. http:// web.uri.edu/sheepngoat/files/Pasture-Management -Internal-Parasite-pub2.pdf.

36. Coffey, Linda. "Managing Internal Parasites: Success Stories." ATTRA, National Center for Appropriate Technology. Accessed August 31, 2017. https://attra .ncat.org/attra-pub/summaries/summary.php?pub=493.

37. Mayer, Ralph, and Tom Olsen. "Estimated Costs for Livestock Fencing." Iowa State University. Accessed September 1, 2017. https://www.extension.iastate.edu /agdm/livestock/html/b1-75.html.

Chapter 4

1. Garrett, H. E., et al. "Hardwood Silvopasture Management in North America." In *New Vistas in Agroforestry*. Dordrecht: Springer Netherlands, 2004, pp. 21–33.

2. Bellassen, Valentin, and Sebastiaan Luyssaert. "Carbon Sequestration: Managing Forests in Uncertain Times." *Nature News* 506, no. 7487 (February 13, 2014): 153. doi:10.1038/506153a.

3. Masterson, Kathleen. "Mimicking Mother Nature, UVM Scientists 'Nudge' Forests Toward Old Growth Conditions." Accessed July 21, 2017. http://digital.vpr .net/post/mimicking-mother-nature-uvm-scientists -nudge-forests-toward-old-growth-conditions.

4. Ford, Sarah E. "Integrating Management for Old-Growth Characteristics with Enhanced Carbon Storage of Northern Hardwood-Conifer Forests." University of Vermont and State Agricultural College, 2016. http:// search.proquest.com/openview/892451a23f03acfe868a 62000d7d9e01/1?pq-origsite=gscholar&cbl=18750&d iss=y.

5. Champion, H. G. "Forest Terminology. (Society of American Foresters. Revised Edition)." (1950):

366-366. http://www.uky.edu/~jmlhot2/courses /for350/Stand%20Descriptions%20and%20 Supporting%20Material_UT%20Clatterbuck.pdf.

6. "Climate Change Indicators: US and Global Precipitation." US Environmental Protection Agency, June 27, 2016. https://www.epa.gov/climate-indicators /climate-change-indicators-us-and-global-precipitation.

7. "Basal Area: A Measure Made for Management: ANR-1371." ACES Publications. Accessed February 11, 2017. http://www.aces.edu/pubs/docs/A/ANR-1371 /index2.tmpl.

8. "Thinning in Mixed Hardwood Forests." USDA Forest Service, Northern Research Station. Accessed September 11, 2017. https://www.nrs.fs.fed.us/sustaining _forests/conserve_enhance/timber/stand_density.

9. Nyland, Ralph D. *Silviculture: Concepts and Applications*. Long Grove, IL: Waveland Press, 2016.

10. Carmean, Willard H., Jerold T. Hahn, and Rodney D. Jacobs. *Site Index Curves for Forest Tree Species in the Eastern United States*, 1989. https://www.treesearch .fs.fed.us/pubs/10192.

11. Chedzoy, Brett. *Silvopasture: Creating Pasture in the Woods*. ForestConnect, 2014. https://www.youtube .com/watch?v=TOsVYP0LRv0.

12. "Site Quality & Stand Density." *Florida Forest Stewardship*, University of Florida Institute of Food and Agricultural Sciences Extension. Accessed July 22, 2017. http://www.sfrc.ufl.edu/Extension /florida_forestry_information/forest_management /site_quality_and_stand_density.html.

13. Southern Group of Foresters. "The 'No. 1 Shelterbelt' Celebrates 75 Years." SouthernForests.org. Accessed September 4, 2017. http://www.southernforests.org /resources/publications/the-southern-perspective /the-southern-perspective-september-2010/update -from-the-states/the-no.-1-shelterbelt-celebrates -75-years.

14. Williams, Gerald W. *The USDA Forest Service: The First Century*. Washington, DC: USDA Forest Service, 2005.

15. Marttila-Losure, Heidi. "Shelterbelts, One of the Great Soil Conservation Measures of the 1930s, Are Being Removed." *Dakotafire*, February 4, 2013. http:// dakotafire.net/land/shelterbelts-one-of-the-great -soil-conservation-measures-of-the-1930s-are-being -removed/3347.

16. Hinsley, S. A., and P. E. Bellamy. "The Influence of Hedge Structure, Management and Landscape Context on the Value of Hedgerows to Birds: A Review." *Journal*

of Environmental Management 60, no. 1 (September 1, 2000): 33–49. doi:10.1006/jema.2000.0360.

17. Mitchell, Wilma A., W. P. Kuvleskey Jr., Donna Burks, and Chester O. Martin. *Hedgerow and Fencerow Management on Corps of Engineers Projects.* No. ERDC-TN-EMRRP-SI-23. Vicksburg, MS: Engineer Research and Development Center, 2001.

18. Burel, Françoise. "Hedgerows and Their Role in Agricultural Landscapes." *Critical Reviews in Plant Sciences* 15, no. 2 (1996): 169–90.

19. Brandle, James R., Laurie Hodges, and Xinhua H. Zhou. "Windbreaks in North American Agricultural Systems." In *New Vistas in Agroforestry.* Dordrecht: Springer Netherlands, 2004, pp. 65–78.

20. Kentucky Department of Fish and Wildlife. "Edge Feathering." Accessed September 11, 2017. https://fw.ky.gov/Wildlife/Documents/edgefeathering.pdf.

21. Sun, Meng. "Glyphosate Listed Effective July 7, 2017, as Known to the State of California to Cause Cancer." *OEHHA*, June 26, 2017. https://oehha.ca.gov/proposition-65/crnr/glyphosate-listed-effective-july-7-2017-known-state-california-cause-cancer.

22. Natural Resource Conservation Service. "NRCS Spec Guide Sheet." Accessed September 5, 2017. https://www.nrcs.usda.gov/Internet/FSE_DOCUMENTS/nrcs144p2_016364.pdf.

23. Russell, Matthew B., et al. "Interactions Between White-Tailed Deer Density and the Composition of Forest Understories in the Northern United States." *Forest Ecology and Management* 384 (January 2017): 26–33. doi:10.1016/j.foreco.2016.10.038.

24. The Nature Conservancy. "The Future of NY Forests." (2010) Accessed September 5, 2017. https://www.nature.org/ourinitiatives/regions/northamerica/unitedstates/newyork/lands-forests/forestregenfactsheet.pdf.

25. Smallidge, Peter. "Low Cost Fence Designs to Limit Deer Impacts." Cornell University. Accessed September 5, 2017. https://blogs.cornell.edu/cceforestconnect/files/2015/12/Fencing-xanc6w.pdf.

26. Donahue, Brian, et al. *A New England Food Vision.* Durham, NH: Food Solutions New England, University of New Hampshire, 2014. Available at http://foodsolutionsne.org/new-england-food-vision.

CHAPTER 5

1. Alemu, Molla Mekonnen. "Ecological Benefits of Trees as Windbreaks and Shelterbelts." *International Journal of Ecosystem* 6, no. 1 (2016): 10–13.

2. Wray, Paul Harland, and Laura Sternweis. *Farmstead Windbreaks: Planning.* Iowa State University Extension, 1997.

3. Kallenbach, R. L. "Integrating Silvopastures into Current Forage-Livestock Systems." In *Agroforestry Comes of Age: Putting Science into Practice* (2009), pp. 455–61.

4. Lin, C. H., R. L. McGraw, M. F. George, and H. E. Garrett. "Shade Effects on Forage Crops with Potential in Temperate Agroforestry Practices." *Agroforestry Systems* 44, no. 2 (1998): 109–19.

5. Ladyman, K. P., et al. "Quality and Quantity Evaluations of Shade Grown Forages." In *Proceedings of the 8th North American Agroforestry Conference* (2003), p. 175.

6. Varella, A. C., et al. "Do Light and Alfalfa Responses to Cloth and Slatted Shade Represent Those Measured Under an Agroforestry System?" *Agroforestry Systems* 81, no. 2 (2011): 157–73.

7. Kallenbach, R. L., M. S. Kerley, and G. J. Bishop-Hurley. "Cumulative Forage Production, Forage Quality and Livestock Performance from an Annual Ryegrass and Cereal Rye Mixture in a Pine Walnut Silvopasture." *Agroforestry Systems* 66, no. 1 (January 2006): 43–53. doi:10.1007/s10457-005-6640-6.

8. Kephart, Kevin D., and Dwayne R. Buxton. "Forage Quality Responses of C3 and C4 Perennial Grasses to Shade." *Crop Science* 33, no. 4 (1993): 831–37.

9. Lin, C. H., M. L. McGraw, M. F. George, and H. E. Garrett. "Nutritive Quality and Morphological Development Under Partial Shade of Some Forage Species with Agroforestry Potential." *Agroforestry Systems* 53, no. 3 (2001): 269–81.

10. Kallenbach, R. L., M. S. Kerley, and G. J. Bishop-Hurley. "Cumulative Forage Production, Forage Quality and Livestock Performance from an Annual Ryegrass and Cereal Rye Mixture in a Pine Walnut Silvopasture." *Agroforestry Systems* 66, no. 1 (January 2006): 43–53. doi:10.1007/s10457-005-6640-6.

11. Feldhake, C. M. "Forage Frost Protection Potential of Conifer Silvopastures." *Agricultural and Forest Meteorology* 112, no. 2 (2002): 123–30.

12. Clason, T. R., and S. H. Sharrow. "Silvopastoral Practices." In *North American Agroforestry: An Integrated Science and Practice.* Madison, WI: ASA, 2000, pp. 119–47.

13. Finch, V. A. "Body Temperature in Beef Cattle: Its Control and Relevance to Production in the Tropics." *Journal of Animal Science* 62, no. 2 (1986): 531–42.

14. Morrison, S. R. "Ruminant Heat Stress: Effect on Production and Means of Alleviation." *Journal of Animal Science* 57, no. 6 (1983): 1594–600.

15. Blackshaw, Judith K., and A. W. Blackshaw. "Heat Stress in Cattle and the Effect of Shade on Production and Behaviour: A Review." *Australian Journal of Experimental Agriculture* 34, no. 2 (1994): 285–95.

16. Karki, Uma, and Mary S. Goodman. "Microclimatic Differences Between Mature Loblolly-Pine Silvopasture and Open-Pasture." *Agroforestry Systems* 89, no. 2 (2015): 319–25.

17. Martsolf Jr., Jay David. "Microclimatic Modification Through Shade Induced Changes in Net Radiation." Dissertation, University of Missouri–Columbia, 1966.

18. Hamilton, Jim. "Silvopasture: Establishment and Management Principles for Pine Forests in the Southeastern United States," 2008. Accessed September 11, 2017. https://www.silvopasture.org/pdf_content/silvopasture_handbook.pdf.

19. Green, Ted. "Tree Hay: A Forgotten Fodder | Agricology," 2016. Accessed August 5, 2017. http://www.agricology.co.uk/tree-hay-forgotten-fodder.

20. Tame, Mike. "Management of Trace Elements and Vitamins in Organic Ruminant Livestock Nutrition in the Context of the Whole Farm System," 2008. Available at http://www.organicresearchcentre.com/manage/authincludes/article_uploads/iota/research-reviews. Specifically review management-of-trace-elements-and-vitamins-in-organic-ruminant-livestock-nutrition.doc and the-role-and-management-of-herbal-pastures-for-animal-health.doc.

21. Heady, Harold F. "Palatability of Herbage and Animal Preference." *Journal of Range Management* 17, no. 2 (1964): 76–82.

22. Wilkinson, Ian. "Herbal Leys." *Cattle Breeder*, 2011. Accessed September 8, 2017. https://www.cotswoldseeds.com/files/cotswoldseeds/leys%20article_1.pdf.

23. "Herbs to Help Your Pasture." *Red Devon USA*, September 20, 2010. http://reddevonusa.com/blog/herbs-to-help-your-pasture.

24. Waller, P. J., et al. "Plants as De-Worming Agents of Livestock in the Nordic Countries: Historical Perspective, Popular Beliefs and Prospects for the Future." *Acta Veterinaria Scandinavica* 42, no. 1 (2001): 31–44. doi:10.1186/1751-0147-42-31.

25. Klebelsberg, Christoph. "Replicating the Practice of Willow Silage as Winter Fodder to Reduce the Need for Grazing Livestock, and Lead to Forest Regeneration—South Asia Pro-Poor Livestock Policy Programme," 2009. Accessed August 6, 2017. http://www.sapplpp.org/informationhub/cpr-livestock/queries/replicating-the-practice-of-willow-silage-as-winter-fodder-to-reduce-the-need-for-grazing-livestock.html#.WiMFC46BLzo.

26. Smith, Jo, et al. "Nutritional and Fermentation Quality of Ensiled Willow from an Integrated Feed and Bioenergy Agroforestry System in UK." *Maataloustieteen Päivät* 8, no. 9.1 (2014): 2014. http://www.smts.fi/MTP_julkaisu_2014/Posterit/064Smith_ym_Nutritional_and_fermentation_quality_of_ensiled_willow.pdf

27. Rytter, Rose-Marie, Lars Rytter, and Lars Högbom. "Carbon Sequestration in Willow (*Salix* spp.) Plantations on Former Arable Land Estimated by Repeated Field Sampling and C Budget Calculation." *Biomass and Bioenergy* 83 (December 2015): 483–92. doi:10.1016/j.biombioe.2015.10.009.

28. McIvor, Ian. "Willows for the Farm." Poplar & Willow Research Trust, 2013. Accessed August 9, 2017. http://www.poplarandwillow.org.nz/documents/brochure-1-willows-for-the-farm.pdf.

29. Hawks Bay Regional Council, "Fodder Willows," 1996. Accessed August 9, 2017. http://www.hbrc.govt.nz/assets/Document-Library/Information-Sheets/Land/fodderwil.pdf.

30. Roder, W. "Experiences with Tree Fodders in Temperate Regions of Bhutan." *Agroforestry Systems* 17, no. 3 (March 1, 1992): 263–70. doi:10.1007/BF00054151.

31. McCabe, Sharon M., and T. N. Barry. "Nutritive Value of Willow (*Salix* sp.) for Sheep, Goats and Deer." *Journal of Agricultural Science* 111, no. 1 (1988): 1–9. doi:10.1017/S0021859600082745.

32. Mupeyo, Bornwell, et al. "Effects of Feeding Willow (*Salix* spp.) upon Death of Established Parasites and Parasite Fecundity." *Fuel and Energy Abstracts* 164 (2011): 8–20. doi:10.1016/j.anifeedsci.2010.11.015.

33. Sanga, U., F. D. Provenza, and J. J. Villalba. 2011. "Transmission of Self-Medicative Behaviour from Mother to Offspring in Sheep." *Animal Behaviour* 82 (2011): 219–27. https://extension.usu.edu/behave/past-projects/pasture-projects/control-parasites.

34. Boutsikaris, Costa. "Episode 6—Willow Coffins." Woodlanders.org video 14:57. Accessed August 9, 2017. http://www.woodlanders.com/blog/2017/4/16/episode-6-willow-coffins.

35. Perttu, Kurth L., and Haworth Continuing Features Submission. "Biomass Production and Nutrient Removal

from Municipal Wastes Using Willow Vegetation Filters." *Journal of Sustainable Forestry* 1, no. 3 (March 28, 1994): 57–70. doi:10.1300/J091v01n03_05.

36. Dalziel, Sarah. "How to Make Willow Charcoal for Artists." May 11, 2016. Accessed August 9, 2017. https://joybileefarm.com/how-to-make-willow -charcoal.

37. Hangs, R. D., H. P. Ahmed, and J. J. Schoenau. "Influence of Willow Biochar Amendment on Soil Nitrogen Availability and Greenhouse Gas Production in Two Fertilized Temperate Prairie Soils." *BioEnergy Research* 9, no. 1 (March 1, 2016): 157–71. doi:10.1007/s12155-015-9671-5.

38. Ehrlich, Steven D. "Willow Bark." University of Maryland Medical Center, 2015. Accessed August 9, 2017. http://www.umm.edu/health/medical/altmed /herb/willow-bark.

39. Caputo, Jesse. "Research Summary: Sequestration of Carbon by Shrub Willow Offsets Greenhouse Gas Emissions." *Extension*, 2016. Accessed August 9, 2017. http://articles.extension.org/pages/70406/research -summary:-sequestration-of-carbon-by-shrub-willow -offsets-greenhouse-gas-emissions.

40. Stone, Katharine R. 2009. "*Robinia pseudoacacia*." In Fire Effects Information System [online]. USDA Forest Service, Rocky Mountain Research Station, Fire Sciences Laboratory (Producer). Available at http:// www.fs.fed.us/database/feis [December 2, 2017].

41. Horton, G. M. J., and D. A. Christensen. "Nutritional Value of Black Locust Tree Leaf Meal (*Robinia pseudo- acacia*) and Alfalfa Meal." *Canadian Journal of Animal Science* 61, no. 2 (1981): 503–06.

42. Barrett, Robert P., Tesfai Mebrahtu, and James W. Hanover. "Black Locust: A Multi-Purpose Tree Species for Temperate Climates." In *Advances in New Crops*. Portland: Timber Press, 1990, pp. 278–83. Accessed August 5, 2017. https://hort.purdue.edu/newcrop /proceedings1990/V1-278.html.

43. Toensmeier, Eric. "Intensive Silvopasture." *Drawdown*, February 7, 2017. http://www.drawdown.org/solutions /coming-attractions/intensive-silvopasture.

44. Cardona, Cuartas, et al. "Contribution of Intensive Silvopastoral Systems to Animal Performance and to Adaptation and Mitigation of Climate Change." *Revista Colombiana de Ciencias Pecuarias* 27, no. 2 (June 2014): 76–94.

45. Calle, Zoraida, et al. "A Strategy for Scaling-Up Intensive Silvopastoral Systems in Colombia." *Journal of Sustainable Forestry* 32, no. 7 (2013): 677–93. http:// elti.fesprojects.net/ISTF%20Conference%202012 /zoraida_calle_istfconference2012.pdf.

46. McLane, Eben. "Planting the Next Generation of Waterproof Lumber." Northern Woodlands, Winter 2004. Accessed August 9, 2017. http://northern woodlands.org/articles/article/planting_the_next _generation_of_waterproof_lumber.

47. Rédei, Károly, Zoltán Osvath-Bujtas, and Irina Veperdi. "Black Locust (*Robinia pseudoacacia* L.) Improvement in Hungary: A Review." *Acta Silvatica et Lignaria Hungarica* 4 (2008): 127–32.

48. Keresztesi, B. "A Breeding and Cultivation of Black Locust, *Robinia pseudoacacia*, in Hungary." *Forest Ecology and Management* 6, no. 3 (1983): 217–44. https://doi.org/10.1016/S0378-1127(83)80004-8.

49. Food and Agriculture Organization of the United Nations. *Mulberry for Animal Production: Proceedings of an Electronic Conference Carried Out Between May and August 2000*. FAO, 2002.

50. Datta, R. K., A. Sarkar, P. R. M. Rao, and N. R. Singhvi. "Utilization of Mulberry as Animal Fodder in India." In *Mulberry for Animal Production: Proceedings of an Electronic Conference Carried Out Between May and August 2000*. FAO, 2002, pp. 183–88. Accessed August 5, 2017. http://www.fao.org/docrep/005/X9895E /x9895e0h.htm.

51. Sánchez, Manuel D. "Mulberry: An Exceptional Forage Available Almost Worldwide," 2002, pp. 271–89. Accessed August 5, 2017. http://www.fao.org /livestock/agap/frg/mulberry/Papers/HTML /Mulbwar2.htm.

52. Sujathamma, P., G. Savithri, and K. Kavyasudha. "Value Addition of Mulberry (*Morus* spp.)." *International Journal of Emerging Technologies in Computational and Applied Sciences* 5, no. 4 (2013): 352–56.

53. Ly, J., Chhay Ty, Chiv Phiny, and T. R. Preston. "Some Aspects of the Nutritive Value of Leaf Meals of *Trichanthera gigantea* and *Morus alba* for Mong Cai Pigs." *Livestock Research for Rural Development* 13, no. 3 (2001); Phiny, Chiv, T. R. Preston, and J. Ly. "Mulberry (*Morus alba*) Leaves as Protein Source for Young Pigs Fed Rice-Based Diets: Digestibility Studies." *Livestock Research for Rural Development* 15, no. 1 (2003).

54. Zhou, Z., B. Zhou, L. Ren, and Q. Meng. "Effect of Ensiled Mulberry Leaves and Sun-Dried Mulberry Fruit Pomace on Finishing Steer Growth Performance, Blood Biochemical Parameters, and Carcass Characteristics." *PLoS One* 9, no. 1 (2014): e85406. https://doi.org /10.1371/journal.pone.0085406.

55. Kemp, P. D., A. D. Mackay, L. A. Matheson, and M. E. Timmins. "The Forage Value of Poplars and Willows." *Proceedings of the New Zealand Grassland Association* 63 (2001), pp. 115–20. Accessed August 9, 2017. https://www.grassland.org.nz/publications/nzgrassland_publication_275.pdf.

56. Charlton, J. F. L., G. B. Douglas, B. J. Wills, and J. E. Prebble. "Farmer Experience with Tree Fodder." *Using Trees on Farms.* Grassland Research and Practice Series, no. 10 (2003): 7–16.

57. McWilliam, E. L., et al. "The Effect of Different Levels of Poplar (*Populus*) Supplementation on the Reproductive Performance of Ewes Grazing Low Quality Drought Pasture During Mating." *Animal Feed Science and Technology* 115, no. 1 (2004): 1–18.

58. Johnson, Jacob W., et al. "Millwood Honeylocust Trees: Seedpod Nutritive Value and Yield Characteristics." *Agroforestry Systems* 87, no. 4 (August 1, 2013): 849–56. doi:10.1007/s10457-013-9601-5.

59. *Honeylocust Research Newsletter* 2. Accessed August 9, 2017. http://faculty.virginia.edu/honeylocust-agro-forestry/agroforestry/Honeylocust%20Research%20Newsletter%20No.1.htm.

60. Dini-Papanastasi, O. "Early Growth and Pod Production of Ten French Honey Locust Varieties in a Semi-Arid Mediterranean Environment of Greece." *Cahiers Options Méditerranéennes* 62 (2004): 323–26.

61. Forwood, J. R., and C. E. Owensby. "Nutritive Value of Tree Leaves in the Kansas Flint Hills." *Journal of Range Management*, 1985, 61–64.

62. Burner, David M., et al. "Yield Components and Nutritive Value of *Robinia pseudoacacia* and *Albizia julibrissin* in Arkansas, USA." *Agroforestry Systems* 72, no. 1 (2008): 51–62.

63. Toensmeier, Eric. "Not All Nitrogren Fixers Are Created Equal." Accessed September 7, 2017. http://www.perennialsolutions.org/all-nitrogen-fixers-are-not-created-equal.

64. Hamilton, Jim. "Silvopasture: Establishment & Management Principles for Pine Forests in the Southeastern United States." Accessed August 9, 2017. https://www.silvopasture.org/pdf_content/silvopasture_handbook.pdf.

65. Sharrow, Steven H., and Rick Fletcher. "Converting a Pasture to a Silvopasture in the Pacific Northwest," 2003. http://digitalcommons.unl.edu/agroforestnotes/26.

66. Olson, Bret, and Karen Launchbaugh. "Managing Herbaceous Broadleaf Weeds with Targeted Grazing." *Targeted Grazing: A Natural Approach to Vegetation Management and Landscape Enhancement.* Centennial, CO: American Sheep Industry Association, 2006, 58–67.

67. Hansen, Melissa. "Less Plum Curculio Damage Was Seen." *Good Fruit Grower.* Accessed August 6, 2017. http://www.goodfruit.com/less-plum-curculio-damage-was-seen; "Potential of Organic Hogs as a Tool for Post-Harvest Orchard Floor Sanitation and Pest Management." *Ceres Trust*, June 25, 2013. https://cerestrust.org/potential-of-organic-hogs.

68. Temel, Suleyman, and Mucahit Pehluvan. "Evaluación de Las Hojas de Huertos Frutales y de Álamo, En Otoño, Como Fuente de Forraje Alternativo Para La Alimentación Del Ganado." *Ciencia e Investigación Agraria* 42, no. 1 (April 2015): 27–33. doi:10.4067/S0718-16202015000100003.

69. Voth, Kathy. "Training Sheep for Vineyard Management." *On Pasture*, February 3, 2015. http://onpasture.com/2015/02/02/training-sheep-for-vineyard-management.

70. Antiquum Farm. "Managing Vineyards with Livestock." Accessed August 10, 2017. https://antiquumfarm.com/practices/husbandry.

71. Hawke's Bay Wine Growers. "A Guide to Using Sheep for Leaf-Plucking in the Vineyard." Accessed August 10, 2017. https://www.premier1supplies.com/img/newsletter/09-05-13-sheep/sheep-for-leaf-plucking-booklet.pdf.

72. Burke, Michael. "Final Report for Integration of Intensive Sheep Grazing with a Vineyard/Orchard Operation." *SARE Reporting System.* Accessed August 10, 2017. https://projects.sare.org/project-reports/fnc99-267.

73. Decouson, J. *Sheep and Orchards: A Promising Association for More Sustainable Cider Apple Production.* 2006. Accessed June 6, 2017. http://www.archiveofciderpomology.co.uk/ArchiveReports/SheepAndOrchards2011.pdf

74. King, Tim. "Grazing Pigs in Orchards Creates Healthy Orchards, Pigs." TheLandOnline.com. Accessed August 10, 2017. http://www.thelandonline.com/news/grazing-pigs-in-orchards-creates-healthy-orchards-pigs/article_59a98680-ce2b-11e6-bc58-1fe8be4cd73c.html.

75. Freeman, Sarah. "'Orchard Pigs' Return Virtue Cider to Its Roots." *Tales of the Cocktail*, October 14, 2015. https://talesofthecocktail.com/in-depth/orchard-pigs-return-virtue-cider-its-roots.

76. Baumgartner, Jo Ann. "A Farmer's Guide to Food Safety and Conservation: Facts, Tips & Frequently Asked

Questions" 2013. Accessed August 6, 2017. http:// www.nyfarmersmarket.com/wp-content/uploads /Farmer-Guide-to-Food-Safety.pdf.

77. Rittenhouse, Thea. "Tip Sheet: Manure in Organic Production Systems." ATTRA, 2015. Accessed August 6, 2017. https://www.ams.usda.gov/sites/default/files /media/Manure%20in%20Organic%20Production%20 Systems_FINAL.pdf.

78. "Standards for the Growing, Harvesting, Packing, and Holding of Produce for Human Consumption." *Federal Register*, November 27, 2015. https://www.federal register.gov/documents/2015/11/27/2015-28159 /standards-for-the-growing-harvesting-packing- and-holding-of-produce-for-human-consumption. Specifically, see the link to the responses to comments: https://www.federalregister.gov/d/2015-28159/p-1231.

79. Marcus, Amling. "*Escherichia Coli* Field Contamination of Pecan Nuts." Accessed August 6, 2017. http:// pubmedcentralcanada.ca/pmcc/articles/PMC379774 /pdf/applmicro00032-0077.pdf.

80. Nunn, L., et al. "Rotationally Grazing Hogs for Orchard Floor Management in Organic Apple Orchards." *Acta Horticulturae* 737 (2007): 71.

81. Hoar, B., et al. "Buffers Between Grazing Sheep and Leafy Crops Augment Food Safety." *California Agriculture* 67, no. 2 (April 1, 2013): 104–09.

82. The Organic Center. "*E. coli* 0157:H7 Frequently Asked Questions." 2006. Accessed August 7, 2017. https:// organic-center.org/reportfiles/EColiFAQReport.pdf.

83. McWilliams, James. "Beware the Myth of Grass-Fed Beef: Cows Raised at Pasture Are Not Immune to Deadly *E. coli* Bacteria." Slate.com, January 2010. Accessed August 7, 2017. http://www.slate.com /articles/health_and_science/green_room/2010/01 /beware_the_myth_of_grassfed_beef.html.

84. Weeda, W. C. "The Effect of Cattle Dung Patches on Pasture Growth, Botanical Composition, and Pasture Utilisation." *New Zealand Journal of Agricultural Research* 10, no. 1 (February 1967): 150–59. doi:10.108 0/00288233.1967.10423087.

85. "Congressman Bishop Heralds Win for Pecan, Tree Nut Industries." *Congressman Sanford Bishop*, May 29, 2014. https://bishop.house.gov/media-center /press-releases/congressman-bishop-heralds-win -for-pecan-tree-nut-industries.

86. Reid, William. "Northern Pecans: Impacts of Grazing Cattle in a Pecan Grove." *Northern Pecans*, June 21, 2014. http://northernpecans.blogspot.com/2014/06 /impacts-of-grazing-cattle-in-pecan-grove.html.

87. "Two Story Agriculture—Pecans and Livestock." Noble Research Institute, July 1996. Accessed August 10, 2017. http://www.noble.org/news/publications /ag-news-and-views/1996/july/two-story-agriculture ---pecans-and-livestock.

88. SARE.org. "Integrating Cattle, Pecans." Accessed August 10, 2017. http://www.sare.org/Learning -Center/SARE-Biennial-Reports/Archives-of-Biennial -Reports-Highlights/2003-Annual-Report/Text -Version/Integrating-Cattle-Pecans.

89. "The Hickory-Pecan Plant Catalog." Badgersett Research. Accessed August 10, 2017. http://www .badgersett.com/plants/orderhickories.html.

90. Garrett, H. E., and W. B. Kurtz. "Silvicultural and Economic Relationships of Integrated Forestry- Farming with Black Walnut." *Agroforestry Systems* 1, no. 3 (September 1, 1983): 245–56. doi:10.1007/ BF00130610.

91. "Toxic Plant Profile: Black Walnut." University of Maryland Extension. Accessed August 10, 2017. https://extension.umd.edu/learn/toxic-plant -profile-black-walnut.

92. Scott, R., and W. C. Sullivan (2007). "A Review of Suitable Companion Crops for Black Walnut." *Agroforestry Systems* 71, no. 3: 185–93. doi:10.1007 /s10457-007-9071-8.

93. Cole, Leslie. "Peerless Pork." OregonLive.com. Accessed August 10, 2017. http://www.oregonlive.com/food day/index.ssf/2009/06/peerless_pork.html.

94. Burnder, David. "Silvopasture Add Value to Christmas Tree Plantation." *AFTA* 11 (July 2003). Accessed August 9, 2017. http://www.aftaweb.org/latest -newsletter/temporate-agroforester/55-2003-vol-11 /july-no-3/31-christmas-tree-plantation.html.

95. Shropshire Sheep Breeders Association. *Two Crops from One Acre: A Comprehensive Guide to Using Shropshire Sheep for Grazing Tree Plantations*, 2008. Accessed August 9, 2017. http://www.shropshire-sheep.co.uk /publications/SSBA%20P16%20Two%20Crops%20 Booklet.pdf.

96. Sharrow, S. H., W. C. Leininger, and K. A. Osman. 1992. "Sheep Grazing Effects on Coastal Douglas Fir Forest Growth: A Ten-Year Perspective." *Forest Ecology and Management* 50: 75–84.

97. Compas, L. 2000. Small farms, big ideas, News for Missouri sustainable agriculture. Accessed April 3, 2003. http://www.aftaweb.org/latest-newsletter /temporate-agroforester/55-2003-vol-11/july -no-3/31-christmas-tree-plantation.html.

CHAPTER 6

1. "Balancing Your Animals with Your Forage." USDA Natural Resources Conservation Service. Accessed September 11, 2017. https://www.nrcs.usda.gov /Internet/FSE_DOCUMENTS/stelprdb1097070.pdf.

2. Voth, Kathy. "The New Grazing Charts Are Here :-) The New Grazing Charts Are Here!" *On Pasture*, January 3, 2017. http://onpasture.com/2017/01/02 /the-new-grazing-charts-are-here-the-new-grazing -charts-are-here.

3. "Livestock Shade Structures—Costs and Benefits." USDA Natural Resources Conservation Service, 2007. Available at www.silvopasturebook.com/resources.

4. Higgins, Stephen, and Sarah Wightman. "Shade Options for Grazing Cattle." University of Kentucky, 1999. Accessed August 11, 2017. http://www2.ca.uky .edu/agcomm/pubs/aen/aen99/aen99.pdf.

5. Orefice, Joeseph. "Final Report for FNE12-762." *SARE Reporting System*. Accessed August 10, 2017. https:// projects.sare.org/project-reports/fne12-762.

6. Grado, S. C., C. H. Hovermale, and D. G. St. Louis. "A Financial Analysis of a Silvopasture System in Southern Mississippi." *Agroforestry Systems* 53, no. 3 (November 1, 2001): 313–22. doi:10.1023/A:1013375426677.

7. Frey, Gregory E., et al. "Adoption and Economics of Silvopasture Systems by Farm Size in Northeastern Argentina." In *When Trees and Crops Get Together*, Proceedings of the 10th North American Agroforestry Conference. Quebec City, Canada, 2007.

8. Stainback, G. Andrew, and Janaki R. R. Alavalapati. "Restoring Longleaf Pine Through Silvopasture Practices: An Economic Analysis." *Forest Policy and Economics* 6, no. 3 (June 1, 2004): 371–78. doi:10.1016/j.forpol.2004.03.012.

9. Clason, T. R. "Economic Implications of Silvipastures on Southern Pine Plantations." *Agroforestry Systems* 29 (1995): 227–38.

10. Dangerfield, C. W., and R. L. Harwell. "An Analysis of a Silvopastoral System for the Marginal Land in the Southeast United States." *Agroforestry Systems* 10 (1990): 187–97.

11. Eldridge, R. W., Kenneth H. Burdine, and Richard Trimble. "The Economics of Rotational Grazing." Agricultural Economics—No 05-02. University of Kentucky, 2005. Accessed September 12, 2017. http:// georgiaforages.caes.uga.edu/events/GS11/11/UKY %20-%20rotational%20grazing%20economics.pdf.

12. Undersander, D. J., et al. *Pastures for Profit: A Guide to Rotational Grazing*. Cooperative Extension Publications, University of Wisconsin, 2002. Available at https://www.nrcs.usda.gov/Internet /FSE_DOCUMENTS/stelprdb1097378.pdf.

13. Parker, W. J., L. D. Muller, and D. R. Buckmaster. "Management and Economic Implications of Intensive Grazing on Dairy Farms in the Northeastern States." *Journal of Dairy Science* 75, no. 9 (September 1, 1992): 2587–97. doi:10.3168/jds.S0022-0302(92)78021-7.

14. Department of the Treasury. "Farmer's Tax Guide." Publication 225, 2017. https://www.irs.gov/pub /irs-pdf/p225.pdf.

15. Godsley, Larry. "Tax Considerations for the Establishment of Agroforestry Practices." University of Missouri Agroforestry Center, 2010. Available at https://extensiondata.missouri.edu/pub/pdf/agguides /agroforestry/af1004.pdf.

16. Taylor, Adam, and David Mercker. "Why It Is Difficult to Write a Crop Budget for Hardwood Timber." University of Tennessee, 2016. Accessed August 14, 2017. https://extension.tennessee.edu/publications/ Documents/W365.pdf.

17. Curtis, Kynda, Shane Feuz, and Nelissa Aybar. "Consumer Willingness to Pay for Specialty Meats," 2012. http://digitalcommons.usu.edu/extension _curall/1016.

18. Tucker, Maggie. "Consumer Prejudices Affect Willingness to Pay for Beef." Texas A&M Animal Science, September 2016. Accessed August 18, 2017. https://animalscience.tamu.edu/2016/09/14/ consumer-prejudices-affect-willingness-to-pay-for-beef.

19. Curtis, K., et al. "Consumer Preferences for Meat Attributes." University of Nevada, 2006. https:// www.unce.unr.edu/publications/files/ag/2008 /fs0811.pdf.

20. Ellis, Stu. "How Should Beef Be Marketed for Consumers to Pay More for Perceived Value?" Farmgateblog.com, May 2011. Accessed August 18, 2017. http://www.farmgateblog.com/article/1360/ how-should-beef-be-marketed-for-consumers-to-pay -more-for-perceived-value.

21. Leroux, Matthew. "Guide to Marketing Channel Selection." Cornell Cooperative Extension, Tompkins County, 2010. Accessed August 18, 2017. http:// ccetompkins.org/resources/guide-to-marketing -channels.

Index

Note: Page numbers in *italics* refer to figures and illustrations.

About the Author

Steve Gabriel, coauthor of *Farming the Woods* (with Ken Mudge), is an ecologist, educator, and forest farmer who has lived most of his life in the Finger Lakes region of New York. He passionately pursues work that reconnects people to the forested landscape and supports them to grow their skills in forest stewardship.

He currently splits his time between working for the Cornell Small Farms Program as Agroforestry Extension Specialist and developing the farm he runs with wife, Elizabeth, Wellspring Forest Farm, which produces shiitake and oyster mushrooms, duck eggs, pastured lamb, nursery trees, and maple syrup.

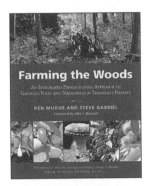
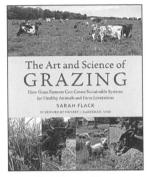
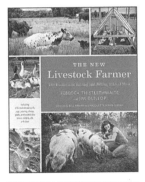
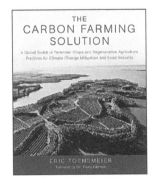

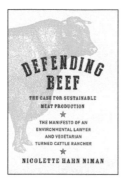